Understanding Risks and Uncertainties in Energy and Climate Policy

Haris Doukas • Alexandros Flamos • Jenny Lieu
Editors

Understanding Risks and Uncertainties in Energy and Climate Policy

Multidisciplinary Methods and Tools for a Low Carbon Society

Springer Open

Editors
Haris Doukas
Electrical & Computer Engineering
National Technical University of Athens
Athens, Greece

Alexandros Flamos
Industrial Management and Technology
University of Piraeus
Piraeus, Greece

Jenny Lieu
Science Policy Research Unit
University of Sussex
Brighton, UK

ISBN 978-3-030-03151-0 ISBN 978-3-030-03152-7 (eBook)
https://doi.org/10.1007/978-3-030-03152-7

Library of Congress Control Number: 2018964075

This Springer imprint is published by the registered company Springer Nature Switzerland AG
The registered company address is: Gewerbestrasse 11, 6330 Cham, Switzerland

Foreword

A dozen years ago I participated in a workshop organised by the University of Graz, Austria. The workshop took place in a castle near Graz, Schloss Seggau, which since 1218 has been home to the Catholic bishop for that region. While I have forgotten most of what was discussed at the workshop itself, one moment has stuck in my mind ever since. I was walking through the castle and found myself in the bishops' portrait gallery, images of 60 bishops spanning eight centuries. Here is what stuck: the first 57 bishops all looked the same, and then the last three were different. After 750 years of following the rules, the artists suddenly started playing with the use of colour and light, with surrealism, with brush strokes and with background shapes. And the bishops started playing too. Finally, too, they were smiling. It was as if, from one year to the next, everyone woke up. And what also stuck in my mind was the fact that awakening coincided, almost exactly, with the year I was born, 1965. I certainly don't claim to have caused it! But the fact is that I and indeed all of us are living in a time that is unlike any that has come before.

And so it isn't particularly surprising that we don't have it all figured out yet. Certainly, we haven't figured out how to deal with climate change yet, and that is a big problem. Climate change has the potential to fundamentally alter the living environment on which our thin veneer of civilisation rests. Mainly, it is a result of industrialisation, this age of using vast amounts of energy to magnify our powers and build and consume huge amounts of stuff. We've been causing this for centuries now, since the industrial revolution, but it was only in the 1960s that we began to wake up to it and to other large-scale environmental issues. About the same time that the bishops of Schloss Seggau woke up to smiling.

And so we are still figuring out how to handle it all. First, it took us a long time to decide that it was a problem worth worrying about. Eventually the natural scientists said yes. The next step was cost–benefit analysis. One of the first models to give us insight was a climate-economy model, DICE, developed by William Nordhaus at Yale University. His model suggested that climate change could have serious economic consequences, but there was also reason to believe that the policies to stop climate change could have big economic consequences as well. He advised that we balance the two. And in fact for about 20 years, academics were primarily concerned with finding the right balance, figuring out how much we should reduce our emissions to maximise human welfare.

But then we moved on. A little over a decade ago, the Intergovernmental Panel and Climate Change (IPCC) published a report that said more or less the same thing as a British sponsored report, written by the economist Nicolas Stern: namely that the risks of climate change were so serious that doing cost–benefit analysis didn't make sense. We were facing an existential threat, and it didn't make sense to use the usual discount rates, or to try to equalise average marginal cost and benefit functions. What defined the next decade of climate policy analysis were the risks and uncertainties of climate change itself. Coming out of this, in 2015, world leaders agreed in Paris to do everything possible to stop climate change, as soon as possible. They agreed to work to limit global warming to less than 2 °C, ideally to 1.5 °C. That issue was settled.

And now we have entered a new stage of thinking about climate change and climate policy. The defining question now is how exactly to achieve the goals that leaders in Paris set for themselves, set for us all. The reality is that they had no idea, and to a large extent we don't either. What define this new stage are not the risks and uncertainties of climate change, but rather the risks and uncertainties inherent in everything we might do to stop it.

Back when William Nordhaus was analysing climate change with a cost–benefit model, it appeared that mitigating it would be costly. Now we simply don't know. It might, depending on how we do it, or it might not cost us anything and might even lead to great economic growth. But we have also discovered that what we don't know about mitigation, the risks and uncertainties, covers a lot more than costs. We don't know the local environmental impacts, like how different energy sources use different amounts of water. Or how people will fear those environmental impacts, or how they will turn their fear into action and then block the construction of new energy infrastructure.

What we do know is that we need to act quickly. To hold global warming to under 2 °C we need to stop new investment into old-style energy infrastructure right about now and stop whatever old stuff remains within the next 20–30 years. To act this quickly, we need to assess the risks and uncertainties of the pathways forward, to make sure we aren't committing ourselves to a new pathway that is as unsustainable—environmentally or socially—as the last.

And that is where this book comes in, and the EU research project, on which it was based. The TRANSrisk project had as its intention to really begin to model the risks and uncertainties of different climate mitigation pathways, generating results that could lead to agreement on rapid change. Part of that task involved developing new modelling tools. Those are the tools described in this book. They are tools to assess the economic risks, the environmental risks, the social risks and uncertainties, of everything we could do to stop climate change. The tools and applications described in this book are tools that we need, tools we need to understand how to use, if we are to act quickly to stop climate change.

We live in a unique age, for better and for worse. The bishops are smiling, and I for one think that is a very good thing. We could be destroying the environment that supports us, a very bad thing. But we also may be discovering and inventing, for the first time, a set of tools to allow ourselves to stop doing so. We don't have this age figured out. This is not the first book proposing new tools to figure out climate change, and it will not, hopefully, be the last. But we won't get to those books in the future unless we read this book today: a book that captures essential progress on a pathway towards better understanding.

Climate Policy, ETH Zürich
Zürich, Switzerland
June 2018

Anthony Patt

Editorial

Addressing climate change and its negative impact on our society and environment is one of the major challenges of the twenty-first century. Strategies for this transition to a low-carbon society can be supported by innovations in technological systems (e.g. energy and water systems), policy instruments and market systems. Societal perceptions and business approaches may also need to be reframed to tackle the long-standing challenges of unsustainable behaviours. In the transition towards a low-carbon future, there can be barriers or risks that prevent the implementation of technological and policy innovations (or other actions). In addition, the innovations or actions themselves can also lead to future risks or unintended negative consequences that are difficult to capture using quantitative or qualitative methods independently.

The TRANSrisk project[1] takes a unique approach to evaluating risks and uncertainties in low-carbon pathways, combining economic computer models with qualitative inputs from a diverse range of stakeholders in the area of study. Quantitative tools include models exploring future climate evolution and its impacts, as well as the costs and benefits associated with different mitigation pathways. These tools can enhance uncertainty analysis and robust decision-making processes, through the quantification of risks and interrelations of climate change mitigation pathways, political opinion and public acceptability. However, they can contain a high degree of uncertainty, and some critical issues or risks that prevent the implementation of low-carbon (technology) options are difficult to quantify, for example public acceptance (or lack thereof). On the other hand, qualitative methods can capture challenges and risk through multi-stakeholder engagement—collective intelligence can help overcome quantitative limitations.

---

[1]TRANSrisk (2018) Transitions pathways and risk analysis for climate change mitigation and adaption strategies. EU Horizon 2020 Programme, GA: 642260. Available at: http://transrisk-project.eu/.

By considering a mix of quantitative and/or qualitative methods within and beyond the TRANSrisk approach, this special edition presents a range of innovative methodologies, tools and new framings to better consider elements of risks and uncertainty in the support of energy and climate policies and strategies.

Organisation of the Book

The issue opens with a study by **Alexandros Nikas et al.**, which provides a simple overview and organising scheme of integrated assessment models. It does this by delving into the characteristics of more than 60 individual models, describing the main ways in which certain classes or groups of climate-economy models differ from one another. This analysis provides an initial understanding of generic model structures and offers descriptions and comparisons of the main classes of models.

Jenny Lieu et al. continue by presenting a consensus building in engagement processes' framework that includes Indigenous consensus, knowledge, interests and rights as a focal point of a consultation in the decision-making process. The consultation process is presented within the context of land use decisions impacting a low-carbon future for oil sands development in Alberta, Canada. The framework aims to help reduce risks resulting from decisions that do not consider the interests and rights of those communities most impacted by resource development or climate mitigation pathways.

The third study is by **Sotiris Papadelis and Alexandros Flamos**. In this chapter, a step-by-step application of calibrating an agent-based model is presented. In particular, an agent-based model for small-scale PV adoption was calibrated on the historical data for small-scale solar PV capacity additions that took place in Greece from January 2010 to February 2013.

Pedro Crespo del Granado et al. explore how long-term scenarios for transmission expansion and decarbonisation policies influence the evolution of the EU power system infrastructure. They use an EU electricity investment model to determine the optimal portfolio of electricity generation technologies and compute their respective costs and emissions achieved towards 2050. Based on the investment model's results, they investigate how these portfolios perform under divergent policy or geopolitical developments. They apply a robust optimisation tool, which selects ideal portfolios by stress testing a particular scenario or policy choice under uncertainty of input parameters.

Hera Neofytou et al. continue and investigate EU countries' scenarios/targets combinations and ambition levels scenarios to tackle climate change. The research focuses on assessment of each alternative climate and energy policy scenario and its socio-economic, environmental and energy impacts with the application of multi-criteria decision analysis.

Bob van der Zwaan et al. inspect the interrelationship between energy and water use in the Middle East. They present results for projected power production, water withdrawal and water consumption levels until 2050 in the Middle East under both baseline and stringent climate policy scenarios. They also analyse how the use of different cooling techniques for the main power production options in the Middle East can yield water withdrawal and consumption savings in the electricity sector in the region.

In the next study, **Evangelos Grigoroudis and Konstantinos Petridis** estimate the environmental efficiency of countries using data envelopment analysis. The model considers national economies as production units, where resources (e.g. labour, energy, capital) are used to produce economic outputs (e.g. income), as well as undesirable outputs (e.g. emissions). In this context, a country is considered efficient if it produces the maximum possible wealth with the minimum harmful emissions while at the same time minimising the necessary resources. Using stochastic variables, the model also considers the uncertainty of collected data.

Delton Chen et al. investigate standard market-based policies for addressing climate change, focusing on those that aim to internalise the social cost of carbon (SCC) into the economy with either carbon taxes or cap-and-trade schemes. For a variety of reasons, standard policies are failing to manage the systemic risk of dangerous-to-catastrophic climate change. In this chapter, they clarify and expand on a market hypothesis that argues for a second externalised cost of carbon, called the risk cost of carbon (RCC), as the appropriate solution to this risk problem.

Theocharis Tsoutsos and Sotiris N. Kamenopoulos explore the social acceptability of renewable energy projects. Understanding the process under which a social licence to operate may be granted to a renewable energy project is important. In order to illustrate how these projects can be assessed from a sustainability point of view, two hypothetical scenarios were constructed. The scenarios were then evaluated by five imaginary stakeholders under specific criteria and sustainability principals, with the use of the multi-criteria decision analysis combined with the multi-attribute utility theory.

Finally, **Eike Blume-Werry et al.** propose a unilateral approach by a state introducing a CO_2 levy that internalises and prices CO_2 at a national level. The suggested climate and supply market model thereby incentivises and rewards production from CO_2-neutral sources during times when this does not cover the targeted share of production. The model is explored further by using Switzerland as an example, showing that a cross-sector carbon price can be implemented at acceptable costs for consumers.

Acknowledgement

The editors would like to express their sincere thanks and gratitude to the authors, who submitted chapters to this book, and the members of the advisory board, Assoc. Professor Kathleen Araujo, Boise State University's School of Public Service; Professor Gordon Mackerron, Science Policy Research Unit, University of Sussex; and Professor Yiannis Siskos, University of Piraeus.

The editors would also like to especially thank the referees, who spent their valuable time on providing their detailed reviews. Without their help, it would be impossible to prepare this book in line with the high standards set from the beginning.

Moreover, the editors would like to thank the members of the Decisions Support Systems (DSS) Laboratory of National Technical University of Athens (NTUA), the Technoeconomics of Energy Systems Laboratory (TEESLab) of the University of Piraeus (UNIPI) and the University of Sussex team for their hard work and coordination of activities within the TRANSrisk project, as well as the project officers, Frederik Accoe and Gema San Bruno, for their dedication and continuous support.

Finally, the editors would like to acknowledge the support from the European Commission (EC), which is financing TRANSrisk under the Horizon 2020 Research and Innovation Programme grant agreement No. 642260. The content of this book is the sole responsibility of its authors and editors and does not necessarily reflect the views of the EC.

As editors of this book we do hope that it will contribute to understanding risks and uncertainties in energy and climate policy towards a low-carbon future.

Athens, Greece
Piraeus, Greece
Brighton, UK
May 31, 2018

Haris Doukas
Alexandros Flamos
Jenny Lieu

Contents

About the Editors

Haris Doukas is an Assistant Professor in the School of Electrical & Computer Engineering (ECE) of the National Technical University of Athens (NTUA). He has a degree in mechanical engineering (Aristotle University of Thessaloniki—AUTH, 2003) and a PhD in the area of decision support systems for the sustainable energy sector's operation (ECE, NTUA, 2009). His scientific and research expertise includes the development of models and decision support systems for energy and environmental policy. He has participated as an expert in the design of energy policies, programmes promoting renewables and energy efficiency and specific energy management actions at a local, national and European level. As an expert, he has also participated in similar initiatives outside of Europe, for example in the countries of the Mediterranean basin and the member states of the Gulf Cooperation Council. In these fields, he has published more than 100 papers in international scientific journals and a book titled *Decision Support Policy Models for Energy and Environmental Systems*. He has also made numerous presentations in international conferences and published several chapters and articles in technical books and magazines. Dr. Doukas is an Associate Editor of *the Operational Research International Journal* (ORIJ) of Springer and participates as a member of the Editorial Board in journals related to the energy policy and operational research. Dr. Doukas has received awards for his work from the State Scholarship Foundation (IKY), the NTUA, the AUTH, the Technical Chamber of Greece (TCG) and the Hellenic Operational Research Society (HELORS).

Alexandros Flamos is an Associate Professor at the University of Piraeus (UNIPI), Director of the Technoeconomics of Energy Systems Laboratory (TEESlab) of UNIPI and Editor in Chief at the *Energy Sources, Part B: Economics, Planning, and Policy* (ESPB) scientific journal (Taylor & Francis). He has a degree in electrical and computer engineering and a PhD in the area of decision support systems (DSS) applied in energy and environmental policy and planning. At UNIPI and at the National Technical University of Athens (NTUA), he teaches energy economics, energy technologies and environment, utilisation of renewable energy sources (RES), management of energy sources and engineering economics and technoeconomics of energy systems at both under- and postgraduate level. He has over 20 years of experience in the areas of DSS, energy management and planning and energy modelling and their applications for analysing energy policy scenarios, intra- and cross-sectoral interactions of energy and climate policies, security of energy supply, business models and energy pricing competitiveness. He is member of the research committee of UNIPI, and he has held the position of scientific coordinator/senior researcher in more than 30 European Commission (EC)-funded projects (EC R&D FP5, FP6, FP7, H2020, IEE, MEDA, ENPI, etc.) related to energy systems management, appraisal and planning. He has also worked as a consultant for projects funded by international donors (e.g. the European Investment Bank and EuropeAid). He has significant experience working as an Editor in Chief, Guest Editor and Editorial Board member for scientific journals. He also has published more than 100 papers in high impact international scientific journals (peer reviewed) and international conferences and has been an invited speaker at major international energy policy events (MENAREC, Euro-Asia meetings, EU-GCC, etc.).

Jenny Lieu is a Research Fellow at SPRU (Science Policy Research Unit) in the University of Sussex (UofS). She is currently the Co-Principal Investigator for TRANSrisk (TR), an EU Horizon 2020 project that evaluates risks and uncertainties in low-carbon transition pathways at national, regional and global levels. She has a diverse research background in energy and climate policy and innovation studies. Some of the research areas she has worked in include policy evaluation and policy mixes (APRAISE, Assessment of Policy Interrelationships and Impacts on Sustainability in Europe, EU 7th Framework Programme); integration of energy, environment and climate modelling with stakeholder engagement; oils sands development and First Nations communities in Alberta, Canada (TR); new urbanism in Calgary (TR), Canada (University of Calgary); policy learning and transfer in the development of China's renewable energy policy (Tsinghua University); fast breeder reactors in the UK (UofS and EDF); gender issues in agriculture (United Nations Conference on Trade and Development); and biogas applications and Clean Development Mechanism (UofS). In addition to her research activities, Jenny has also provided consultancy services, producing strategic recommendations for energy and environmental infrastructure firms and conducting policy analysis for international institutions. She has also been a Co-Editor of a Routledge Energy Transition books series and is guest editor for the journal *Environmental Innovation and Societal Transitions*.

A Detailed Overview and Consistent Classification of Climate-Economy Models

Alexandros Nikas, Haris Doukas, and Andreas Papandreou

Abstract The proliferation and growing variety of climate-economy models and what are known as integrated assessment models (IAMs) can make it difficult for someone interested in following the debate to place any specific model, or the discussion about the merits of one or another, into a broader context. The literature related to climate-economy modelling is already vast: apart from a very large number of models and an even larger number of applications, there already exist many good surveys comparing—inter alia—modelling frameworks, model assumptions and model results. The objective of this chapter is to provide a simple overview and organising scheme of this modelling world by delving into the characteristics of more than 60 individual IAMs towards describing the main ways in which certain classes or groups of climate-economy models differ from one another. In contrast to other more detailed or narrowly focused "overviews" and literature reviews, this analysis takes less for granted and aims at providing an initial understanding of generic model structures. After briefly discussing some principles of classification that can help organise this often daunting modelling world, the chapter offers descriptions and comparisons of the main classes of models.

Keywords Climate policy · Integrated assessment models · Equilibrium · Macroeconometric · Energy systems · Climate-economy modelling · Optimal growth · Uncertainty

A. Nikas (✉) · H. Doukas
Decision Support Systems Laboratory, School of Electrical and Computer Engineering, National Technical University of Athens, Athens, Greece
e-mail: anikas@epu.ntua.gr; h_doukas@epu.ntua.gr

A. Papandreou
Department of Economics, University of Athens, Athens, Greece
e-mail: aap@econ.uoa.gr

H. Doukas et al. (eds.), *Understanding Risks and Uncertainties in Energy and Climate Policy*, https://doi.org/10.1007/978-3-030-03152-7_1

1 Introduction

Many modelling frameworks have been developed to provide an understanding of the drivers of climate change and to assist policy formation (Flamos 2016). When climate change emerged as a serious issue in the 1970s, there were no theoretical tools that could provide a more integrated understanding of the phenomenon or provide richer insights into policy response. Models of physical dimensions of the climate system (mostly ecosystem models) were extended to consider the processes by which greenhouse gas emissions were generated and could be limited. General circulation models that dealt with atmospheric parts of the climate system were being linked to ocean models. Economists were modifying global energy-economy analysis to project greenhouse gas emissions, considering ways to reduce them and incorporating aggregated physical dimensions of the climate system. Scientists from different disciplines were linking models and analyses to provide a more integrated understanding of different facets of a highly complex interrelated phenomenon (Weyant 2009).

At a broad level, we can see the following interlinked chain of interactions. Human-induced climate change results from an increase in GHG emissions and their levels of concentration in the atmosphere. Climate science tells us how different concentration levels of GHGs may affect the temperature, precipitation, cloud formation, wind and sea level rise. These changes in turn result in various physical, environmental and social impacts like change in crop yields, water supply, species loss and migration. These impacts can then be translated into monetary terms, or processed through a model of the economy, to give a single measure of the economic cost of climate change. As these changes take place over time, models attempt to project parts or the whole dynamic process of increasing emissions, temperature changes, physical impacts and economic damages. The economy is not only affected by climate change, but it is also the perpetrator of climate change as growth in production and consumption gives rise to more GHG emissions. The most important part of the economy that determines the rate of emissions is the energy system or the forms and uses of energy. Each part of this climate-economy interaction is characterised by uncertainty (Papadelis et al. 2013) and some degree of scientific disagreement.

Various ways of climate-economy modelling can to a large extent be understood by the different ways in which they model parts of this highly interconnected process. Figure 1 below provides a depiction of climate-economy dynamics, identifying four key modules of climate-economy modelling. The climate module describes the link between GHG emission, atmospheric concentrations and the resulting variation in temperature and other climatic changes (precipitation, cloud cover, extreme weather events, climate discontinuities, etc.). The impacts module

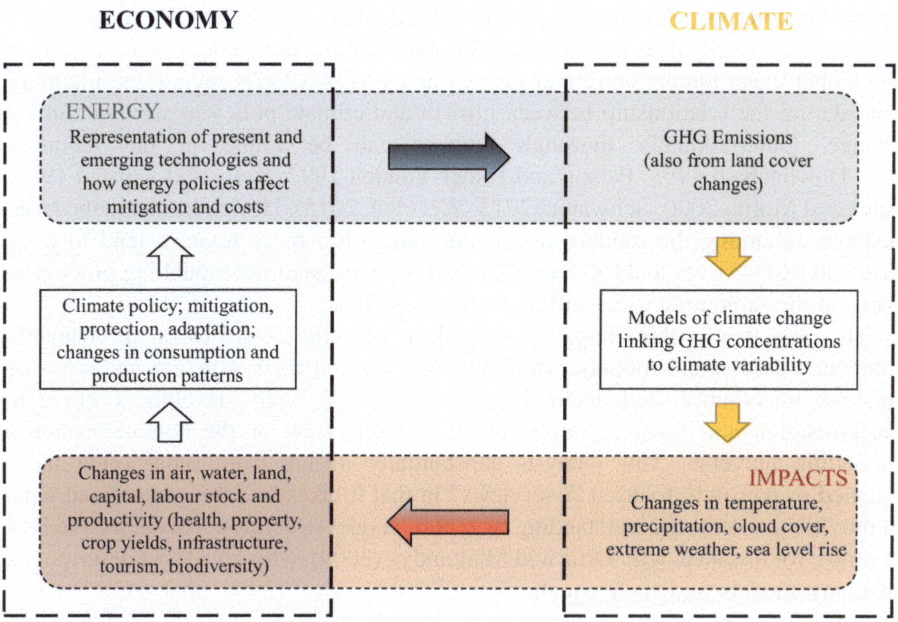

Fig. 1 Climate-economy dynamics with four modules: Economy, climate, impacts, and energy

(or damage function) expresses physical or environmental outcomes as a function of climate variables. For instance, a model might have an agricultural damage function relating variability in temperature, precipitation and cloud cover to crop yields. An economy module may describe the dynamics or growth of an economy, how emissions vary with growth and climate policies and how climate-induced physical and environmental changes might affect parts or all of an economy. The economy model is often augmented with a more detailed energy module that describes the factors determining the uses of different sources of energy and the cost of emission reductions.

The great variety of climate-economy models reflect in part the range of underlying scientific disciplines influencing their development, alternative methodologies and assumptions, as well as the different questions or issues they address. The large and growing number of models and their relative complexity can make it bewildering to distinguish them or understand their unique attributes. There already exist many good reviews of different categories of integrated assessment or climate-economy models in the literature: Füssel (2009) provides general reviews; a special issue of *The Energy Journal* provides more detailed and technical comparisons of IAMs (Weyant 1999); Tol and Fankhauser (1998) and Yohe (1999) review the modelling and representation of impacts in IAMs; Hitz and Smith (2004) review the way different IAMs deal with the global impacts of climate change as a function of the global mean temperature; and Lecocq and Shalizi (2007) review the literature considering the relationship between growth and climate policy as well as climate change. Other equally thorough reviews can be found in the literature (e.g. Dowlatabadi 1995; Parson and Fisher-Vanden 1997; Kelly and Kolstad 1999; Rana and Morita 2000; Schwanitz 2013; Wei et al. 2015). However, due to the large differences among the models and model categories, these reviews tend to have different perspectives and focus on very specific aspects of the modelling processes, while at the same time using different categorisations.

The objective of this chapter is to look at key characteristics of the available integrated assessment models, while also focusing on their structure and ways of treating uncertainty and technology, in order to help develop a concrete categorisation and form a simple and useful overview of the climate-economy modelling universe. This analysis substantially differs from other often more detailed or narrowly focused "overviews" in that it takes less for granted and aims at providing an initial understanding of generic model structures. This objective is in contrast, for instance, with Ortiz and Markandya (2009), who give short descriptions of many models and their equations, or Stanton et al. (2009) that focus on key assumptions affecting model outcomes or Füssel (2010) that focuses on how adaptation to climate change is incorporated in models. In this respect, this simple and brief overview is meant to complement these other less generic discussions and act as an initial guide to this vast terrain. Section 2 provides an overview of the six classes in which we categorise the existing integrated assessment modelling frameworks. Sections 3–7 present the unique features of each one of the model categories, as well as key characteristics of a large number of representative IAMs. Finally, Sect. 9 concludes the analysis and discusses some key remarks.

It should be noted that there exist in the literature different criteria for considering a modelling framework as an IAM. According to most reviews and from a strict point of view, only models with a close loop between economy and environment effects can be classified as IAMs; thus, most partial equilibrium models cannot be considered alone as IAMs, but can certainly be used as part of an IAM modelling suite. In this research, all models that include separate modules for climate, economy and energy are considered to be IAMs. Exceptions include certain energy system models (Sect. 6) that may not explicitly include a climate module but may rather abstract from climate by including emissions (without climate change or damages), which have also been included in other reviews (e.g. Stanton et al. 2009).

2 Classifying Climate-Economy Models

There exist a large number of various classifications in the literature, which do not fully align with each other or with the one presented in this chapter (e.g. Füssel 2010; Stanton et al. 2009; Ortiz and Markandya 2009; Söderholm 2007). In particular, Füssel (2010) following an older tradition divides them according to the kind of decision analytical frameworks to which they are applied, Stanton et al. (2009) divide them according to model structures and Ortiz and Markandya (2009) classify integrated assessment models by whether all four modules (climate, impacts, economy, energy) are used and how they are combined.

Drawing on these classifications and a detailed literature review of applications, six general modelling structures or approaches are presented. These are distinguished primarily by how the economy is modelled and the way the other three modules (climate, impacts, energy) are integrated. Of course, the nature of these models slightly hinders their consistent classification, given that certain IAMs may inevitably be sorted into more than one class. The six model classes that are presented in the following sections are briefly introduced below:

1. **Optimal growth (or welfare optimisation)** IAMs represent the economy as a single all-encompassing sector. They are designed to determine the climate policy and investment levels that maximise welfare (future against present consumption) over time, by identifying the emission abatement levels for each time step. They tend to be fairly simple, highly aggregated and transparent models that capture the trajectory of an economy and its interaction with climate in a fully integrated fashion, meaning that all modules are represented and endogenously determined.
2. **General equilibrium (or usually referred to as computable general equilibrium—CGE)** models have a more detailed representation of the economy with multiple sectors and often include higher resolution of energy technologies and regional detail. Rather than seeking optimal policies, they consider the impacts of specific policies on economic, social and environmental parameters. The richer representation of the economy comes at a cost in that the growth of the economy is harder to model and its structure more complex.
3. **Partial equilibrium models** provide a detailed analysis of the interaction between environmental impacts and a particular sector of the economy. These are usually used to assess potential climate-induced damages to a specific sector of the economy and are often linked to computable general equilibrium models.

    (a) **Energy system models** can be considered as a subcategory of partial equilibrium models that provide a detailed account of the energy sector, i.e. energy technologies and their associated costs. These are used, inter alia, to determine the least-cost ways of attaining GHG emission reductions or the costs of alternative climate policies. They are often linked with computable general equilibrium or macroeconometric models in order to add the desired level of insight to top-down approaches.

4. **Macroeconometric models**, like computable general equilibrium models, can be quite detailed in terms of energy technologies and geographic scope and are also used to evaluate alternative climate policies, but they differ in that they do not assume that consumers and producers behave optimally or that markets clear and reach equilibrium in the short term. Instead, they use historical data and econometrically estimated parameters and relations to dynamically and more realistically simulate the behaviour of the economy.
5. **Other integrated assessment models** refer to models that may have little in common except that they do not fit neatly into any of the previous well-known groups. A key departure is that they model the economy in a highly "reduced form" or simply use exogenous growth scenarios (no model at all). Although they significantly differ from one another, they all tend to be more policy-oriented than models of the other five classes.

Table 1 provides an overview of characteristics of the different approaches. The first column labels the overall approach, while the remaining columns describe how each approach varies in the way the four different modules are modelled. The table, acting as a reference point and organising principle, also aligns with the descriptions of the six approaches presented in the remainder of the paper so that the different elements in the boxes can be further explained.

The classification scheme presented in Table 1 is not meant to be exhaustive or comprehensive, and models will often not fit neatly into one of these approaches, while many combine elements of different categories. For instance, Füssel (2010) introduces a separate category referred to as "policy guidance models" and represented by ICLIPS (Toth 2005), which integrates the first four approaches into one model. Furthermore, combining models of different categories towards adding the desired level of detail is not uncommon in the climate policy literature; for example, CGE and macroeconometric models are often combined with energy system models. The connection between the different model categories is an important aspect in the modelling literature, given that certain models focus on specific sectors often neglecting the impacts on the other sectors. At the same time, IAMs that cover the energy, economy and climate are criticised for sacrificing the necessary model granularity for the sake of simplicity. From a modelling perspective, this linkage is complex and understudied in the literature (e.g. Karkatsoulis et al. 2017). Table 2 attempts to provide an overview of the six categories of integrated assessment models with some of their most prominent modelling frameworks, along with a short description and set of indicative applications. Sixty-one modelling frameworks have been reviewed and assessed for the purpose of this study.

Table 1 Key characteristics and classification of climate-economy models

Category	Economy	Impacts	Energy	Climate
Optimal growth models	Neoclassical growth Highly aggregated Tracks long-term trajectory of economy (dynamic) Single production function Single representative agent Policy optimisation Global	Highly aggregated monetary damage function that translates temperature change to loss of GDP All damages in monetary terms	Aggregated energy sector (top-down)	Reduced-form equations linking emissions to temperature
General equilibrium models	Multisectoral (multiple production functions) Single representative agent Optimising behaviour of producers and consumers Global, regional, national, local More difficult to incorporate dynamics	Shocks introduced into production functions, e.g. reduction in crop yields reflected in agricultural production function leading to a shift in supply Shocks can be based on expert judgements or drawn from biophysical or statistical models of impacts	Often detailed description of energy system (multiple energy sources explicitly modelled) Often linked to bottom-up energy system models	Climate scenarios are exogenous and used to drive climate variables and impacts
Partial equilibrium models	Detailed modelling of a single sector, e.g. agriculture Static and dynamic Often combined or linked with other top-down models	(a) Detailed biophysical model of impacts to specific sector relating key climate variables to impacts (b) Statistical analysis leading directly to monetary value of impacts	Little or no representation of energy sector	Exogenous climate scenarios used to drive impacts

(continued)

Table 1 (continued)

Category	Economy	Impacts	Energy	Climate
Energy system models	Can be considered as a subcategory of partial equilibrium models (with some instances including a macro feedback) Often linked to top-down models (making them "hybrid")	Used to assess costs of reducing emissions; no need to represent impacts	Detailed representation of alternative energy technologies and mitigation opportunities and policies	Focus on emissions rather than climate change
Macroeconometric models	Input-output econometric (multisectoral) Macroeconomy components Simulation: agents not assumed to feature optimising behaviour "Keynesian" Dynamic	Used primarily to evaluate mitigation and adaptation policies rather than evaluate damages arising from climate change	Detailed description of energy system	Exogenous scenarios
Other models	Various simple representations of economy or exogenous scenarios of economic growth	Variety of detailed representations of impacts and damages both physical and monetary	Various aggregated and detailed energy models	Reduced-form equations linking emissions to temperature and other variables

Table 2 Classification of the 61 reviewed models, with a short description and a set of indicative applications

Category	Model	Introduced by	Short description	Indicative applications
Optimal growth models	AD-FAIR	Hof et al. (2009)	A combination of FAIR (analysis of environmental costs of mitigation) and AD-RICE (adaptation cost)	Hof et al. (2012)
	AIM/Dynamic Global	Masui et al. (2006a)	A global dynamic optimisation model for assessing CO_2 reductions and economic impacts by considering energy saving investments for Japan	Xu and Masui (2009)
	AIM/Enduse	Kainuma et al. (2011)	An optimisation model for Japan, designed to assess energy saving investments	Kainuma et al. (2003)
	CETA-M	Peck and Teisberg (1993)	A growth model representing world economic growth, energy consumption, energy technology choice, global warming and global warming costs	Peck and Teisberg (1995, 1999)
	DEMETER-1 and DEMETER-1CCS	Gerlagh (2006)	A growth model with learning-by-doing for fossil fuels and non-carbon energy	Gerlagh (2007)
	DICE	Nordhaus and Yang (1996)	A modified Ramsey-style optimal economic growth model	Nordhaus (2010)
	DICE-2007	Nordhaus (2008)	An extension of DICE that comprises a modified Ramsey-style optimal economic growth model	Ackerman et al. (2010)
	ENTICE	Popp (2004)	An extension of DICE, including endogenous technological change	Popp (2006a)
	FAIR 2.1	Den Elzen (2005)	A top-down model comprising a socio-economic scenario module, an emission pathway module, a climate model, an abatement costs module, a damage module and a macroeconomic growth module	Hof et al. (2008)

(continued)

Table 2 (continued)

Category	Model	Introduced by	Short description	Indicative applications
	FEEM-RICE	Buonanno et al. (2003)	A global climate-economy model; extension of RICE and DICE that includes endogenous technological change	Buchner and Carraro (2005)
	GRAPE	Kurosawa et al. (1999)	A growth model consisting of five modules: energy, climate, land use, macroeconomics and environmental impacts	Kurosawa (2004, 2006)
	MERGE	Manne and Richels (2005)	A climate market-oriented, fully integrated assessment model	Kypreos (2007, 2008)
	MIND	Edenhofer et al. (2005b)	An endogenous growth model focusing on the energy sector	Edenhofer et al. (2005a)
	RICE	Nordhaus (1994)	A multiregional extension of DICE	Rosendahl (2004), Schultz and Kasting (1997)
	RICE-99	Nordhaus and Boyer (2000)	A Ramsey-Koopmans single-sector optimal growth model suitably extended to incorporate the interactions between economic activities and climate	Bosetti et al. (2005)
	WITCH	Bosetti et al. (2007)	A top-down Ramsey-type neoclassical optimal growth model with an energy input specification operating as a bottom-up model	Bosetti et al. (2009)
General equilibrium models	AIM	Kainuma et al. (1999)	A global CGE model with recursive dynamics that estimates GHG emissions absorption in the Asia-Pacific region and its impact on environment, society and economy	Dai et al. (2011), Fujino et al. (2006)
	AIM/Material	Masui et al. (2003)	A country-based CGE model with recursive dynamics	Masui (2005)
	Dynamic GTAP	Walmsley et al. (2006a)	A dynamic CGE model	Golub et al. (2009)

	G-CUBED	McKibbin and Wilcoxen (1999)	A multi-country, multi-sector, intertemporal model applied in the study of GHG policy, trade liberalisation, tax and microeconomic policy	McKibbin et al. (2004)
	GEM-E3	Van Regemorter (2005)	A static CGE model evaluating energy, climate and environmental policies.	Nilsson (1999)
	GREEN	Burniaux et al. (1992)	A dynamic CGE model studying the economic effects of policies aiming to reduce CO_2 emissions in Europe	Nicoletti and Oliveira-Martins (1993)
	GTAP-E	Burniaux and Truong (2002)	A static CGE model	Kremers et al. (2000)
	GTEM	Pant (2007)	A dynamic model of the world economy developed to address global change issues, including those related to climate change policy	Jakeman and Fisher (2006)
	ICES	Bosello et al. (2009)	A recursive dynamic, multiregional CGE model of the world economy, extension of GTAP-E	Bosello et al. (2010), Parrado and De Dian (2014)
	IGEM	Goettle et al. (2007)	A dynamic model of the US economy describing growth due to capital accumulation, technical change and population change	Goettle and Fawcett (2009)
	IMACLIM-R	Crassous et al. (2006)	A multi-sector multi-region recursive CGE model projecting the world economy on a yearly basis	Crassous et al. (2006)
	LINKAGE	Van der Mensbrugghe (2005)	A global/multi-region, multi-sector, dynamic applied general equilibrium model	Laborde et al. (2016)
	MEMO	Bukowski and Kowal (2010)	A large-scale, multi-sector dynamic stochastic general equilibrium model for Poland	

(continued)

Table 2 (continued)

Category	Model	Introduced by	Short description	Indicative applications
	MIRAGE	Bchir et al. (2002)	A multi-region, multi-sector CGE model devoted to trade policy analysis	Zaki (2011)
	MIT EPPA	Paltsev et al. (2005)	A recursive-dynamic multiregional CGE model of the world economy that is built on the GTAP dataset and additional data for urban GHGs	Viguier et al. (2003)
	MS-MRT	Bernstein et al. (1999b)	A multi-sector, multi-region CGE model that has been used to analyse the global impacts of the Kyoto Protocol with emphasis on the international trade aspects of climate policy	Bernstein et al. (1999a)
	SGM 2004	Edmonds et al. (2004)	A CGE model analysing issues related to energy, economy and GHG emissions	Schumacher and Sands (2006)
	WIAGEM	Kemfert (2001)	An economic approach focusing on the international energy market that integrates climate interrelations via temperature changes and sea level variations	Kemfert (2005), Kemfert et al. (2006), Kemfert and Truong (2007)
	WORLDSCAN	Lejour et al. (2006)	A recursive dynamic CGE for long-term issues in international economics	Bollen and Gielen (1999), Bollen (2015)
Partial equilibrium models	GIM	Mendelsohn et al. (2000)	An adaptation partial equilibrium model, predicting the market impacts across climatic scenarios.	Mendelsohn and Williams (2004)
	MiniCAM/GCAM	Edmonds and Reiley (1985)	An IAM of moderate complexity focused on the power and agriculture sectors	Scott et al. (1999)
	TIAM-ECN	Keppo and van der Zwaan (2012)	A version of the TIAM model broadly used for studying energy technology and climate policy scenarios, comprising a global bottom-up energy system model	van der Zwaan et al. (2013)

Energy system models	DNE21+	Sano et al. (2005)	A detailed energy module, rich in terms of energy sources and technologies	Rout et al. (2008), Oda et al. (2009), Wada et al. (2012)
	Calliope	Pfenninger (2015)	A multi-scale energy systems (MUSES) modelling framework, for developing energy system models	Redondo and van Vliet (2015)
	EFOM	Finon (1976)	An energy system model for France	Van der Voort (1982)
	ERIS	Turton and Barreto (2004a)	An energy system model, also covering several non-electric sectors (transportation and thermal needs, and corresponding technologies)	Barreto and Kypreos (2004), Turton and Barreto (2004b)
	GENIE	Mattsson and Wene (1997)	An energy system model that focuses on the implementation of new technologies (primarily photovoltaics and fuel cells) globally	Mattson (2002)
	GET-LFL	Hedenus et al. (2006)	A cost minimisation model, designed to compare the effect of introducing induced technological change in an energy system	Edenhofer et al. (2006)
	MARKAL/TIMES	Fishbone and Abiock (1981)	A model that extensively covers energy production and consumption technologies, as well as linkages to other economic sectors (through exogenous specification of useful energy demand)	Rafaj and Kypreos (2007), Seebregts et al. (2000)
	MEDEE 2	Lapillonne (1978)	A bottom-up demand forecasting model that enables the assessment of the impact of energy efficiency policies at the national level	Lapillonne (1980)
	MESSAGE	Messner (1997)	A cost minimisation, dynamic, linear programming model of the overall energy system	Hainoun et al. (2010), Sullivan et al. (2013)
	NEMS	Gabriel et al. (2001)	A large-scale energy-economy model that computes equilibrium fuel prices and quantities in the US energy sector	Yu (2008)

(continued)

Table 2 (continued)

Category	Model	Introduced by	Short description	Indicative applications
	POLES	Criqui et al. (1998)	A world energy-economy partial equilibrium simulation model of the energy sector, modelling from upstream production to final user demand and greenhouse gas emissions	Criqui et al. (1999), Kouvaritakis et al. (2000a, b), Kitous et al. (2010)
	PRIMES	Capros et al. (1998)	An energy model focusing on market mechanisms aimed at explicitly projecting prices influencing the evolution of energy demand and supply as well as technology progress	Capros et al. (2007, 2016)
	WEM	International Energy Agency (2010)	A large-scale mathematical model designed by IEA to replicate how energy markets function.	Kesicki and Yanagisawa (2015)
Macroeconometric models	E3ME	Barker and Zagame (1995)	A non-CGE energy-environment economy model of the world's economic and energy systems and the environment	Barker (1998, 1999), Barker and Rosendahl (2000), Ščasný et al. (2009)
	E3MG	Barker et al. (2006)	A large-scale macroeconomic simulation model of the global economy with detailed, integrated treatments of energy demand and the consequent atmospheric emissions. It follows E3ME's approach except that at the global level various markets are closed, allowing for imbalances	Barker et al. (2008), Barker and Scrieciu (2010), Dagoumas and Barker (2010)
	MDM-E3	Junankar et al. (2007)	A multisectoral dynamic model of the UK for assessing energy-economy-environment issues and other policies	Ekins and Etheridge (2006)
	Oxford Global Macroeconomic and Energy Model	Cooper et al. (1999)	A macroeconometric model designed for assessing the impact of policies towards controlling carbon emissions	Barker and Ekins (2001)

Other models	CIAS	Warren et al. (2008)	A multi-institutional modular IAM for modelling climate change	Warren et al. (2012)
	FUND	Tol (1997)	A non-CGE policy optimisation model that supports policy makers in understanding what an optimal policy looks like, rather than evaluating the consequences of proposed policies	Link and Tol (2004), Ackerman and Munitz (2012)
	ICAM-3	Dowlatabadi (1998)	A simulation model designed to assess the mitigation cost regarding technical change	Dowlatabadi (2000)
	IGSM2	Sokolov et al. (2005)	An economic model for analysis of GHG and aerosol precursor emissions and mitigation proposals	Reilly et al. (2006)
	IMAGE 2.4	Bouwman et al. (2006)	A policy evaluation model	Stehfest et al. (2009)
	PAGE2002	Hope (2006)	A version of the PAGE model that incorporates the five IPCC reasons for concern	Hope (2008, 2009)
	PAGE09	Hope (2011)	An updated version of PAGE2002 that takes into account IPCC 4th Assessment Report	Hope (2013)

3 Optimal Growth Models

Optimal growth or welfare optimisation IAMs tend to be more transparent because they are relatively simple, compared to models of other categories with more complex structures. They have solid microeconomic foundations and focus explicitly on the development of the economy over time. Social welfare is often defined as the utility of a representative agent, and the overall objective is to maximise aggregated welfare over time. In neoclassical economic growth models, economies make investments in capital, education and technologies. These enhance future consumption by sacrificing some present consumption, the objective being to find the right balance between present consumption and investment in future consumption so as to maximise overall welfare. IAMs of this class extend the neoclassical growth models by including the "natural capital" of the climate system as an additional kind of capital (Nordhaus 2014). Increased emissions of GHGs effectively deplete natural capital, while abatement investments augment it. In addition to standard investments, natural capital used today enhances present consumption versus expending resources in order to protect the climate system, or to avoid damage from climate change, for future welfare. In terms of climate policy, these models compare alternative paths of emissions over time (abatement) in order to find the policy that maximises overall social welfare.

Table 3 presents a set of key models falling under this class, along with information regarding the perspective of the model, the number of regions and forecasting period it can cover and the damage function by means of which the damages are translated into monetary terms. The model perspective describes the overall approach of the modelling framework: a top-down approach looks at the system under examination as a whole and uses reduced form behavioural relationships with econometrical validation, while bottom-up approaches are developed from an engineering perspective and start from the sector of interest in detail before expanding the focus onto the whole system. Other settings are more flexible and can be described as being developed from a hybrid perspective, i.e. combine different levels of detail for specific sectors or the system—a particular class of hybrid models are economic engineering models, which combine microeconomic foundations of behaviour with explicit engineering and technology details (see, e.g. Sect. 6).

Most of the modelling frameworks in this category are top-down approaches, with the exception of AIM/Enduse, which can also be considered as a non-integrated assessment model, since no economic module is included. CETA-M, WITCH and MERGE feature a hybrid model perspective, while DEMETER-1(CCS) is a classic top-down model incorporating insights from the bottom-up literature regarding learning-by-doing effects (Ortiz and Markandya 2009).

The DICE (dynamic integrated climate economy) global model (Nordhaus and Yang 1996) is selected in this study as a representative model of this category. In DICE, countries are aggregated into a single level of output, capital stock, technology and emissions (in a regional setting, RICE is a multi-region version of DICE). The social welfare function represents the world's well-defined set of preferences

Table 3 Overview of optimal growth integrated assessment models

Model	Model perspective	Regions	Forecasting period	Damages: Function	Damages: Source
AD-FAIR	Top-down	Regional (17)	2010–2100	$\frac{GDr,t}{Yr,t} = a1, r\Delta Tt + a2, r\Delta Tt^{a3,r}$ where $a1, a2, a3$ are parameters, t is period, r is region, ΔT is temperature change, Y is GDP	Hof (2010)
AIM/ Dynamic Global	Top-down	Global (6)	1995–2100	Sectoral models	
AIM/ Enduse	Bottom-up	National (Japan)	2005–2050	No economic module incorporated	
CETA-M	Hybrid	Global (OECD, ROW)	2000–2150	$Ct = aLtT^{\lambda}$, T is temp rice above pre-industrial age, λ is parameter, t is period, C is annual warming cost, a is scaling constant	Peck and Teisberg (1993)
DEMETER-1 and DEMETER-1CCS	Top-down	Global	150 years	No damage treatment	Ortiz and Markandya (2009)
DICE	Top-down	Global	1985–2105	$R = \frac{1}{1 + 0.00284 \cdot T^2}$ where T is global temperature	Ackerman and Stanton (2012)
DICE-2007	Top-down	Regional (17)	Until 2200	$\frac{D(t)}{Y} = 1 - \frac{1}{1 + \eta \cdot T^B}$ where $\frac{D(t)}{Y}$ is fractional reduction of production, $B = 2$, $\eta = \eta$ non-catastrophic $+ \eta$ catastrophic $= 28\%$ T is global warming relative to pre-industrial	Kopp and Mignone (2012)

(continued)

Table 3 (continued)

Model	Model perspective	Regions	Forecasting period	Damages	
				Function	Source
ENTICE	Top-down	Global (6)	2000–2300	$\Omega(t) = \frac{1}{1 + a1 + Tt + a2Tt^2}$ where t is period, T is global temperature, $a1, a2$ are parameters	Popp (2006b)
FAIR2.1	Top-down	National (Japan)	2005–2250	$Yt = \mathrm{A} \cdot Kt^a \cdot Lt^{(1-a)}, Kt + 1 = Kt - \eta Kt + It$ where Y : GDP, K : capital, L : labor, a : capital elasticity of production, I : investment, η : depreciation, t : period	Hof et al. (2008)
FEEM-RICE	Top-down	Global (OECD, ROW)	Until 2200	$\Omega(n,t) = \frac{1 - b1, n \cdot \mu(n,t)^{b2,n}}{1 + \theta 1 \frac{1}{\exp(\mathrm{SAD}(n,t))} \cdot \frac{T(t)^{\theta_2}}{2.5}}$ where T is global temp, $\theta 1, \theta 2, b1, b2, n$ are Nordhaus coefficients, n is region, t is time, SAD is stock of capital devoted to adaptation, μ is abatement cost	Bosello (2010)
GRAPE	Top-down	Global	2000–2100	$R = \frac{1}{1 + 0.00284 \cdot T^2}$ where T is global temperature	Koji et al. (2009)
MERGE	Hybrid	Global	2000–2150	Market Damages : $Dt, r = d1, r \cdot \Delta ATt, r^{d2,n} \cdot \mathrm{GDP}t, r$ where ΔAT is variation of temperature, t is period, r is region Non-market damages : $\mathrm{WTP}t, r = \frac{d3, r \cdot \Delta ATt, r^{d4,r}}{1 + 100\exp\left(\frac{-0.23\mathrm{GDP}t, r}{\mathrm{Pop}t, r}\right)}$ where Pop is capital, and $d1, d2, d3, d4$ are damage parameters	Manne et al. (1995)

MIND	Top-down	Global	1995–2300	No damage treatment	Ortiz and Markandya (2009)
RICE-1994	Top-down	Global	1990–2200	$\Omega(t) = \frac{1 - b1, i \cdot \mu i(t)^{b2}}{1 + \theta 1 \cdot iT(t)^{\theta 2}}$ where T is temperature, t is period, i is region, μ is rate of emissions reduction, $b1, b2$ are parameters of emissions reduction cost $\theta 1, \theta 2$ are parameters of climate damage function	Nordhaus and Yang (1996)
RICE-99	Top-down	Regional (7)	1995–2105	$\Omega(t) = \frac{1 - b1 \cdot \mu(t)^2}{1 + D(\mathrm{t})}, D(t) = \theta 1 T(t) + \theta 2 T(t)^2$ T is global temp, μ is control rate, $\theta 1, \theta 2$ parameters of damage function b is parameter of emissions reduction cost function, $t = 1990, 2000, \ldots$	Nordhaus (2002)
WITCH	Hybrid	Regional (10)	2000–2150	$\Omega t = \frac{1}{1 + \theta 1, n \cdot Tt + \theta 2, n \cdot Tt^2}$ where T is temperature, t is period, n is region, $\theta 1, n, \theta 2, n$ are damage function parameters	Bosetti et al. (2007)

and accordingly ranks different consumption paths. Welfare is increasing in per capita consumption for each generation but with diminishing marginal utility of consumption. The more you consume (or the wealthier you are), the less valuable an additional unit of consumption is. If we expect future generations to be wealthier, then additional consumption for them is less valuable than it is for us. Two central normative parameters determine the relative importance given to different generations. The pure rate of time preferences is a subjective weighting of consumption at different times, and a positive value means that immediate consumption is valued more than future consumption. Higher values increase the bias towards present consumption. The elasticity of the marginal utility of consumption is a measure of how many additional units of consumption fall in value in terms of utility. For a given growth in consumption over time, higher elasticity means that an additional unit of consumption is more valuable today than in the future.

The overall savings rate for physical capital and the rate of control of emissions of greenhouse gases are the two key decision variables for the economy. A single commodity can be either consumed or invested. Consumption viewed broadly includes food and shelter as well as environmental amenities and services. The production of output is represented by a Cobb-Douglas production function in capital, labour and energy. Energy can come from carbon-based fuels or non-carbon-based technologies. Technological change can come from economy-wide technological change or carbon-saving technological change.

The key feature that turns the neoclassical growth model into a fully integrated assessment model is the linking of certain geophysical relationships affecting climate change to the economy: the carbon cycle, a radiative forcing equation, climate change equations and a climate-damage relationship. In the DICE-2007 model, industrial CO_2 emissions constitute the only GHG that can be controlled and vary with total output, a time-varying emissions-output ratio and a rate of control of emissions. The cost of tougher climate policies will be reflected in reductions in output. A radiative forcing equation calculates the impact of GHG accumulation on the radiation balance of the globe. The climate equations that draw from general circulation models calculate the mean surface temperature of the globe and the average temperature of the deep oceans. The climate-damage relationship translates climate change into economic damages by drawing on estimates of economic impacts from other work.

Regarding technology, all optimal growth models among those reviewed assume exogenous (induced) technological change (ITC), while most also incorporate parameters that are endogenously approached (endogenous technological change—ETC); the only exception appears to be the WITCH model, in which technology is exclusively changed endogenously (Table 4). In contrast to DICE, in the latest DICE-2007 model (Nordhaus 2008), both forms of technological change are exogenous, which can be perceived as a serious limitation, especially as changes in carbon prices would be expected to induce carbon-saving technological change.

As suggested in Table 4, uncertainty in optimal growth models is usually treated deterministically, in their original design; only DICE applications have treated uncertainty in a probabilistic manner, by means of Monte Carlo analysis

Table 4 Technological change and uncertainty treatment in optimal growth models

Model	Technological change		Uncertainty treatment		Uncertainty factors
	Endogenous	Exogenous	Deterministic	Stochastic	
AD-FAIR		✓	Sensitivity analysis		Future damage costs, adaptation costs (Hof et al. 2009, 2012)
AIM/Dynamic Global	✓	✓	Scenario analysis; sensitivity analysis		Capital depreciation rate (Xu and Masui 2009), energy efficiency improvements (Masui et al. 2006a)
AIM/Enduse		✓	Scenario analysis; sensitivity analysis		Demographic trends, economic growth rate, industrial structure and goods shipments, required office space, transportation (Kainuma et al. 2000), availability, performance and cost of new technologies (Akashi et al. 2012)
CETA-M		✓	Scenario analysis; sensitivity analysis		Autonomous energy efficiency improvement rate, fossil fuel resources, electric and non-electric backstop costs, warming per emissions, adaptation costs, exogenous emissions, carbon cycle, labour growth rate, discount rate (Peck and Teisberg 1993)
DEMETER-1 and DEMETER-1CCS	✓	✓	Scenario analysis; sensitivity analysis		CCS costs (Gerlagh and van der Zwaan 2006)
DICE	✓	✓	Sensitivity analysis	Monte Carlo analysis	All model parameters (e.g. rate of capital depreciation, elasticity of output, rate of CO_2 transfer, pure rate of social time preference, etc.) (Nordhaus and Yang 1996); climate sensitivity (Ackerman et al. 2010)
DICE-2007		✓		Monte Carlo analysis	Total factor of productivity, rate of decarburisation, population growth, cost of backstop technology, damage-output coefficient, atmospheric retention fraction of carbon dioxide,

(continued)

Table 4 (continued)

Model	Technological change		Uncertainty treatment		Uncertainty factors
	Endogenous	Exogenous	Deterministic	Stochastic	
					temperature sensitivity coefficient, and total availability of fossil fuels (Nordhaus 2007)
ENTICE	✓	✓	Scenario analysis; sensitivity analysis		Opportunity cost of R&D, deviation between the private and social rates of return to R&D, decay rate of knowledge, return to energy R&D, elasticity of R&D, exogenous reduction of carbon intensity (Popp 2004)
FAIR2.1		✓	Scenario analysis		Climate sensitivity, abatement costs, damage costs, discount rate, time horizon (Hof et al. 2008)
FEEM-RICE	✓	✓	Sensitivity analysis		Discount rate, climate change damage (Bosello 2010)
GRAPE		✓	Scenario analysis		CCS (carbon transport and storage) costs (Kurosawa 2004)
MERGE		✓	Scenario analysis (2005 version); sensitivity analysis	Monte Carlo analysis (2008 version); sensitivity analysis	Global efficiency price of carbon, abatement strategy (Manne and Richels 2005); climate sensitivity, carbon value (Kypreos 2008)
MIND	✓	✓	Sensitivity analysis		Resource costs, labour R&D, energy R&D, learning rate, resource extraction learning rate (Edenhofer et al. 2005b)
RICE-1994		✓	Sensitivity analysis		All model parameters (e.g. rate of capital depreciation, elasticity of output, rate of CO_2 transfer, pure rate of social time preference, etc.) (Nordhaus and Yang 1996)
RICE-99		✓	Scenario analysis		Population growth, total factor productivity, energy efficiency, changes in land use, climate sensitivity, discount rate (Von Below and Persson 2008)
WITCH	✓		Scenario analysis (2005 version)	Monte Carlo analysis (2008 version)	Abatement costs (Bosetti et al. 2008)

(e.g. Nordhaus 2007; Ackerman et al. 2010). Furthermore, the MERGE and WITCH integrated assessment models were upgraded in their 2008 versions so as to include Monte Carlo analysis in treating uncertainty.

4 Computable General Equilibrium Models

This section draws heavily on Wing's lucid presentation of the basic structure of all computable general equilibrium models used for economy-environment interactions (Wing 2011). CGE models are an algebraic representation of the intricate functioning of a market economy and are based on the core abstract theoretical foundation of how a decentralised price system works. Demand and supply for goods arise from consumers maximising utility and producers maximising profits, with prices bringing about equilibrium in the markets. What makes them "computable" models is the use of economic data to derive numerical parameters that will simulate the real world when solved for equilibrium prices, demand and supply levels of goods. National Accounts for a specific year provide the requisite information on expenditures for goods and services by production sectors and households, as well as the division of factors of production across producers. Data is also required to determine the price elasticities of demands and supplies and factor substitution. This data is usually derived from other empirical work analysing how the behaviour of agents responds to price changes. With the use of National Accounts and elasticity data, the parameters of CGE equations are "calibrated" so that the equilibrium solution of the numerical model precisely reproduces the data of a real economy for a given year.

The general way that CGE models are used for policy analysis is to change one or more of the exogenous parameters of the economy and compute the new equilibrium. Comparing the new counterfactual equilibrium to the initial equilibrium vectors of prices and activity levels as well as the level of utility of the representative household provides insights about the effect of a "shock" on the economy.

By modelling the linkages between the different sectors of an economy, CGE models are able to capture not only the direct impacts of a policy on one sector of the economy but to trace its full (or general equilibrium) impact on the interdependent sectors of an economy and ultimately the change in consumption (or utility) of the representative agent, which is a measure of the welfare impact. A carbon tax will not just increase the cost of certain forms of energy but will also affect the demand and supply of other goods. This is one of the main advantages of these models relative to partial equilibrium models that focus on a single sector or other models that do not have a detailed multi-sector representation of the economy. For example, neoclassical growth models that model the economy as a single sector cannot capture these general equilibrium effects, despite focusing on a broader understanding of long-term dynamics. An overview of the reviewed CGE models can be found in Table 5.

CGE models have traditionally focused on evaluating the costs of emission reductions, alternative mitigation policies and the damages resulting from climate change. Increasing attention is also given to considering the costs and benefits of

Table 5 Overview of computable general equilibrium models

Model	Model perspective	Regions	Forecasting period
AIM	Top-down	Regional (21)	1990–2100
AIM/Material	Top-down	Japan	1995–2010
Dynamic GTAP	Top-down	Global	1997–2025
G-CUBED	Top-down	Regional (8)	2000–2100
GEM-E3	Top-down	Regional (Europe)	1996–2020
GREEN	Top-down	Global	1985–2050
GTAP-E	Top-down	Regional (5)	
GTEM	Top-down	Global	1997–2100
ICES	Top-down	Regional (14)	2001–2050
IGEM	Top-down	Regional (USA)	2000–2060
IMACLIM-R	Top-down	Regional (5)	1997–2100
LINKAGE	Top-down	Global	2004–2080
MEMO	Bottom-up	National	2010–2030
MIRAGE	Top-down	Regional	2004–2020
MIT EPPA	Top-down	Regional (16)	2000–2100
MS-MRT	Hybrid	Regional (10)	2000–2030
SGM 2004	Hybrid	Global (14)	2000–2050
WIAGEM	Top-down	Regional (25)	2000–2050
WORLDSCAN	Top-down	Regional (16)	2000–2050

adaptation. A standard exercise is to examine the effect that a carbon tax will have on an economy's output and emissions of greenhouse gases. A multi-region model with international trade could examine how carbon tax policies would perform if different countries apply different tax rates (Elliott et al. 2010). A typical way of capturing the impacts of climate change on an economy in a CGE model is to model the shocks through several possible channels. Rising temperatures can lead to changes in consumer expenditure patterns such as an increase in demand for air conditioning in the summer or a drop in demand for heating in the winter. By introducing a shock parameter into the representative agent's expenditure function, this influence can be captured. To the extent that climate change reduces the productivity of space conditioning, the shock parameter rises, leading to increases in expenditure required for a given level of space conditioning and thus ultimately negatively impacting the households' welfare. In a similar fashion, shock parameters can be introduced to account for changes in the productivity of primary factors in various industries. If climate change reduces (or increases) the yield of certain crops, then a shock parameter can be changed to capture the reduced productivity in the agricultural sector. Comparing a benchmark equilibrium where the shock parameter has a unit value with a new equilibrium resulting from changed parameter values on agricultural productivity will give a measure of the welfare loss from this impact of climate change. Shock parameters can also be introduced to capture reductions in the aggregate endowments of capital and labour such as those arising from damage to property or from increased morbidity or mortality. By appropriately incorporating a full set of possible climate change impacts through relevant shock parameters, the computable general equilibrium model is able to tally the loss of welfare to the representative agent resulting from the several productivity shocks. This method, however, does not capture impacts that do not register in markets such as loss in biodiversity or an increase in risk of morbidity and mortality.

Technological change and uncertainty treatment are presented in Table 6. In this domain, CGEs appear similar to optimal growth models, in that they mostly feature induced technological change, although GEM-E3, IGEM, IMACLIM-R and MIT EPPA also have technological aspects endogenously determined, and in that they all treat uncertainty by means of scenario or deterministic sensitivity analyses. The only exception in CGE modelling applications in the literature appears to be the MIT EPPA model, in which anthropogenic emissions of greenhouse gases and precursors were sampled with Monte Carlo analysis (Webster et al. 2002, 2003).

Since the economy in CGE models is always at an optimum, any restrictions on GHGs necessarily lead to costs or losses in output. No regrets or double dividends are possible in these models; this contrasts with some macroeconometric models discussed in Sect. 7. DeCanio (2003) carries out a detailed, fundamental critique on the underlying theoretical basis of CGE models.

5 Partial Equilibrium Models

Partial equilibrium analysis differs from general equilibrium modelling primarily by focusing on a specific market or sector and assuming that prices (or conditions) in the rest of the economy remain constant or unchanged. It is usually justified theoretically when the changes being considered affect primarily one market and are expected to have a relatively small impact on the rest of the economy. Partial equilibrium analysis is used extensively to estimate the impacts of climate change in different sectors of the economy. Although partial equilibrium analysis is unable to capture the broader implications that climate impacts or mitigation will have as sector changes reverberate through all economic sectors, it has the advantage of providing a more detailed understanding compared to a general equilibrium appraisal, which can be very useful in designing policy.

One of the early applications of partial equilibrium analysis to assess the impacts of climate change was on the agricultural sector. Two distinct ways to measure the impact of climate change on agriculture have emerged: a statistical approach and a biophysical approach. Mendelsohn et al. (1999) used the biophysical approach to estimate damages to the agricultural sector in the USA: simulation models were used to predict changes in yield from crop (damage function), and then these provided the inputs for a spatial partial equilibrium model of the US agricultural sector. Much like their general equilibrium counterparts, this shift in the parameter of a production function that would result from climate change brings about a new equilibrium, and the difference in welfare is a measure of the economic damage caused.

Table 6 Technological change and uncertainty treatment in CGE models

Model	Technological change		Uncertainty treatment		Uncertainty factors
	Endogenous	Exogenous	Deterministic	Stochastic	
AIM		✓	Scenario analysis		Population growth, economic growth, climate sensitivity (Matsuoka et al. 1995); industry production structure, product composition, technology popularising rate external policy environment aspects (Wen et al. 2014)
AIM/Material		✓	Scenario analysis		Labour productivity, non-energy material input reduction, energy efficiency improvement, energy mix (Masui et al. 2006b)
Dynamic GTAP		✓	Scenario analysis		Policy mix (Walmsley et al. 2006b); total factor productivity (Ianchovichina et al. 2001); demographic trends (Tyers and Shi 2012)
G-CUBED		✓	Scenario analysis		Emission permit allocation regimes (McKibbin and Wilcoxen 2009)
GEM-E3	✓		Scenario analysis		Policy mix (Nilsson 1999)
GREEN		✓	Scenario analysis		Market structure (Burniaux 1998)
GTAP-E		✓	Scenario analysis		Emission permit allocation regimes (Nijkamp et al. 2005)
GTEM		✓	Scenario analysis		Supply of labour, land growth, natural resources, population growth (Pant et al. 2007); emission trading regimes, economic growth, consumption growth (Tulpulé et al. 1998)
ICES		✓	Scenario analysis		Temperature increase (Bosello et al. 2009); carbon credit allocation regime (Bosello et al. 2010)

IGEM	✓	✓	Scenario analysis		Carbon offset mechanism structure (Fawcett 2011)
IMACLIM-R	✓		Scenario analysis		Economic growth, technological change, policy mix (Waisman et al. 2012)
LINKAGE		✓	Scenario analysis		Agricultural productivity growth (Zhai and Zhuang 2012)
MEMO		✓	Scenario analysis; sensitivity analysis		Policy mix, technological mix (Bukowski and Kowal 2010)
MIRAGE		✓	Scenario analysis		Market structure (Bchir et al. 2003)
MIT EPPA	✓	✓	Scenario analysis	Monte Carlo analysis; statistical analysis	Burden sharing scheme (Viguier et al. 2003); anthropogenic emissions for all GHGs and aerosol and GHG precursors (Webster et al. 2002, 2003)
MS-MRT		✓	Scenario analysis		Emissions trading system structure (Bernstein et al. 1999b)
SGM 2004		✓	Scenario analysis; sensitivity analysis		Technology readiness, carbon prices, interest rates (Schumacher and Sands 2006)
WIAGEM		✓	Scenario analysis; sensitivity analysis		Emission stabilisation levels (Kemfert and Truong 2007); radiative forcing constraints, set of GHGs, emission abatement elasticities (Kemfert et al. 2006)
WORLDSCAN		✓	Scenario analysis		Emission stabilisation levels (Bollen and Gieen 1999)

Producers' costs are a function of the outputs of the goods they produce, factor prices and exogenous environmental inputs like climate, soil quality, air quality and water quality. In the statistical approach, land is separated from other production inputs and considered to be heterogeneous with different environmental characteristics. Perfect competition for land is assumed which implies that entry and exit will drive pure profits to zero. In this case land value should be equal to the net revenue from the land, which is linked to the value of land. This "Ricardian" model allows one to measure the impact of a change in an environmental variable through the impact on the value of land. A change in an environmental factor that damages production will lead to a fall in the stream of future of land rents and in the value of land. If prices for all markets (except that of land) are assumed constant, the value of the environment change is captured by the change in aggregate land values. In order to discern the impact of climate variation (as well as mean temperature) on property values, Mendelsohn et al. (1999) used cross-sectional data on aggregate land values in different counties along with 12 climatic variables (of temperature and precipitation). They then regressed aggregate farm values on climate, soil and economic variables in order to determine the marginal impact of each climate variable. Essentially, differences in temperature and weather patterns at a point in time across various regions were used to project climate change impacts in the future.

Both biophysical and statistical partial equilibrium approaches have been used to estimate impacts of climate change across multiple sectors of an economy. With consistent assumptions across sectoral analyses, these are often added up to provide a value of total impact to an economy. This total value does not capture non-market impacts such as health and biodiversity.

The idea of drawing on the literature of biophysical models to find the physical climate impacts for different sectors and then translate these into monetary values by various methods (first order calculations, partial equilibrium models, and guesstimates) and add them up to find a total monetary value of "damages" from climate change is also known as the "enumerative approach". This is probably one of the easiest methods to grasp conceptually as the move from the physical to the monetary valuation is more transparent. Estimates of "physical effects" of climate change on specific sectors of the economy or environmental services are obtained from natural science work (climate models, impact models and laboratory experiments). Economic valuation methods are then used to place a monetary value on the physical impact, and this can either provide an estimate of damage to a specific sector like agriculture, or the values of damages to the different sectors (tourism, agriculture, forestry, biodiversity, etc.) can be added up to give an estimate of the total damages of climate change to a region. For instance, engineering estimates can be used to find the physical effect that a rise in sea level has on coastal protection and loss of land. Economic estimates of the cost of coastal protection and the value of lost land or land protection follow. For goods that are not traded in markets, like climate impacts to health and biodiversity, other economic techniques are needed to estimate monetary values. Physical loss to health could be translated into monetary terms by considering medical expenses, productivity loss or citizens' willingness to pay to avoid risk of health damages (Tol 2010).

Theory does not suggest that adding up the separate economic sectoral impacts will lead to the same result as evaluating total climate change impacts with a computable general equilibrium or macroeconomic model that incorporate all market interactions. A particularly stark example of the importance of capturing market interactions can be found in the literature on what is known as the "rebound effect" (Sorrell 2009). Improvement in energy efficiency might be expected to directly lead to reductions in energy consumption and GHG emissions, and many policies to improve energy efficiency have that aim. An analysis that looks just at the immediate impact of energy efficiency or savings policy might lead to that conclusion. However, empirical and theoretical literature show that there can often be a rebound effect such that any gains in energy efficiency are countered by increases in energy consumption. This could happen because the improved energy efficiency brings about a fall in the price of energy leading to increased use of energy or the development of new energy-using products.

The more detailed analysis and quantification of damage functions that are part of the partial equilibrium studies or enumerative approaches have also been used as inputs into general equilibrium models; for example, Jorgenson et al. (2004) use sectoral damage functions as inputs into their Intertemporal General Equilibrium Model (IGEM). This distinguished their approach from many computable general equilibrium models that did not rely on damage functions derived from more detailed empirical sectoral studies. A damage function is used to describe how unit costs or the supply of input factors changes as a result of climate variation. For instance, for the sectors of crop agriculture, forestry, energy and water, a damage function is used in order to relate the percentage change in unit production costs to changes in temperature and precipitation. The difference between the unit price of producing a given quantity before and after the impact of climate change is used in the model to reflect the change in productivity or inputs required to produce the same amount of a good. This productivity change is incorporated in the relevant sector of the CGE to model the full economic implications of the climate impacts.

The three partial equilibrium models reviewed are presented in Table 7. It is obvious that technological change differs across the three models, all of which are of global coverage. Only the GCAM model (formerly known as MiniCAM) features both endogenous and induced technological progress as well as characteristics of both bottom-up and top-down approaches (Urban et al. 2007), and, contrary to GIM and TIAM-ECN, there have been applications featuring uncertain parameters treated stochastically by means of Monte Carlo analysis (Scott et al. 1999).

One of the advantages of a sectoral approach is that it allows a much more detailed understanding of the various climate impacts and a more refined estimation of specific impacts on parts of the economy. One recent study known as PESETA (Ciscar et al. 2009) combined sectoral and computable general equilibrium models to produce a Europe-wide analysis of climate impacts for five categories: agriculture, tourism, river floods, coastal systems and health. The study pointed to the disadvantages of other regional integrated assessment studies that rely on reduced-form damage functions relating global temperature to GDP (as most wealth maximising IAMs do). Specifically, these reduced-form damage functions are often based on

Table 7 Overview of partial equilibrium models, including uncertainty treatment and technological change

Model	GIM	MiniCAM/GCAM	TIAM-ECN
Model perspective	Top-down	Hybrid	Bottom-up
Regions	Global (178)	Global (14)	Global (36)
Forecasting period	1990–2100	1990–2100	2010–2100
Technological change	Endogenous	Endogenous; exogenous	Exogenous
Uncertainty treatment	Deterministic (scenario analysis)	Deterministic (scenario analysis); stochastic (Monte Carlo analysis)	Deterministic (scenario analysis)
Uncertainty factors	Climate prediction models (Mendelsohn and Williams 2004)	Policy mix, emissions, atmospheric concentration, radiative forcing, global mean temperature and climate sensitivity, damages for the no intervention case, cost of stabilising emissions, cost of stabilising the atmosphere, cost of stabilising the climate (Scott et al. 1999)	Radiative forcing levels (van der Zwaan et al. 2013)

literature that draws on different and possibly inconsistent climate scenarios; only average temperature and precipitation are used rather than a fuller set of climate variables at an appropriate time-space resolution, resulting in estimates of impacts not having a detailed enough geographical resolution. In contrast, PESETA followed an enumerative (bottom-up) approach, which means that the impact assessment was based on much more detailed sectoral models deriving from the regions under study. In order to meaningfully add these impacts up, common climate scenarios were used at a high time-space resolution. Finally, the impacts derived from each sector were fed into a computable general equilibrium model (GEM-E3) allowing for the assessment of impacts, after market interactions had been incorporated. Another recent example of a national climate change assessment combining a sectoral approach with a top-down CGE can be found in the Garnaut Climate Change Review (Garnaut 2008).

6 Energy System Models

While much of the discussion on partial equilibrium models (Sect. 5) focused on ways of estimating damages that climate change may cause, energy system models focus on the key sector determining GHG emissions and costs arising from emission reduction policies. Numerous models have been developed over the years to provide energy policy guidance and that have evolved into integrated assessment models or components of IAMs. The analysis of energy and environmental policy often demands a level of technological explicitness and detail that goes beyond macroeconomic models that do not differentiate technology stocks. A class of technology-oriented models known as "bottom-up" models emerged in the 1970s (following the first oil crisis) and are still being developed, for the purpose of addressing this need for detail. While these were developed for energy resource planning purposes and began with simple, single-sector accounting tools, they soon evolved into complex and dynamic optimisation and simulation frameworks for energy and climate policy appraisal at local, national or international levels (Greening and Bataille 2009). As Mundaca et al. (2010) note, these models are disaggregated representations of the energy-economy system, entailing detailed characterisations of existing and new energy technologies, and can simulate alternative technological pathways. Besides considering least-cost means of achieving emission targets, these models are employed towards identifying a number of climate-energy issues, including best technology opportunities, costs of alternative mitigation policies and the potential for greater energy efficiency.

Energy system models can broadly be classified as optimisation models or simulation models. Optimisation models use information on costs and constraints of technology characteristics to identify the "best", "least-cost" or "optimal" technology. The consumer is assumed to be rational, and energy supplies are allocated to energy demands, based on minimum lifecycle technology costs. By incorporating a constraint on emissions, an optimisation model can estimate the least costs of achieving a target. Simulation models are designed to capture technological and economic dynamics as realistically as possible. Rather than seeking to find a least-cost solution, they model the most probable responses to policy shocks. Producers and consumers may carry out production and consumption activities with different objectives in contrast to optimisation models that usually operate from the perspective of a single optimising decision-maker. There are many simulation models of various degrees of sophistication. A sequential iterative simulation process is used to find an equilibrium set of prices and demands. A policy application affects prices, and the iterations continue until a new equilibrium is found. Outcomes are very sensitive to the dynamics and technologies assumed. A simulation, for instance, of a GHG policy will lead to very different results if carbon capture and storage or other backstop technologies are included. Table 8 presents an overview of the reviewed energy system models, along with their system, geographical and time coverage, mathematical structure and perspective.

The technological explicitness and detail of energy system models allow them to consider such issues as how policies can promote technology commercialisation and diffusion, but they have been criticised for lacking microeconomic (or behavioural) realism and "macroeconomic completeness" (or feedbacks). In terms of behavioural realism, they have been criticised for being too optimistic on the profitability of attaining energy efficiency from the diffusion of low-emission or inexpensive technologies. Part of the problem is that bottom-up models focus mostly on the financial costs while not taking into account such key factors as greater risks, intangible costs and longer payback periods associated with investments in energy efficiency. For instance, two light bulbs that may appear to provide the same service in terms of lumens may differ in risk of premature failure, payback period, shape, hue of light or time it takes for a bulb to reach full intensity; similarly, public transit and single-occupancy vehicles may provide the same personal transportation service, but evidence suggests that some consumers perceive public transportation as being of lower convenience, status and comfort level. By incorporating more parameters to gauge for consumers preferences, like time preferences or perception of risks, models will be better able to explain and predict the potential uptake and diffusion of new technologies. Mundaca et al. (2010) review models that attempt to capture more of this behavioural realism for the analysis of energy efficiency policies.

Being essentially partial equilibrium models focusing on energy consumption, energy system models tend to find relatively low mitigation costs because they only consider the impact of emission reduction strategies on energy system costs usually ignoring feedback loops and interactions with other sectors of the economy; exceptions of such models including feedback loops, however, can be found in the literature (e.g. Karkatsoulis et al. 2017). For instance, these models assume that investments within the energy sector can be funded at a constant rate of interest. An ambitious climate policy, however, would lead to a depreciation of capital stocks in certain sectors and accordingly change the return on investment in the energy sector as well as a concomitant reallocation of investments across sectors. These investment dynamics are a critical determinant of macroeconomic costs missed by the partial equilibrium analysis. For the same reason, most energy system models tend to neglect the potentially significant rebound effects and crowding-out implications of investments (Edenhofer et al. 2006).

As Table 8 suggests, all energy system models are bottom-up; exceptions include hybrid models (Greening and Bataille 2009) that also feature characteristics of top-down approaches, like MESSAGE and WEM (Urban et al. 2007), and economic engineering models in particular, combining microeconomic foundations of behaviour with technology details (such as DNE21+, NEMS, POLES and PRIMES). In a broader perspective, there are numerous ways that bottom-up models have added macroeconomic components, whether these derive from optimal growth models, macroeconomic models or computable general equilibrium models. Because of the technological explicitness of bottom-up models, the top-down feedbacks have focused on direct adjustment effects on the demand for energy-using goods and services in response to changes in the cost of delivery, but do not capture the secondary macroeconomic effects like change to wages, cost of capital, exchange rates and government budgets resulting from energy price changes. When governmental energy policies are moderate in scope, they are unlikely to have substantial macroeconomic implications, but as policies become more ambitious—as would be required for a rapid reduction in GHG emissions or a big shift towards renewable energy sources—the macroeconomic consequences constitute an important factor in assessing the policy outcomes (Greening and Bataille 2009).

Just as bottom-up models have been trying to overcome their weaknesses by combining elements of top-down models, there have been many attempts by

Table 8 Overview of energy system models

Model	System coverage	Mathematical structure	Model perspective	Regions	Forecasting period
Calliope	Partial equilibrium	Optimisation	Bottom-up	National	
DNE21+	Limited macro feedback	Optimisation	Economic engineering	Regional (77)	2000–2100
EFOM	Partial equilibrium	Optimisation	Bottom-up	National	1974–2020
ERIS	Partial equilibrium	Optimisation	Bottom-up	Regional (12)	1990–2050
GENIE	Limited macro feedback	Optimisation	Bottom-up	Regional (4)	1995–2050
GET-LFL	Partial equilibrium	Optimisation	Bottom-up	Global	2000–2050
MARKAL/ TIMES	Partial equilibrium	Optimisation	Bottom-up	Regional (5)	1990–2050
MEDEE 2	Partial equilibrium	Simulation	Bottom-up	National	1990–2040
MESSAGE	Partial equilibrium	Optimisation	Hybrid	Regional (11)	2005–2100
NEMS	Limited macro feedback	Market equilibrium	Economic engineering	National (USA)	2000–2030
POLES	Partial equilibrium	Market equilibrium	Economic engineering	Regional (18)	1980–2100
PRIMES	Partial equilibrium	Market equilibrium	Economic engineering	Regional (Europe)	2005–2050
WEM	Limited macro feedback	Optimisation	Hybrid	Regional (25)	2015–2040

top-down modellers to enhance their technological explicitness by incorporating elements of bottom-up models. Technological change has generally been captured in top-down models with the use of two key parameters: elasticity of substitution (ESUB) and the autonomous energy efficiency index (AEEI). ESUBs are used to capture the degree to which a relative price change will affect the substitution between any two pairs of aggregate inputs (capital, labour, energy, materials) and between different forms of final energy. In general, the easier it is to substitute capital for energy or one form of energy for another, the lower the cost of reducing energy use or GHG emissions. AEEI gives the rate at which price-independent technological evolution improves energy productivity, and itself depends on changes in technology and capital stock turnover. A higher AEEI means that the economy becomes energy-efficient faster. ESUB and AEEI are often estimated from aggregate, historical data, but these may not be good indicators for future values under different policy regimes. A policy focus on low to zero GHG emissions may have a substantial impact on AEEI and ESUB that is not captured by looking into the past. This inadequacy of top-down models partly explains the push towards greater technological detail and making technological change endogenous. Another reason is that top-down models represent technological change as an abstract, aggregate phenomenon which may be adequate to assess economy-wide instruments like taxes and tradable permits but are unable to consider technology-focused policies. There are several different ways of incorporating elements of bottom-up models into top-down models, but there are limitations to how much technological detail can be incorporated without running into computational and other difficulties.

Technological change along with uncertainty treatment in the reviewed energy system models is presented in Table 9. It is evident that there exists a balance between endogenous and exogenous technological change among energy system models. Uncertainty treatment, however, is again treated mostly by means of deterministic approaches, i.e. scenario and sensitivity analysis, with the most prominent exception being that of the MARKAL/TIMES model, which also features Monte Carlo analysis according to Seebregts et al. (2002).

There exist a large number of general reviews of existing energy system models (e.g. Worrell et al. 2004; Jebaraj and Iniyan 2006), while Mundaca et al. (2010) provide a review of models with a specific focus on energy efficiency, also identifying a separate category of models called "accounting models".

It should be noted that energy system models can be perceived as a cross-cutting category of models based on this classification, ranging from partial equilibrium to neoclassical/optimal growth models. In essence, they are partial equilibrium models, assuming equilibrium in one particular sector, i.e. the power sector. However, this category also includes models with features from other modelling approaches and structures, as shown in Table 8. For example, GET-LFL is an energy system model that can be classified as a cost minimisation IAM (Wei et al. 2015). The same applies for DNE21+, MIND and MESSAGE (Stanton et al. 2009). Furthermore, it should be mentioned that energy system models do note solely focus on the power sector but all economic sectors that are consumers of energy. For example, they are also used in studies oriented on the transport sector (e.g. Siskos et al. 2015). The latter is responsible for around 25% of energy-related GHG emissions today and is widely acknowledged to be the most inflexible sector of the energy system with regard to deep emissions reduction in the future (e.g. Hickman et al. 2010).

7 Macroeconometric Models

Environmental policy issues around 1990 pushed the development of computable general equilibrium models like the GREEN model of OECD, while in Europe, there was a parallel development of the CGE model GEM-E3 and the input-output econometric (or macroeconometric) model E3ME, which integrated energy and emissions in the economic model. E3ME stands for energy-environment-economy (E3) multisectoral model at the European level and, along with its variants, remains one of the most prominent macroeconometric models for appraisal of climate policy and climate-economy interactions. E3MG is a similar model that focuses on the

Table 9 Technological change and uncertainty treatment in energy system models

Model	Technological change		Uncertainty treatment		Uncertainty factors
	Endogenous	Exogenous	Deterministic	Stochastic	
DNE21+	✓	✓	Scenario analysis		Costs of emission reduction potential (Akimoto et al. 2010); policy in effect, carbon tax, stabilisation levels (Akinoto et al. 2004); energy demand (Yamaji et al. 2000)
EFOM		✓	Scenario analysis		Cost minimisation strategies (Van der Voort 1982); growth of electricity demand (Plinke et al. 1990)
ERIS	✓		Scenario analysis; sensitivity analysis	Subjective probability	Technological learning (Kypreos et al. 2000); depreciation rate (Barreto and Kypraios 2004)
GENIE	✓		Scenario analysis		Technological investment profiles, mitigation policy in effect (Mattsson and Wene 1997)
GET-LFL	✓		Scenario analysis		Emissions stabilisation levels, technological change (Hedenus et al. 2006)
MARKAL/ TIMES	✓	✓	Scenario analysis; sensitivity analysis	Stochastic programming; Monte Carlo analysis (Seebregts et al. 2002)	Discount rate, fossil fuel prices, emission constraints (Seebregts et al. 2000); cost externalities (Rafaj and Kypreos 2007)
MEDEE 2		✓	Scenario analysis		Policy mix (Lapillonne 1980)
MESSAGE		✓	Scenario analysis; sensitivity analysis		Discount rate, operation time of wind turbines, overnight cost of nuclear power plants (Hainoun et al. 2010); electricity demand, GHG constraints (Sullivan et al. 2013)
NEMS		✓	Scenario analysis		Electric vehicle load profiles (Yu 2008)
POLES	✓		Scenario analysis		Climate change (Mima and Criqui 2009); emission stabilisation (Kitous et al. 2010)
PRIMES		✓	Scenario analysis		CO_2 emission reduction (Capros et al. 1999); scale of climate action, delay in climate action, technological diffusion delays (Capros et al. 2012)
WEM		✓	Scenario analysis		Policy mix (Kesicki and Yanagisawa 2015)

global level, as does the Oxford Global Macroeconomic and Energy Model, while the MDM-E3 model focuses on the UK economy (Table 10).

Macroeconometric models are "integrated" or hybrid in that they combine top-down macro models with bottom-up energy system models. The goal of integrating the two types of models is to provide a dynamic, non-linear picture of economic change in a detailed interindustry framework (West 2002). While static models can measure long-run comparative equilibrium solutions, the macroeconometric models are able to track the time path of the economy through

Table 10 Overview of macroeconometric models, including uncertainty treatment and technological change

Model	E3ME	E3MG	MDM-E3	Oxford global macroeconomic and energy model
Model perspective	Hybrid (Cambridge Econometrics 2014): top-down (energy); bottom-up (electricity supply)	Hybrid (Barker and Scrieciu 2010): top-down (interactions, feedback and spillover effects between the required investments and outcomes and the rest of the economy); bottom-up (energy technology model)	Hybrid (Barker et al. 2007): top-down; bottom-up (less detailed electricity sub-model, energy technology model)	Top-down
Regions	Global (59)	Global (20)	National	Global (22)
Forecasting period	1990–2100	1971–2100	Until 2030	2005–2020
Technological change	Endogenous	Endogenous	Exogenous	Exogenous
Uncertainty treatment	Deterministic (scenario analysis); stochastic (probabilistic analysis)	Deterministic (scenario analysis); stochastic (probabilistic analysis)	Deterministic (scenario analysis)	Deterministic (scenario analysis)
Uncertainty factors	All model parameters (Mercure et al. 2017); level of coordination in fiscal policies (Barker 1998, 1999)	Policy mix, carbon prices, new investments, emissions trading system structure (Barker et al. 2012); technology penetration levels (Dagoumas and Barker 2010)	Policy mix (Barker et al. 2007; Ekins and Etheridge 2006)	Policy mix, permit trading system structure (Cooper et al. 1999)

short-run disequilibrium adjustments. Macroeconometric models incorporate equations that trace the trajectory of national economic aggregates as well as related components of economic activity like labour, savings and consumption. These equations are estimated econometrically. Aggregate potential output is usually simulated as a function of aggregate inputs of capital and labour, and sometimes energy and materials. Transactions among economic sectors are described by input-output models. More or less aggregated sectoral demand functions are estimated by means of historical data, e.g. for energy services and food, allowing for projections of future trends in response to a carbon tax or other climate policies. The accuracy of these forecasts depends on the extent to which historical changes, e.g. technology changes induced by past price changes, are likely to be a good indicator for future changes.

Unlike CGE and optimal growth models, macroeconometric models do not assume that markets clear in the short and medium run, and demand and supply do not derive from optimising behaviour of consumers and producers. They are disequilibrium models with demand and supply approaching equilibrium in the long run. Because of the fact that they do not posit optimising behaviour on the part of agents or a "central planner", they are characterised as simulation models, representing as closely as possible the dynamics of the real world. The economy and energy system are described by a set of rules that need not lead to full equilibrium. Some macroeconometric models also allow for structural unemployment resulting from inadequate demand for labour in the long run (Hourcade et al. 1996). CGE models usually assume that there is no unemployment or that the labour market clears. The Post-Keynesian E3ME macroeconometric model estimates labour with various disaggregated equations, e.g. working hours are estimated for men and women, different ages and sectors, thus allowing the model to forecast full-time and part-time workers. Disequilibrium in the labour market or unemployment is therefore a feature of this model. For this reason, it is sometimes argued that CGE models are more suitable for describing long-run steady-state behaviour, while macroeconometric models are more suitable for forecasting short-term outcomes. The parallel development of these "very different models" has given rise to an ongoing, often-heated discussion and conflicting positions between the input-output econometric and the CGE community (e.g. Grassini 2009). Robinson (2006) provides a nice account of the historical tension between CGE and macroeconometric models and the remaining theoretical difficulties of reconciling the approaches. Kratena and Streicher (2009) on the other hand attempt to better identify the key features differentiating the two approaches and suggest that the distance between them is much smaller than usually assumed.

One of the model outcomes that has often set the E3ME model apart from other top-down models is that it can give rise to negative costs, i.e. the imposition of climate policies can actually lead to increases in employment and output. Since structural unemployment is possible in this model, a transition to a low carbon economy can potentially enhance effective demand for labour reducing the lost output. In the E3MG model where the labour market and other markets may not clear, part of the impact of induced technological change arising from climate policies is to raise growth through increased transfer of labour from traditional to modern sectors (Edenhofer et al. 2006).

As many CGE models assume the economy is always at an optimum, including full employment, any "constraints" resulting from climate policies can only result in additional costs to the economy. In general, first best models assume perfect markets and optimal policy implementation so that no-regret options are impossible. Second best models essentially allow for the possibility that climate policies can reduce market imperfections as a side benefit. This way, the costs of climate protection can be diminished or even become negative. Although imperfections or second best modelling is a fundamental feature of most macroeconometric models, since many of these draw heavily on the Keynesian tradition, different kinds of imperfections are often incorporated in GCE models and are usually implicitly assumed in most bottom-up (engineering) energy sector models (Sect. 6). Double dividends arise because climate policy redresses one imperfection (missing market or other institutions for climate protection) while also potentially reducing other market imperfections, e.g. barriers that prevent uptake of new technologies. When incorporating side benefits, for example, from reducing "distortionary" taxes with revenue from a carbon tax, in second best models, it would help to consider whether climate policies are the best way of dealing with many market imperfections and the extent to which these benefits should be attributed to climate policy per se.

8 Other Integrated Assessment Models

Optimal growth and CGE models are both based on a specific long-standing theoretical foundation so that most of these models can be understood as variations (though sometimes substantial) on a theme and comparability seems to be more straightforward. This section presents models that are hard to classify into any of the previously discussed models and delves into one well-known non-CGE model, the PAGE2002 model, as indicative of the kind of possible departures from standard neoclassical growth, CGE and macroeconometric models. These models are presented in Table 11. It should be noted that FUND should not be considered as a hybrid model, but it can run different optimisation modes, including top-down or bottom-up, cooperative or non-cooperative and with or without interregional capital transfer.

The PAGE2002 model attracted much attention recently due to it being the top-down model that the Stern team relied on for many of the much publicised aggregate climate change damages. One of the features that made it attractive to the Stern team was the model's central focus on taking account of uncertainty in many of the climate-economy parameters. PAGE was developed as a computer simulation model in 1992 for use in decision-making within the European Commission. It was explicitly designed to be comprehensive but "accessible to policy makers" with the "simplest credible functional forms" (Hope et al. 1993), so that it remains transparent and able to run fast and repeatedly using a random sample of uncertain input

Table 11 Overview of models not falling into any of the previously discussed groups

Model	Model perspective	Regions	Forecasting period	Damages	
				Damage function	Source
CIAS	Top-down	Global (20)	2000–2100	No damage treatment; holistic assessment of damages due to climate change	Schellnhuber et al. (2003)
FUND	Bottom-up and top-down	Global (16)	1950–2300	Damage in FUND is distinguished between market and non-market effects; the former affect investment (economic growth) and consumption (welfare). The damage cost module per sector is presented in detail in (Ortiz and Markandya 2009), while other damage functions are used in the literature (Anthoff et al. 2009; Narita et al. 2009; Tol 2002)	Ortiz and Markandya (2009), Anthoff et al. (2009), Narita et al. (2009), Tol (2002)
ICAM-3	Bottom-up	Regional (11)	1975–2100	$D = a \cdot A + \gamma \cdot G + \lambda \cdot L(a, \gamma)$ α, β, γ comprise the policy choice, G is the vector of the costs of various geoengineering activities, L is the vector of losses (e.g.GDP), and λ the projection of the losses.	Dowlatabadi (1998)
IGSM2	Top-down	Global	2000–2100	No damage treatment; IGSM2, along with the human systems component, comprise the MIT EPPA general equilibrium model	
IMAGE 2.4	Top-down	Global (26)	2000–2100	$Yt = A \cdot Kt^{a} \cdot Lt^{1-a}$ Y is GDP, K is capital, L is labor, a is elasticity of production, $Kt + 1 = Kt - \eta \cdot Kt + It, \eta$ is depreciation, I is investment	Bouwman et al. (2006)
PAGE2002	Top-down	Regional (8)	2000–2200	$Dt = \frac{Tpret^{pow}}{2.5} \cdot W + (Tpret - Tdis)Pdis \cdot Wdis$ Dt is damage as a precentage of GDP, $Tpre$ is global temp above the pre-industrial age, $Tdis$ is global temperature where is chance of discontinuity, $Pdis$ is parameter positively related of a discontinuity, W is GDP lost for a 2.5°C warming	Hof et al. (2008)
PAGE09	Top-down	Regional (8)	2000–2200	$D = a \cdot \frac{Tact^{\beta}}{Tcal}$ where D is damage, a is damage at calibration temp, $Tcal$ is calibration temp, $Tact$ is actual temp rise, β is damage exponent	Pycroft et al. (2011)

parameters (Plambeck et al. 1997). A common criticism of IAMs is that they are so complex and opaque ("black box") that it is hard to see how the various underlying assumptions affect their outcomes. The use of simple equations to capture complex climatic and economic phenomena is, according to Hope (2006), justified because the results approximate those of the most complex climate simulations and all aspects of climate change are subject to profound uncertainty. Uncertainty is a central focus of the model that builds up probability distributions of the results by representing the key inputs to the marginal impacts with probability distributions (stochastic treatment). An approximate probability distribution is generated for the model outputs of temperature rise, climate change damages and costs of adaptation and prevention. This is meant to help decision-makers perform a risk analysis so that they can select a policy that balances the cost of intervention against the benefits of mitigating potential climate change impacts (Plambeck et al. 1997).

A full description of the PAGE2002 model and all of its equations can be found in Hope (2006). A number of equations are focused on determining the temperature rise from excess concentrations of each of the greenhouse gases caused by human activities. There is no module of an economy. The economic side of the model is limited to a few equations that link market and non-market damages to temperature increases and calculate costs of avoiding or diminishing these damages through adaptation and/or emission reductions. In estimating damages arising from climate change, an "enumerative approach" is taken, which means that total damages are a simple aggregation of damages in individual sectors. There is thus no general equilibrium type of accounting for the many possible interactions between sectors. Although it has been assumed that this will lead to a lower estimate of total damages than that from a model that captures interactions, it is difficult to understand the magnitude of the difference. Climate change impacts are assumed to occur if temperature rises at a rate above some tolerable rate of change or level of temperature. These rates and levels vary with regions, and a regional multiplier captures these differences. Adaptation policies in any given year can increase both the tolerable rate of change and the level of temperature rise. The regional impact of climate change is therefore a function of the regional temperature rise and how much this is in excess of the regional tolerable rate of change or level of temperature that also depends on adaptation policies. A weighted index translates the regional temperature rise into monetary damage by multiplying the excess regional temperature rise by a regionally weighted percentage loss of GDP (based on estimates) times the regions' estimated GDP. This is done for all eight regions in the model, for the market and non-market sectors and for every time period. By adding together market and non-market sector damages, the model finds aggregate damages per region per period, and this can then be discounted with regional and time variable discount rates. To get the net present value of global climate change impacts, the model aggregates the net present values of all regional damages.

Adaptation costs depend on the change in the rate and level of tolerable temperature rise that can be brought about by adaptation policy within each region. With appropriate weighting and use of uncertainty parameters, the net present value of costs can be estimated for different regional adaptation strategies. The costs of preventing climate change are based on estimates of mitigating emissions below business as usual levels and are also weighted by region and discounted per time period. Only the direct costs of preventing greenhouse gas emissions are included in the model, although secondary benefits (like the concomitant reduction in atmospheric pollution) can be incorporated by reducing the preventative cost parameters. In many of the model's equations, parameters are used to capture uncertainty like that relating to the equilibrium warming from a doubling of CO_2 concentration, or the uncertainty about future growth or policy. In PAGE2002, there are about 80 uncertainty parameters depending on the regions and impact sectors used for a given run.

Although little can be obtained from a comparative analysis of models that fall into this "other models" class, Table 12 presents how uncertainty is treated and how technological change is introduced in these seven models.

A key difference between these models and models of other categories lies in the treatment of uncertain parameters. With the exception of the Community Integrated Assessment System (CIAS) model, all other models feature the capacity to approach uncertainty stochastically, by means of Monte Carlo analysis in FUND (Ackerman and Munitz 2012), IGSM2 (Webster et al. 2003, 2012) and IMAGE (Van Vuuren 2007) and probability distributions in ICAM-3 (e.g. Dowlatabadi 1998). PAGE, as already discussed, incorporates probability distributions for treating a large number of uncertain parameters, as well as Latin hypercube sampling, which was preferred over Monte Carlo analysis.

9 Concluding Remarks

This book chapter has tried to convey the broad outlines and main distinguishing features of alternative climate-economy model frameworks. The main objective has been to provide a simple overview and organising scheme into what can be a daunting wealth of different climate-economy or integrated assessment models. Rather than attempting to provide brief descriptions of a large sample of climate-economy models, the paper has tried to portray the main features of a small number of different classes of models, while delving into some key aspects of the models' perspective, structure, coverage and ways of treating uncertainty and technological advancement. Furthermore, no attempt was made to consider or compare results from alternative models; there are many surveys that compare model outcomes and consider how these differences can be explained either by the features of the model or by the specific assumptions embedded in these. Some suggest that the model framework can have substantial implications on the outcomes. Lanz and Rausch (2011) show systematic differences in outcomes from general equilibrium and energy system models. In contrast, one survey by Edenhofer et al. (2006) suggests that the underlying differences in outcomes lie not necessarily in the model type per se but the assumptions commonly made by the researchers working with different model types. The distinction between "model type" and "assumptions" may be

Table 12 Uncertainty and technological change in models not falling into any of the identified categories

Model	Technological change		Uncertainty treatment		Uncertainty factors
	Endogenous	Exogenous	Deterministic	Stochastic	
CIAS	✓		Scenario analysis (Warren et al. 2008)		Emission stabilisation levels (Warren et al. 2012)
FUND		✓	Scenario analysis	Monte Carlo analysis	Emission reduction gas mix (Tol 2006); climate damages (Ackerman and Munitz 2012)
ICAM-3	✓	✓	Scenario analysis; sensitivity analysis	Probability distributions	Mitigation action delay, experience with control technologies; more than 2000 stochastic parameters regarding paths followed, demographics, economic activity, resources and prices of fossil and non-fossil fuels, the energy intensity of economic activity, short-run budget adjustments when energy prices change and long-run lagged inter-fuel substitution elasticities (Dowlatabadi 1998)
IGSM2		✓	Scenario analysis	Monte Carlo analysis	Emission reduction gas mix, policy mix (Reilly et al. 2006); climate parameters, anthropogenic emissions (Webster et al. 2003), climate responses (Webster et al. 2012)
IMAGE 2.4		✓	Scenario analysis; sensitivity analysis	Monte Carlo analysis	Dietary variants; variations of CO_2 fertilisation, recovery period of natural vegetation, potential feedbacks of decreasing demand on intensification in the agricultural system (Stehfest et al. 2009); large number of input parameters (Van Vuuren 2007)
PAGE2002	✓		Scenario analysis	Probability distributions; Latin hypercube sampling	Emission stabilisation levels (Hope 2008); more than 80 input parameters, depending on the number of sectors and regions selected (Hope 2006)
PAGE09	✓		Scenario analysis; sensitivity analysis	Probability distributions; Latin hypercube sampling	Emission stabilisation levels (Hope 2011); climate sensitivity, discount rate (Pycroft et al. 2011)

somewhat vague, but certainly a deeper understanding of climate-economy modelling really requires a detailed understanding of the role of the many assumptions explicitly or implicitly incorporated in these. Several climate-economy surveys do focus, inter alia, on just such a comparison of the role of assumptions (e.g. Stanton et al. 2009; Söderholm 2007).

It should be noted that many climate-economy models would not fall easily into any of the broad classifications presented here. Moreover, with the modellers' tendency to continuously develop and improve their models, there is much "cross-fertilisation" further blurring distinctions. The diversity of models has many sources. Model selection or design may be driven by different underlying research questions, like whether the focus is on calculating the cost of emission reductions or the long-term damages of climate change or whether one is comparing specific policies or trying to determine an overall optimal global climate policy target. If the focus is on understanding the impacts of climate change to a national economy, a more detailed multisectoral model may be more appropriate than a wealth maximising model that treats the economy as a single sector but captures the long-term global trajectory of the global economy. The differences may reflect deeper philosophical controversies, like whether it is meaningful to assume perfectly functioning markets when the object is to model climate change, which is the grandest instance of market failure extending to so many parts of the economy and with an unprecedented scale in time and space. This is why recent advancements or perspectives call for the use of modelling ensembles that highlight and make use of these differences in structure, design and theoretical foundations, in order to gain better insights (Doukas et al. 2018) and meaningfully inform policy processes; and, even such approaches may miss fundamental aspects that can only be explored with the help of stakeholders (Nikas et al. 2017). These differences may also result from the need to capture one particularly salient feature of climate change like uncertainty. The latter is associated with so many aspects of climate change and respective policy and is therefore, at least to some extent, being treated by means of deterministic or stochastic approaches embedded in the different modelling frameworks.

No doubt these and many other sources of model diversity will continue to drive the development of new and refinement of old models. Although this paper has barely scratched the surface of climate-economy modelling, the following quote seems like an apt closing for the great analytical challenges raised by our need to better understand the required policy response to climate change: "it is difficult to conceive an integrated model that will be able to provide the best answers to all questions. Instead, [. . .] the relative strengths and weaknesses of the different frameworks ensure that the combined contributions rather than individual models provide really valuable policy insights, to which new approaches and new frameworks for coupling economic and climate models can contribute" (Toth 2005).

Acknowledgements The most important part of this research is based on the H2020 European Commission Project "Transitions pathways and risk analysis for climate change mitigation and adaptation strategies—TRANSrisk" under grant agreement No. 642260. The content of this chapter, the sole responsibility for which lies with the authors, does not necessarily reflect the opinion of the European Commission.

References

Ackerman, F., & Munitz, C. (2012). Climate damages in the FUND model: A disaggregated analysis. *Ecological Economics, 77*, 219–224.

Ackerman, F., & Stanton, E. (2012). *Climate risks and carbon prices: Revising the social cost of carbon*.

Ackerman, F., Stanton, E. A., & Bueno, R. (2010). Fat tails, exponents, extreme uncertainty: Simulating catastrophe in DICE. *Ecological Economics, 69*(8), 1657–1665.

Akashi, O., Hijioka, Y., Masui, T., Hanaoka, T., & Kainuma, M. (2012). GHG emission scenarios in Asia and the world: The key technologies for significant reduction. *Energy Economics, 34*, S346–S358.

Akimoto, K., Sano, F., Homma, T., Oda, J., Nagashima, M., & Kii, M. (2010). Estimates of GHG emission reduction potential by country, sector, and cost. *Energy Policy, 38*(7), 3384–3393.

Akimoto, K., Tomoda, T., Fujii, Y., & Yamaji, K. (2004). Assessment of global warming mitigation options with integrated assessment model DNE21. *Energy Economics, 26*(4), 635–653.

Anthoff, D., Hepburn, C., & Tol, R. S. (2009). Equity weighting and the marginal damage costs of climate change. *Ecological Economics, 68*(3), 836–849.

Barker, T. (1998). The effects on competitiveness of coordinated versus unilateral fiscal policies reducing GHG emissions in the EU: An assessment of a 10% reduction by 2010 using the E3ME model. *Energy Policy, 26*(14), 1083–1098.

Barker, T. (1999). Achieving a 10% cut in Europe's carbon dioxide emissions using additional excise duties: Coordinated, uncoordinated and unilateral action using the econometric model E3ME. *Economic Systems Research, 11*(4), 401–422.

Barker, T., & Ekins, P. (2001). *How high are the costs of Kyoto for the US Economy* (No. 4). Tyndall Centre working paper.

Barker, T., & Rosendahl, K. E. (2000). *Ancillary benefits of GHG mitigation in Europe: SO2, NOx and PM10 reductions from policies to meet Kyoto targets using the E3ME model and Externe valuations. Ancillary benefits and costs of greenhouse gas mitigation* (pp. 413–450).

Barker, T., & Scrieciu, S. Ş. (2010). Modeling low climate stabilization with E3MG: Towards a 'new economics' approach to simulating energy-environment-economy system dynamics. *The Energy Journal, 31*, 137–164.

Barker, T., & Zagame, P. (1995). *E3ME: An energy-environment-economy model for Europe*. Brüssel: European Commission.

Barker, T., Anger, A., Chewpreecha, U., & Pollitt, H. (2012). A new economics approach to modelling policies to achieve global 2020 targets for climate stabilisation. *International Review of Applied Economics, 26*(2), 205–221.

Barker, T., Ekins, P., & Foxon, T. (2007). The macro-economic rebound effect and the UK economy. *Energy Policy, 35*(10), 4935–4946.

Barker, T., Pan, H., Köhler, J., Warren, R., & Winne, S. (2006). Decarbonizing the global economy with induced technological change: Scenarios to 2100 using E3MG. *The Energy Journal, 27*, 241–258.

Barker, T., Scrieciu, S. S., & Foxon, T. (2008). Achieving the G8 50% target: Modelling induced and accelerated technological change using the macro-econometric model E3MG. *Climate Policy, 8*(Suppl 1), S30–S45.

Barreto, L., & Kypreos, S. (2004). Endogenizing R&D and market experience in the "bottom-up" energy-systems ERIS model. *Technovation, 24*(8), 615–629.

Bchir, H., Decreux, Y., Guérin, J. L., & Jean, S. (2002). *MIRAGE, a computable general equilibrium model for trade policy analysis* (Vol. 17). CEPII Working paper.

Bchir, M. H., Fontagné, L., & Zanghieri, P. (2003). *The impact of EU enlargement on Member States: A CGE approach* (Vol. 10). CEPII.

Bernstein, P. M., Montgomery, W. D., & Rutherford, T. F. (1999a). Global impacts of the Kyoto agreement: Results from the MS-MRT model. *Resource and Energy Economics, 21*(3), 375–413.

Bernstein, P. M., Montgomery, W. D., Rutherford, T. F., & Yang, G. F. (1999b). Effects of restrictions on international permit trading: The MS-MRT model. *The Energy Journal, 20*, 221–256.

Bollen, J. (2015). The value of air pollution co-benefits of climate policies: Analysis with a global sector-trade CGE model called WorldScan. *Technological Forecasting and Social Change, 90*, 178–191.

Bollen, J. C., & Gielen, A. M. (1999). Economic impacts of multilateral emission reduction policies: Simulations with WorldScan. In *International environmental agreements on climate change* (pp. 155–167). Dordrecht: Springer Netherlands.

Bosello, F. (2010). *Adaptation, mitigation and 'green' R&D to combat global climate change. Insights from an empirical integrated assessment exercise.*

Bosello, F., De Cian, E., Eboli, F., & Parrado, R. (2009). *Macro economic assessment of climate change impacts: A regional and sectoral perspective. Impacts of climate change and biodiversity effects*. Final report of the CLIBIO project, European Investment Bank, University Research Sponsorship Programme.

Bosello, F., Eboli, F., Parrado, R., & Rosa, R. (2010). *REDD in the carbon market: A general equilibrium analysis.*

Bosetti, V., Carraro, C., & Galeotti, M. (2005). *The dynamics of carbon and energy intensity in a model of endogenous technical change.*

Bosetti, V., Carraro, C., Duval, R., Sgobbi, A., & Tavoni, M. (2009). *The role of R&D and technology diffusion in climate change mitigation: New perspectives using the WITCH model.*

Bosetti, V., Golub, A., Markandya, A., Massetti, E., & Tavoni, M. (2008). Abatement cost uncertainty and policy instrument selection. A dynamic analysis. In *Modelling environment-improving technological innovations under uncertainty* (pp. 127–157). Oxon: Routledge.

Bosetti, V., Massetti, E., & Tavoni, M. (2007). *The WITCH model: Structure, baseline, solutions.*

Bouwman, A. F., Kram, T., & Klein Goldewijk, K. (2006). *Integrated modelling of global environmental change. An overview of IMAGE 2.4.*

Buchner, B., & Carraro, C. (2005). Modelling climate policy: Perspectives on future negotiations. *Journal of Policy Modeling, 27*(6), 711–732.

Bukowski, M., & Kowal, P. (2010). *Large scale, multi-sector DSGE model as a climate policy assessment tool.* Working Paper 3, Instytut Badań Strukturalnych, Warszawa.

Buonanno, P., Carraro, C., & Galeotti, M. (2003). Endogenous induced technical change and the costs of Kyoto. *Resource and Energy Economics, 25*(1), 11–34.

Burniaux, J. M. (1998). *How important is market power in achieving Kyoto? An assessment based on the GREEN model* (pp. 17–18). Proceedings of the Workshop on Economic Modelling and Climate Change.

Burniaux, J. M., & Truong, T. P. (2002). *GTAP-E: An energy-environmental version of the GTAP model.* GTAP Technical Papers, 18.

Burniaux, J. M., Martin, J. P., Nicoletti, G., & Martins, J. O. (1992). *GREEN a multi-sector, multi-region general equilibrium model for quantifying the costs of curbing CO2 emissions.*

Cambridge Econometrics. (2014, April). *E3ME technical manual, version 6.0.* University of Cambridge, Cambridge.

Capros, P., De Vita, A., Tasios, N., Siskos, P., Kannavou, M., Petropoulos, A., Evangelopoulou, S., Zampara, M., Papadopoulos, D., Nakos, C., & Paroussos, L. (2016). *EU Reference Scenario 2016-Energy, transport and GHG emissions Trends to 2050.*

Capros, P., Mantzos, L., Kolokotsas, D., Ioannou, N., Georgakopoulos, T., Filippopoulitis, A., & Antoniou, Y. (1998). *The PRIMES energy system model–reference manual.* National Technical University of Athens.

Capros, P., Mantzos, L., Papandreou, V., Tasios, N., & Mantzaras, A. (2007). *Energy systems analysis of CCS Technology; PRIMES model scenarios.* E3ME-lab/ICCS/National Technical University of Athens, Athens. Final report to DG ENV.

Capros, P., Mantzos, L., Vouyoukas, L., & Petrellis, D. (1999). European energy and CO2 emissions trends to 2020: PRIMES model v. 2. *Bulletin of Science, Technology & Society, 19* (6), 474–492.

Capros, P., Tasios, N., De Vita, A., Mantzos, L., & Paroussos, L. (2012). Model-based analysis of decarbonising the EU economy in the time horizon to 2050. *Energy strategy reviews, 1*(2), 76–84.
Ciscar, J. C., Soria, A., Goodess, C. M., Christensen, O. B., Iglesias, A., Garrote, L., Moneo, M., Quiroga, S., Feyen, L., Dankers, R., & Nicholls, R. (2009). *Climate change impacts in Europe*. Final report of the PESETA research project (No. JRC55391). Directorate Growth & Innovation and JRC-Seville, Joint Research Centre.
Cooper, A., Livermore, S., Rossi, V., Wilson, A., & Walker, J. (1999). The economic implications of reducing carbon emissions: A cross-country quantitative investigation using the Oxford Global Macroeconomic and Energy Model. *The Energy Journal, 20*, 335–365.
Crassous, R., Hourcade, J. C., & Sassi, O. (2006). Endogenous structural change and climate targets modeling experiments with Imaclim-R. *The Energy Journal, 27*, 259–276.
Criqui, P., Cattier, F., Kouvaritakis, N., & Thonet, C. (1998). *Technological scenarios, climate change and emission trading: EC-IEA study on energy technology and climate change simulations using the POLES world model*. Sevilla: IPTS.
Criqui, P., Mima, S., & Viguier, L. (1999). Marginal abatement costs of CO2 emission reductions, geographical flexibility and concrete ceilings: An assessment using the POLES model. *Energy Policy, 27*(10), 585–601.
Dagoumas, A. S., & Barker, T. S. (2010). Pathways to a low-carbon economy for the UK with the macro-econometric E3MG model. *Energy Policy, 38*(6), 3067–3077.
Dai, H., Masui, T., Matsuoka, Y., & Fujimori, S. (2011). Assessment of China's climate commitment and non-fossil energy plan towards 2020 using hybrid AIM/CGE model. *Energy Policy, 39*(5), 2875–2887.
DeCanio, S. (2003). *Economic models of climate change: A critique*. Berlin: Springer.
Den Elzen, M. G. J. (2005). *Analysis of future commitments and costs of countries for the "South-North Dialogue" proposal using the FAIR 2.1 world model*. No. MNP-report, 728001032.
Doukas, H., Nikas, A., González-Eguino, M., Arto, I., & Anger-Kraavi, A. (2018). From integrated to integrative: Delivering on the Paris Agreement. *Sustainability, 10*(7).
Dowlatabadi, H. (1995). Integrated assessment models of climate change: An incomplete overview. *Energy Policy, 23*(4), 289–296.
Dowlatabadi, H. (1998). Sensitivity of climate change mitigation estimates to assumptions about technical change. *Energy Economics, 20*(5), 473–493.
Dowlatabadi, H. (2000). Bumping against a gas ceiling. *Climatic Change, 46*(3), 391–407.
Edenhofer, O., Bauer, N., & Kriegler, E. (2005a). The impact of technological change on climate protection and welfare: Insights from the model MIND. *Ecological Economics, 54*(2), 277–292.
Edenhofer, O., Lessmann, K., & Bauer, N. (2005b). *Mitigation strategies and costs of climate protection: The effects of ETC in the hybrid model MIND*.
Edenhofer, O., Lessmann, K., Kemfert, C., Grubb, M., & Köhler, J. (2006). Induced technological change: Exploring its implications for the economics of atmospheric stabilization: Synthesis report from the innovation modeling comparison project. *The Energy Journal, 27*, 57–107.
Edmonds, J., & Reiley, J. M. (1985). *Global energy-assessing the future*.
Edmonds, J., Pitcher, H., & Sands, R. (2004). Second generation model 2004: An overview. *Special Issue on Endogenous, 30*, 425–448.
Ekins, P., & Etheridge, B. (2006). The environmental and economic impacts of the UK climate change agreements. *Energy Policy, 34*(15), 2071–2086.
Elliott, J., Foster, I., Kortum, S., Munson, T., Cervantes, F. P., & Weisbach, D. (2010). Trade and carbon taxes. *The American Economic Review, 100*(2), 465–469.
Fawcett, A. A. (2011). International offsets usage in proposed US climate change legislation. In *Modeling the economics of greenhouse gas mitigation: Summary of a workshop*.
Finon, D. (1976). *Un modele energetique pour la France* (Vol. 3). Centre national de la recherche scientifique.
Fishbone, L. G., & Abilock, H. (1981). Markal, a linear-programming model for energy systems analysis: Technical description of the bnl version. *International Journal of Energy Research, 5* (4), 353–375.

Flamos, A. (2016). A sectoral micro-economic approach to scenario selection and development: The case of the Greek power sector. *Energies, 9*(2), 77.
Fujino, J., Nair, R., Kainuma, M., Masui, T., & Matsuoka, Y. (2006). Multi-gas mitigation analysis on stabilization scenarios using AIM global model. *The Energy Journal, 27*, 343–353.
Füssel, H. M. (2009). An updated assessment of the risks from climate change based on research published since the IPCC Fourth Assessment Report. *Climatic Change, 97*(3), 469–482.
Füssel, H. M. (2010). Modeling impacts and adaptation in global IAMs. *Wiley Interdisciplinary Reviews: Climate Change, 1*(2), 288–303.
Gabriel, S. A., Kydes, A. S., & Whitman, P. (2001). The National Energy Modeling System: A large-scale energy-economic equilibrium model. *Operations Research, 49*(1), 14–25.
Garnaut, R. (2008). *The Garnaut climate change review: Final report.* Cambridge: Cambridge University Press.
Gerlagh, R. (2006). ITC in a global growth-climate model with CCS: The value of induced technical change for climate stabilization. *The Energy Journal, 27*, 223–240.
Gerlagh, R. (2007). Measuring the value of induced technological change. *Energy Policy, 35*(11), 5287–5297.
Gerlagh, R., & Van der Zwaan, B. (2006). Options and instruments for a deep cut in CO_2 emissions: Carbon dioxide capture or renewables, taxes or subsidies? *The Energy Journal, 27*, 25–48.
Goettle, R. J., & Fawcett, A. A. (2009). The structural effects of cap and trade climate policy. *Energy Economics, 31*, S244–S253.
Goettle, R. J., Ho, M. S., Jorgenson, D. W., Slesnick, D. T., Wilcoxen, P. J., & EP-W, E. C. (2007). *IGEM, an inter-temporal general equilibrium model of the US economy with emphasis on growth, energy and the environment.* Prepared for the US Environmental Protection Agency (EPA), Office of Atmospheric Programs, Climate Change Division, EPA Contract EP-W-05-035.
Golub, A., Hertel, T. W., & Sohngen, B. (2009). Land use modelling in a recursively dynamic GTAP framework. In *Economic analysis of land use in global climate change policy* (Vol. 14, p. 235).
Grassini, M. (2009). Rowing along the computable general equilibrium modelling mainstream. *Studies on Russian Economic Development, 20*(2), 134–146.
Greening, L. A., & Bataille, C. (2009). *Bottom-up models of energy: Across the spectrum.*
Hainoun, A., Aldin, M. S., & Almoustafa, S. (2010). Formulating an optimal long-term energy supply strategy for Syria using MESSAGE model. *Energy Policy, 38*(4), 1701–1714.
Hedenus, F., Azar, C., & Lindgren, K. (2006). Induced technological change in a limited foresight optimization model. *The Energy Journal, 27*, 109–122.
Hickman, R., Ashiru, O., & Banister, D. (2010). Transport and climate change: Simulating the options for carbon reduction in London. *Transport Policy, 17*(2), 110–125.
Hitz, S., & Smith, J. (2004). Estimating global impacts from climate change. *Global Environmental Change, 14*(3), 201–218.
Hof, A. F. (2010). *International mitigation and adaptation strategies for climate change policy: An integrated assessment approach.*
Hof, A. F., de Bruin, K. C., Dellink, R. B., den Elzen, M. G., & van Vuuren, D. P. (2009). The effect of different mitigation strategies on international financing of adaptation. *Environmental Science & Policy, 12*(7), 832–843.
Hof, A. F., De Bruin, K., Dellink, R., den Elzen, M. G. J., & van Vuuren, D. P. (2012). Costs, benefits and inter-linkages between adaptation and mitigation. Global Climate Governance Beyond.
Hof, A. F., den Elzen, M. G., & van Vuuren, D. P. (2008). Analysing the costs and benefits of climate policy: Value judgements and scientific uncertainties. *Global Environmental Change, 18*(3), 412–424.
Hope, C. (2006). The marginal impact of CO2 from PAGE2002: An integrated assessment model incorporating the IPCC's five reasons for concern. *Integrated Assessment, 6*(1), 19–56.
Hope, C. (2009). How deep should the deep cuts be? Optimal CO2 emissions over time under uncertainty. *Climate Policy, 9*(1), 3–8.
Hope, C. (2011). *The social cost of CO2 from the PAGE09 model.*

Hope, C. (2013). Critical issues for the calculation of the social cost of CO2: Why the estimates from PAGE09 are higher than those from PAGE2002. *Climatic Change, 117*(3), 531–543.
Hope, C. W. (2008). Optimal carbon emissions and the social cost of carbon over time under uncertainty. *Integrated Assessment Journal, 8*(1), 107–122.
Hope, C., Anderson, J., & Wenman, P. (1993). Policy analysis of the greenhouse effect: An application of the PAGE model. *Energy Policy, 21*(3), 327–338.
Hourcade, J. C., Richels, R., Robinson, J., Chandler, W., Davidson, O., Finon, D., Grubb, M. J., Halsneas, K., Hogan, K., Jaccard, M., & Krause, F. (1996). *Estimating the costs of mitigating greenhouse gases* (pp. 263–296). Climate Change 1995, Economic and Social Dimensions of Climate Change, Contribution of Working Group II.
Ianchovichina, E., Darwin, R., & Shoemaker, R. (2001). Resource use and technological progress in agriculture: A dynamic general equilibrium analysis. *Ecological Economics, 38*(2), 275–291.
International Energy Agency. (2010). *World Energy Outlook 2010*. Paris: OECD/International Energy Agency.
Jakeman, G., & Fisher, B. S. (2006). Benefits of multi-gas mitigation: An application of the Global Trade and Environment Model (GTEM). *The Energy Journal, 27*, 323–342.
Jebaraj, S., & Iniyan, S. (2006). A review of energy models. *Renewable and Sustainable Energy Reviews, 10*(4), 281–311.
Jorgenson, D. W., Goettle, R. J., Hurd, B. H., Smith, J. B., Chestnut, L. G., & Mills, D. M. (2004). *US market consequences of global climate change*. Washington, DC: Pew Center on Global Climate Change.
Junankar, S., Lofsnaes, O., & Summerton, P. (2007). *MDM-E3. A short technical description*. Cambridge: Cambridge Econometrics.
Kainuma, M., Matsuoka, Y., & Morita, T. (1999). Analysis of post-Kyoto scenarios: The Asian-Pacific integrated model. *The Energy Journal, 27*, 207–220.
Kainuma, M., Matsuoka, Y., & Morita, T. (2000). The AIM/end-use model and its application to forecast Japanese carbon dioxide emissions. *European Journal of Operational Research, 122* (2), 416–425.
Kainuma, M., Matsuoka, Y., & Morita, T. (Eds.). (2011). *Climate policy assessment: Asia-Pacific integrated modeling*. Tokyo: Springer Science & Business Media.
Kainuma, M., Matsuoka, Y., Hibino, G., Shimada, K., Ishii, H., Matsui, S., & Morita, T. (2003). Application of AIM/Enduse model to Japan. In *Climate policy assessment* (pp. 155–176). Tokyo: Springer Japan.
Karkatsoulis, P., Siskos, P., Paroussos, L., & Capros, P. (2017). Simulating deep CO2 emission reduction in transport in a general equilibrium framework: The GEM-E3T model. *Transportation Research Part D: Transport and Environment, 55*, 343–358.
Kelly, D. L., & Kolstad, C. D. (1999). Integrated assessment models for climate change control. *International Yearbook of Environmental and Resource Economics, 2000*, 171–197.
Kemfert, C. (2001). *Economy-energy-climate interaction: The model wiagem*.
Kemfert, C. (2005). Induced technological change in a multi-regional, multi-sectoral, integrated assessment model (WIAGEM): Impact assessment of climate policy strategies. *Ecological Economics, 54*(2), 293–305.
Kemfert, C., & Truong, T. (2007). Impact assessment of emissions stabilization scenarios with and without induced technological change. *Energy Policy, 35*(11), 5337–5345.
Kemfert, C., Truong, T. P., & Bruckner, T. (2006). Economic impact assessment of climate change—A multi-gas investigation with WIAGEM-GTAPEL-ICM. *The Energy Journal, 27*, 441–460.
Keppo, I., & van der Zwaan, B. (2012). The impact of uncertainty in climate targets and CO2 storage availability on long-term emissions abatement. *Environmental Modeling & Assessment, 17*(1–2), 177–191.
Kesicki, F., & Yanagisawa, A. (2015). Modelling the potential for industrial energy efficiency in IEA's World Energy Outlook. *Energy Efficiency, 8*(1), 155–169.
Kitous, A., Criqui, P., Bellevrat, E., & Chateau, B. (2010). Transformation patterns of the world-wide energy system-scenarios for the century with the POLES model. *The Energy Journal, 27*, 49–82.

Koji, T., Takanobu, K., Atsushi, K., Norihiro, I., & Masaji, S. (2009). *Measuring weak sustainability for the future: Calculating genuine saving with population change by an integrated assessment model.*

Kopp, R. E., & Mignone, B. K. (2012). *The US Government's social cost of carbon estimates after their first two years: Pathways for improvement.*

Kouvaritakis, N., Soria, A., & Isoard, S. (2000a). Modelling energy technology dynamics: Methodology for adaptive expectations models with learning by doing and learning by searching. *International Journal of Global Energy Issues, 14*(1–4), 104–115.

Kouvaritakis, N., Soria, A., Isoard, S., & Thonet, C. (2000b). Endogenous learning in world post-Kyoto scenarios: Application of the POLES model under adaptive expectations. *International Journal of Global Energy Issues, 14*(1–4), 222–248.

Kratena, K., & Streicher, G. (2009). *Macroeconomic input-output modelling: Structures, functional forms and closure rules*. International Input-Output Association Working Paper WPIOX, 09-009.

Kremers, H., Nijkamp, P., & Wang, S. (2000). *Mailing issues on climate change policies—A discussion of the GTAP-E Model* (No. 00-099/3). Tinbergen Institute Discussion Paper.

Kurosawa, A. (2004). Carbon concentration target and technological choice. *Energy Economics, 26* (4), 675–684.

Kurosawa, A. (2006). Multigas mitigation: An economic analysis using GRAPE model. *The Energy Journal, 27*, 275–288.

Kurosawa, A., Yagita, H., Zhou, W., Tokimatsu, K., & Yanagisawa, Y. (1999). Analysis of carbon emission stabilization targets and adaptation by integrated assessment model. *The Energy Journal, 27*, 157–175.

Kypreos, S. (2007). A MERGE model with endogenous technological change and the cost of carbon stabilization. *Energy Policy, 35*(11), 5327–5336.

Kypreos, S. (2008). Stabilizing global temperature change below thresholds: Monte Carlo analyses with MERGE. *Computational Management Science, 5*(1), 141–170.

Kypreos, S., Barreto, L., Capros, P., & Messner, S. (2000). ERIS: A model prototype with endogenous technological change. *International Journal of Global Energy Issues, 14*(1–4), 347–397.

Laborde, D., Martin, W., & Van der Mensbrugghe, D. (2016). Measuring the impacts of global trade reform with optimal aggregators of distortions. *Review of International Economics, 25*, 403–425.

Lanz, B., & Rausch, S. (2011). General equilibrium, electricity generation technologies and the cost of carbon abatement: A structural sensitivity analysis. *Energy Economics, 33*(5), 1035–1047.

Lapillonne, B. (1978). *MEDEE 2: A model for long-term energy demand evaluation.*

Lapillonne, B. (1980). Long term perspectives of the US energy demand: Application of the MEDEE 2 model to the US. *Energy, 5*(3), 231–257.

Lecocq, F., & Shalizi, Z. (2007). *How might climate change affect economic growth in developing countries? A review of the growth literature with a climate lens*. Policy Research Working Papers, 4315.

Lejour, A., Veenendaal, P., Verweij, G., & van Leeuwen, N. (2006). *WorldScan; a model for international economic policy analysis* (No. 111). CPB Netherlands Bureau for Economic Policy Analysis.

Link, P. M., & Tol, R. S. (2004). Possible economic impacts of a shutdown of the thermohaline circulation: An application of FUND. *Portuguese Economic Journal, 3*(2), 99–114.

Manne, A. S., & Richels, R. G. (2005). MERGE: An integrated assessment model for global climate change. In *Energy and environment* (pp. 175–189). Springer US.

Manne, A., Mendelsohn, R., & Richels, R. (1995). MERGE: A model for evaluating regional and global effects of GHG reduction policies. *Energy Policy, 23*(1), 17–34.

Masui, T. (2005). Policy evaluations under environmental constraints using a computable general equilibrium model. *European Journal of Operational Research, 166*(3), 843–855.

Masui, T., Hanaoka, T., Hikita, S., & Kainuma, M. (2006a). Assessment of CO_3 reductions and economic impacts considering energy-saving investments. *The Energy Journal, 27*, 175–190.

Masui, T., Matsuoka, Y., & Kainuma, M. (2006b). Long-term CO2 emission reduction scenarios in Japan. *Environmental Economics and Policy Studies, 7*(3), 347–366.

Masui, T., Rana, A., & Matsuoka, Y. (2003). AIM/material model. In *Climate policy assessment* (pp. 177–196). Springer Japan.

Matsuoka, Y., Kainuma, M., & Morita, T. (1995). Scenario analysis of global warming using the Asian Pacific Integrated Model (AIM). *Energy Policy, 23*(4–5), 357–371.

Mattsson, N. (2002). Introducing uncertain learning in an energy system model: A pilot study using GENIE. *International Journal of Global Energy Issues, 18*(2–4), 253–265.

Mattsson, N., & Wene, C. O. (1997). Assessing new energy technologies using an energy system model with endogenized experience curves. *International Journal of Energy Research, 21*(4), 385–393.

McKibbin, W. J., & Wilcoxen, P. J. (1999). The theoretical and empirical structure of the G-Cubed model. *Economic Modelling, 16*(1), 123–148.

McKibbin, W. J., & Wilcoxen, P. J. (2009). Uncertainty and climate change policy design. *Journal of Policy Modeling, 31*(3), 463–477.

McKibbin, W. J., Lee, J. W., & Cheong, I. (2004). A dynamic analysis of the Korea–Japan free trade area: Simulations with the G-cubed Asia–Pacific model. *International Economic Journal, 18*(1), 3–32.

Mendelsohn, R., & Williams, L. (2004). Comparing forecasts of the global impacts of climate change. *Mitigation and Adaptation Strategies for Global Change, 9*(4), 315–333.

Mendelsohn, R., Morrison, W., Schlesinger, M. E., & Andronova, N. G. (2000). Country-specific market impacts of climate change. *Climatic Change, 45*(3–4), 553–569.

Mendelsohn, R., Nordhaus, W., & Shaw, D. (1999). *The impact of climate variation on US agriculture* (pp. 55–74). Cambridge: Cambridge University Press.

Mercure, J. F., Pollitt, H., Edwards, N. R., Holden, P. B., Chewpreecha, U., Salas, P., Lam, A., Knobloch, F., & Vinuales, J. (2017). Environmental impact assessment for climate change policy with the simulation-based integrated assessment model E3ME-FTT-GENIE. *Energy Strategy Reviews, 20*, 195–208.

Messner, S. (1997). Endogenized technological learning in an energy systems model. *Journal of Evolutionary Economics, 7*(3), 291–313.

Mima, S., & Criqui, P. (2009). Assessment of the impacts under future climate change on the energy systems with the POLES model. In *International energy workshop*.

Mundaca, L., Neij, L., Worrell, E., & McNeil, M. (2010). Evaluating energy efficiency policies with energy-economy models. *Annual Review of Environment and Resources, 35*, 305–344.

Narita, D., Tol, R. S., & Anthoff, D. (2009). Damage costs of climate change through intensification of tropical cyclone activities: An application of FUND. *Climate Research, 39*(2), 87–97.

Nicoletti, G., & Oliveira-Martins, J. (1993). Global effects of the European carbon tax. In *The European carbon tax: An economic assessment* (pp. 15–48). Springer Netherlands.

Nijkamp, P., Wang, S., & Kremers, H. (2005). Modeling the impacts of international climate change policies in a CGE context: The use of the GTAP-E model. *Economic Modelling, 22*(6), 955–974.

Nikas, A., Doukas, H., Lieu, J., Alvarez Tinoco, R., Charisopoulos, V., & van der Gaast, W. (2017). Managing stakeholder knowledge for the evaluation of innovation systems in the face of climate change. *Journal of Knowledge Management, 21*(5), 1013–1034.

Nilsson, C. (1999). *A unilateral versus a multilateral carbon dioxide tax—A numerical analysis with the European Model GEM-E3* (No. 66).

Nordhaus, W. (2007). *Accompanying notes and documentation on development of DICE-2007 model: Notes on DICE-2007.* delta. v8 as of September 21, 2007. Miscellaneous publication, Yale University, New Haven, NE.

Nordhaus, W. D. (1994). *Managing the global commons: The economics of climate change* (Vol. 31). Cambridge, MA: MIT Press.

Nordhaus, W. D. (2002). Modeling induced innovation in climate-change policy. *Technological Change and the Environment, 9*, 259–290.

Nordhaus, W. D. (2008). *A question of balance: Weighing the options on global warming policies.* New Haven, CT: Yale University Press.

Nordhaus, W. D. (2010). Economic aspects of global warming in a post-Copenhagen environment. *Proceedings of the National Academy of Sciences, 107*(26), 11721–11726.

Nordhaus, W. D. (2014). *A question of balance: Weighing the options on global warming policies*. New Haven, CT: Yale University Press.

Nordhaus, W. D., & Boyer, J. (2000). *Warming the world: Economic models of global warming*. Cambridge, MA: MIT Press.

Nordhaus, W. D., & Yang, Z. (1996). A regional dynamic general-equilibrium model of alternative climate-change strategies. *The American Economic Review, 86*, 741–765.

Oda, J., Akimoto, K., Sano, F., & Homma, T. (2009). Diffusion of CCS and energy efficient technologies in power and iron & steel sectors. *Energy Procedia, 1*(1), 155–161.

Ortiz, R. A., & Markandya, A. (2009). *Integrated impact assessment models of climate change with an emphasis on damage functions: A literature review*. Bilbao, Spain.

Paltsev, S., Reilly, J. M., Jacoby, H. D., Eckaus, R. S., McFarland, J. R., Sarofim, M. C., Asadoorian, M. O., & Babiker, M. H. (2005). *The MIT emissions prediction and policy analysis (EPPA) model: Version 4*. MIT Joint Program on the Science and Policy of Global Change.

Pant, H. M. (2007). *GTEM draft: Global trade and environmental model*. Australian Bureau of Agricultural and Resource Economics.

Pant, H., Tulpulé, V., & Fisher, B. S. (2007). *Global trade and environment model*. Department of Agriculture: Australian Bureau of Agricultural and Resource Economics and Sciences, Canberra, Australia.

Papadelis, S., Flamos, A., & Psarras, J. (2013). A framework to address uncertainties in energy policy formulation. *Energy Sources, Part B: Economics, Planning, and Policy, 8*(3), 270–278.

Parrado, R., & De Cian, E. (2014). Technology spillovers embodied in international trade: Intertemporal, regional and sectoral effects in a global CGE framework. *Energy Economics, 41*, 76–89.

Parson, E. A., & Fisher-Vanden, A. K. (1997). Integrated assessment models of global climate change. *Annual Review of Energy and the Environment, 22*(1), 589–628.

Peck, S. C., & Teisberg, T. J. (1993). Global warming uncertainties and the value of information: An analysis using CETA. *Resource and Energy Economics, 15*(1), 71–97.

Peck, S. C., & Teisberg, T. J. (1995). International CO2 emissions control: An analysis using CETA. *Energy Policy, 23*(4–5), 297–308.

Peck, S. C., & Teisberg, T. J. (1999). CO_2 emissions control agreements: Incentives for regional participation. *The Energy Journal, 27*, 367–390.

Pfenninger, S. (2015). *Calliope: A multi-scale energy systems (MUSES) modeling framework*. Release v0, 3.

Plambeck, E. L., Hope, C., & Anderson, J. (1997). The Page95 model: Integrating the science and economics of global warming. *Energy Economics, 19*(1), 77–101.

Plinke, E., Haasis, H. D., Rentz, O., & Sivrioglu, M. (1990). Analysis of energy and environmental problems in Turkey by using a decision support model. *Ambio, 19*, 75–81.

Popp, D. (2004). ENTICE: Endogenous technological change in the DICE model of global warming. *Journal of Environmental Economics and Management, 48*(1), 742–768.

Popp, D. (2006a). Comparison of climate policies in the ENTICE-BR model. *The Energy Journal, 27*, 163–174.

Popp, D. (2006b). ENTICE-BR: The effects of backstop technology R&D on climate policy models. *Energy Economics, 28*(2), 188–222.

Pycroft, J., Vergano, L., Hope, C., Paci, D., & Ciscar, J. C. (2011). *A tale of tails: Uncertainty and the social cost of carbon dioxide*.

Rafaj, P., & Kypreos, S. (2007). Internalisation of external cost in the power generation sector: Analysis with Global Multi-regional MARKAL model. *Energy Policy, 35*(2), 828–843.

Rana, A., & Morita, T. (2000). Scenarios for greenhouse gas emission mitigation: A review of modeling of strategies and policies in integrated assessment models. *Environmental Economics and Policy Studies, 3*(2), 267–289.

Redondo, P. D., & van Vliet, O. (2015). Modelling the energy future of Switzerland after the phase out of nuclear power plants. *Energy Procedia, 76*, 49–58.

Reilly, J., Sarofim, M., Paltsev, S., & Prinn, R. (2006). The role of non-CO_3 GHGs in climate policy: Analysis using the MIT IGSM. *The Energy Journal, 27*, 503–520.
Robinson, S. (2006). Macro models and multipliers: Leontief, Stone, Keynes, and CGE models. In *Poverty, inequality and development* (pp. 205–232).
Rosendahl, K. E. (2004). Cost-effective environmental policy: Implications of induced technological change. *Journal of Environmental Economics and Management, 48*(3), 1099–1121.
Rout, U. K., Akimoto, K., Sano, F., Oda, J., Homma, T., & Tomoda, T. (2008). Impact assessment of the increase in fossil fuel prices on the global energy system, with and without CO2 concentration stabilization. *Energy Policy, 36*(9), 3477–3484.
Sano, F., Akimoto, K., Homma, T., & Tomoda, T. (2005). *Analysis of technological portfolios for CO2 stabilizations and effects of technological changes.*
Ščasný, M., Píša, V., Pollitt, H., & Chewpreecha, U. (2009). Analyzing macroeconomic effects of environmental taxation in the Czech Republic with the econometric E3ME model. *Finance a Uver: Czech Journal of Economics & Finance, 59*(5), 460–491.
Schultz, P. A., & Kasting, J. F. (1997). Optimal reductions in CO2 emissions. *Energy Policy, 25*(5), 491–500.
Schumacher, K., & Sands, R. D. (2006). Innovative energy technologies and climate policy in Germany. *Energy Policy, 34*(18), 3929–3941.
Schwanitz, V. J. (2013). Evaluating integrated assessment models of global climate change. *Environmental Modelling & Software, 50*, 120–131.
Schellnhuber, H. J., Warren, R., Haxeltine, A., & Naylor, L. (2003). *Developments in integrated assessment: The co-productive approach*. In Proceedings of OECD Workshop on the Benefits of Climate Policy.
Scott, M. J., Sands, R. D., Edmonds, J., Liebetrau, A. M., & Engel, D. W. (1999). Uncertainty in integrated assessment models: Modeling with MiniCAM 1.0. *Energy Policy, 27*(14), 855–879.
Seebregts, A. J., Goldstein, G. A., & Smekens, K. (2002). Energy/environmental modeling with the MARKAL family of models. In *Operations research proceedings 2001* (pp. 75–82). Berlin: Springer.
Seebregts, A., Kram, T., Schaeffer, G. J., & Bos, A. (2000). Endogenous learning and technology clustering: Analysis with MARKAL model of the Western European energy system. *International Journal of Global Energy Issues, 14*(1–4), 289–319.
Siskos, P., Capros, P., & De Vita, A. (2015). CO2 and energy efficiency car standards in the EU in the context of a decarbonisation strategy: A model-based policy assessment. *Energy Policy, 84*, 22–34.
Söderholm, P. (2007). *Modelling the economic costs of climate policy*. Luleå University of Technology.
Sokolov, A. P., Schlosser, C. A., Dutkiewicz, S., Paltsev, S., Kicklighter, D. W., Jacoby, H. D., Prinn, R. G., Forest, C. E., Reilly, J. M., Wang, C., & Felzer, B. S. (2005). *MIT integrated global system model (IGSM) version 2: Model description and baseline evaluation*. MIT Joint Program on the Science and Policy of Global Change.
Sorrell, S. (2009). Jevons' Paradox revisited: The evidence for backfire from improved energy efficiency. *Energy policy, 37*(4), 1456–1469.
Stanton, E. A., Ackerman, F., & Kartha, S. (2009). Inside the integrated assessment models: Four issues in climate economics. *Climate and Development, 1*(2), 166–184.
Stehfest, E., Bouwman, L., Van Vuuren, D. P., Den Elzen, M. G., Eickhout, B., & Kabat, P. (2009). Climate benefits of changing diet. *Climatic Change, 95*(1–2), 83–102.
Sullivan, P., Krey, V., & Riahi, K. (2013). Impacts of considering electric sector variability and reliability in the MESSAGE model. *Energy Strategy Reviews, 1*(3), 157–163.
Tol, R. S. (1997). On the optimal control of carbon dioxide emissions: An application of FUND. *Environmental Modeling and Assessment, 2*(3), 151–163.
Tol, R. S. (2002). Welfare specifications and optimal control of climate change: An application of fund. *Energy Economics, 24*(4), 367–376.
Tol, R. S. (2006). Multi-gas emission reduction for climate change policy: An application of FUND. *The Energy Journal, 27*, 235–250.

Tol, R. S. (2010). Carbon dioxide mitigation. In B. Lomborg (Ed.), *Smart solutions to climate change: Comparing costs and benefits* (pp. 74–105). New York: Cambridge University Press.
Tol, R. S., & Fankhauser, S. (1998). On the representation of impact in integrated assessment models of climate change. *Environmental Modeling and Assessment, 3*(1), 63–74.
Toth, F. L. (2005). Coupling climate and economic dynamics: Recent achievements and unresolved problems. In *The coupling of climate and economic dynamics* (pp. 35–68). Springer Netherlands.
Tulpulé, V., Brown, S., Lim, J., Polidano, C., Pant, H., & Fisher, B. S. (1998). *An economic assessment of the Kyoto Protocol using the Global Trade and Environment Model* (pp. 17–18). Proceedings of the Workshop on Economic Modelling and Climate Change.
Turton, H., & Barreto, L. (2004a). *Cars, hydrogen and climate change: A long-term analysis with the ERIS model*. 6th IAEE European Conference.
Turton, H., & Barreto, L. (2004b). *The extended energy-systems ERIS model: An overview.*
Tyers, R., & Shi, Q. (2012). *Global demographic change, labour force growth and economic performance* (Chapter 13, pp. 342–375).
Urban, F. R. M. J., Benders, R. M. J., & Moll, H. C. (2007). Modelling energy systems for developing countries. *Energy Policy, 35*(6), 3473–3482.
Van der Mensbrugghe, D. (2005). *Linkage technical reference document*. Development Prospects Group, The World Bank.
Van der Voort, E. (1982). The EFOM 12C energy supply model within the EC modelling system. *Omega, 10*(5), 507–523.
van der Zwaan, B., Keppo, I., & Johnsson, F. (2013). How to decarbonize the transport sector? *Energy Policy, 61*, 562–573.
Van Regemorter, D. (2005). *GEM-E3. Computable General equilibrium model for studying economy-energy-environment interactions for Europe and the world.*
Van Vuuren, D. P. (2007). *Energy systems and climate policy-long-term scenarios for an uncertain future*. Doctoral dissertation, Utrecht University.
Viguier, L. L., Babiker, M. H., & Reilly, J. M. (2003). The costs of the Kyoto Protocol in the European Union. *Energy Policy, 31*(5), 459–481.
Von Below, D., & Persson, T. (2008). *Uncertainty, climate change and the global economy* (No. w14426). National Bureau of Economic Research.
Wada, K., Sano, F., Akimoto, K., & Homma, T. (2012). Assessment of Copenhagen pledges with long-term implications. *Energy Economics, 34*, S481–S486.
Waisman, H., Guivarch, C., Grazi, F., & Hourcade, J. C. (2012). The IMACLIM-R model: Infrastructures, technical inertia and the costs of low carbon futures under imperfect foresight. *Climatic Change, 114*(1), 101–120.
Walmsley, T. L., Dimaranan, B. V., & McDougall, R. A. (2006a). A baseline scenario for the dynamic GTAP model. In *Dynamic modeling and applications for global economic analysis* (p. 136).
Walmsley, T. L., Hertel, T. W., & Ianchovichina, E. (2006b). Assessing the impact of China's WTO accession on investment. *Pacific Economic Review, 11*(3), 315–339.
Warren, R., De La Nava Santos, S., Arnell, N. W., Bane, M., Barker, T., Barton, C., Ford, R., Füssel, H. M., Hankin, R. K., Klein, R., & Linstead, C. (2008). Development and illustrative outputs of the Community Integrated Assessment System (CIAS), a multi-institutional modular integrated assessment approach for modelling climate change. *Environmental Modelling & Software, 23*(5), 592–610.
Warren, R., Yu, R. M. S., Osborn, T. J., & de la Nava Santos, S. (2012). European drought regimes under mitigated and unmitigated climate change: Application of the Community Integrated Assessment System (CIAS). *Climate Research, 51*(2), 105–123.
Webster, M. D., Babiker, M., Mayer, M., Reilly, J. M., Harnisch, J., Hyman, R., Sarofim, M. C., & Wang, C. (2002). Uncertainty in emissions projections for climate models. *Atmospheric Environment, 36*(22), 3659–3670.
Webster, M., Forest, C., Reilly, J., Babiker, M., Kicklighter, D., Mayer, M., Prinn, R., Sarofim, M., Sokolov, A., Stone, P., & Wang, C. (2003). Uncertainty analysis of climate change and policy response. *Climatic Change, 61*(3), 295–320.

Webster, M., Sokolov, A. P., Reilly, J. M., Forest, C. E., Paltsev, S., Schlosser, A., Wang, C., Kicklighter, D., Sarofim, M., Melillo, J., & Prinn, R. G. (2012). Analysis of climate policy targets under uncertainty. *Climatic Change, 112*(3–4), 569–583.

Wei, Y. M., Mi, Z. F., & Huang, Z. (2015). Climate policy modeling: An online SCI-E and SSCI based literature review. *Omega, 57*, 70–84.

Wen, Z., Meng, F., & Chen, M. (2014). Estimates of the potential for energy conservation and CO2 emissions mitigation based on Asian-Pacific Integrated Model (AIM): The case of the iron and steel industry in China. *Journal of Cleaner Production, 65*, 120–130.

West, G. R. (2002). Modeling structural linkages in dynamic and spatial interindustry systems. In *Trade, networks and hierarchies* (pp. 225–250). Berlin: Springer.

Weyant, J. P. (1999). The cost of the Kyoto Protocol: A multi-model evaluation. *The Energy Journal*.

Weyant, J. P. (2009). A perspective on integrated assessment. *Climatic Change, 95*(3), 317–323.

Wing, I. S. (2011). Computable general equilibrium models for the analysis of economy-environment interactions. In *Research tools in natural resource and environmental economics* (p. 255).

Worrell, E., Ramesohl, S., & Boyd, G. (2004). Advances in energy forecasting models based on engineering economics. *Annual Review of Environment and Resources, 29*, 345–381.

Xu, Y., & Masui, T. (2009). Local air pollutant emission reduction and ancillary carbon benefits of SO2 control policies: Application of AIM/CGE model to China. *European Journal of Operational Research, 198*(1), 315–325.

Yamaji, K., Fujino, J., & Osada, K. (2000). Global energy system to maintain atmospheric CO2 concentration at 550 ppm. *Environmental Economics and Policy Studies, 3*(2), 159–171.

Yohe, G. W. (1999). The tolerable windows approach: Lessons and limitations. *Climatic Change, 41*(3), 283–295.

Yu, X. (2008). Impacts assessment of PHEV charge profiles on generation expansion using national energy modeling system. In *Power and energy society general meeting-conversion and delivery of electrical energy in the 21st century* (pp. 1–5). IEEE.

Zaki, C. (2011). *Assessing the global effect of trade facilitation: Evidence from the MIRAGE model*. Economic Research Forum Working Paper Series (No. 659).

Zhai, F., & Zhuang, J. (2012). Agricultural impact of climate change: A general equilibrium analysis with special reference to Southeast Asia. In *Climate change in Asia and the Pacific: How can countries adapt* (pp. 17–35).

"Consensus Building in Engagement Processes" for Reducing Risks in Developing Sustainable Pathways: Indigenous Interest as Core Elements of Engagement

Jenny Lieu, Luis D. Virla, Ryan Abel, and Cecilia Fitzpatrick

Abstract Canada is one of the top ten greenhouse gas (GHG) emitters in the world, and of the nation's total, the province of Alberta was the biggest emitter primarily due to the fossil fuel industry and power generation. Alberta is currently facing a challenge to reduce GHG emissions in line with Canada's obligations to meet Paris Agreement goals. Additionally, the oil sand deposits in Alberta are located on the traditional land of Indigenous communities; therefore, the development, regulation and consultation of this sector have a direct impact not only on emissions but also on the socioeconomic welfare of Indigenous communities. Thus, the transition towards a low-carbon pathway in the oil sand industry is closely connected to upholding the rights of Indigenous peoples. Meaningfully consulting with Indigenous peoples is essential when developing low-carbon pathways that impact the environment and wellbeing of the community. Failure to consult proposed changes with Indigenous communities can lead to risks for the government, the industry and the communities themselves. These risks have been prevalent in the current consultation process in Alberta which has led to litigation, creating mistrust between the government, industry and Indigenous community. Given this background, we present a Consensus Building in Engagement Processes framework that includes Indigenous consensus, knowledge, interests and rights as a focal point of a consultation process required for decision-making. The consultation process is presented within the

J. Lieu (✉)
Science Technology Policy Research Unit (SPRU), University of Sussex, Brighton, UK
e-mail: J.Lieu@sussex.ac.uk

L. D. Virla
Department of Chemical and Petroleum Engineering, University of Calgary, Calgary, AB, Canada

R. Abel
Sustainability Department, Fort McKay First Nation, Wood Buffalo, AB, Canada

C. Fitzpatrick
Fort McKay First Nation, Wood Buffalo, AB, Canada

H. Doukas et al. (eds.), *Understanding Risks and Uncertainties in Energy and Climate Policy*, https://doi.org/10.1007/978-3-030-03152-7_2

context of land use decisions impacting a low-carbon future for oil sand development. The framework is based on seeking consensus from all parties involved and aims to help to reduce risks resulting from decisions that do not consider the interests and rights of communities most impacted by resource development or climate mitigation pathways.

Keywords Sustainability · Indigenous knowledge · Consultation · Engagement · Risk · Pathways

1 Introduction

The Paris Agreement, developed within the United Nations Framework Convention on Climate Change (UNFCCC), aims to reinforce the global action to the menace of climate change. The Agreement proposes a framework that integrates economy, policy, technology and society in order to keep global temperature rise in this century below 2 °C above pre-industrial levels (UNFCCC 2017). Due to its geographic location and landscape, Canada faces challenges when addressing climate change. Extreme temperatures, vast land mass, growing population, a diversified growing economy and abundant natural resources are some of the factors impacting GHG emissions in Canada (UNFCCC 2015). By 2014, Canada's total GHG emissions ascended to 732 megatonnes (Mt) CO2 equivalent, excluding land use, land use change and forestry estimates.

Canada contributes 1.6% of global emissions and is one of the top ten emitters (both total and per capita) in the world (World Resources Institute 2016). Fossil fuel production corresponded to the biggest contributors to this total, comprising 27% of total emissions in Canada (Natural Resources Canada 2016). Of the nation's total, the province of Alberta contributed 37.4% in 2014, representing the biggest emitter among all provinces in Canada (Natural Resources Canada 2016). The total per capital emissions in Alberta amounted to 66.6 tonne per capita in 2013, compared to the country average of 20.7 tonne per capita, and are expected to increase further if no measures are taken (Boothe and Boudreault 2016). The close relationship of Alberta with the fossil fuel industry sector and power generation has steadily increased emissions. The oil and gas sector contributed to 39.4% of the GHG emissions in the overall Albertan economy in 1990 and has increased to 48.3% in 2015 (Environment Canada 2017).

As part of the Paris Agreement, in 2016 Canada agreed to decrease GHG emission by 30% below 2005 levels by 2030. The mechanism in place to achieve this goal mainly involves the development of regulatory measures, such as changes in the transportation and electricity sectors, controlling landfill emissions and promoting clean energy technologies. Specifically for the province of Alberta, policies around carbon pricing, emission capping, electricity generation from renewables and increased efficiency in energy usage have been developed to decrease emission growth and maintain similar levels of today (Leach et al. 2015). Prior to the Paris Agreement in 2015, the Government of Alberta released a Climate Leadership Plan for the province which proposed emission caps, an emission trading system and a carbon tax for facilities that exceed the 100,000 CO2 tonnes/year. The Plan set grounds for the Oil Sands Emissions Limit Act, which created a legal obligation for the oil sand sector to limit emissions to 100 Mt per year (Leach et al. 2015). To guarantee the implementation and acceptance of those policies, the concerns of stakeholders including industry, government and society must be properly addressed.

The oil sand deposits in Alberta—the world's third-largest proven oil reserves (Energy Resources Conservation Board 2010)—are located on traditional land of approximately 24 Indigenous[1] communities. For example, the Athabasca region in Alberta overlaps with traditional territories of five Indigenous communities distributed across the Regional Municipality of Wood Buffalo (RMWB): Mikisew Cree First Nation, Athabasca Chipewyan First Nation, Fort McKay First Nation, Fort McMurray No. 468 First Nation and Chipewyan Prairie Dene First Nation. Therefore, the development, regulation and consultation of this sector have a direct impact not only on the environment but also on the socioeconomic welfare of these communities.

Currently, the low-carbon emission pathway is defined at the national and provincial levels in terms of emission reductions. The details of how the low-carbon pathway will be implemented in the oil sand sector are currently being defined. An important factor to consider in a low-carbon future for Alberta is the inclusion of interests and needs of Indigenous communities affected by oil sand development. Indigenous communities must be consulted in the development of a pathway that could significantly impact their future. In fact, the Paris Agreement stipulates that when addressing climate change, all nations in the Paris Agreement are expected to "respect, promote and consider their respective obligations on human rights; the right to health; the rights of Indigenous peoples, local communities, migrants, children, persons with disabilities and people in vulnerable situations; and the right to development, as well as gender equality, empowerment of women and intergenerational equity" (UNFCCC 2017, p. 21).

Studying the rights and interests of the Indigenous population in the context of Canada is important from a sustainability and climate justice perspective. The Alberta government's policy agenda of transitioning towards a low-carbon pathway in the oil sand sector—100 Mt of CO2 cap (Leach et al. 2015)—cannot be disconnected from the rights of the Indigenous people who reside on the lands where oil sands are extracted. Thus, we argue that, when Indigenous rights and values are protected, the lands and the environment overall are also protected, and this supports

[1]Within the Canadian contexts, Indigenous peoples are referred to First Nations, Aboriginal peoples, Indigenous and Native. When discussing constitutional rights, the term "Aboriginal peoples" is used, while other terms are used interchangeably in literature (University of Alberta 2015).

pathways to a low-carbon future (Baker 2013). Therefore, developing a fair consultation process that fully includes Indigenous rights and interests can help contribute towards a more sustainable low-carbon economy.

This chapter presents a framework that includes Indigenous knowledge, interests and rights as a focal point of a consultation process required for decision-making. This framework is based on discussions and interviews with Indigenous community members, industry, academics and government representation over the period of 2016–2017.[2] The consultation process is presented within the context of land use decisions impacting a low-carbon future for oil sand development in Alberta. The consultation involves three major groups including Indigenous communities, industry and government. This chapter will also explore risks in the current resource development consultation process. We define "risk" as barriers to effective consultation or as negative outcomes/impacts that may result from the consultation process being one-sided. For instance, exploring the negative outcomes of land use decisions is crucial, as decisions to develop the oil sands not only increase emissions (and reduce the ability of forests to take up atmospheric carbon due to deforestation in the boreal forest to make way for large projects) but also significantly impact the ability of Indigenous communities to exercise their Aboriginal and Treaty rights and practise their traditional way of life to which Canada has already formally committed.

This chapter is organised into seven sections. The first section begins by presenting an overview of studies that link the Indigenous knowledge and climate change. The next part provides a starting ground for understanding inclusion by recognising Indigenous rights to free, prior and informed consent in consultation. This leads to the third section where we provide a broad perspective of Indigenous rights and consultation in the Canadian context. In the fourth section, we identify risks in the current consultation process in Alberta. We then introduce Indigenous ways of knowing in part 5 and attempt to recentre the focus of consultation on the Indigenous values of respect, relevance, reciprocity and responsibility. This sets the stage for Sect. 6 where we introduce the framing for consensus building in consultation processes, which can be applied to developing future sustainable pathways (or to address broader resource development activities). In the final section, we provide a summary and suggest steps forward to applying the framing.

[2]List interviewees: Interview 1, academic (A1); Interview 2, academic (A2); Interview 3, Indigenous community member (CM1); Interview 4, Indigenous community member (CM2); Interview 5, Indigenous community member (CM3); Interview 6, expert (E1); Interview 7, expert (E2); Interview 8, Indigenous member (IM1); Interview 9, industry player (IP1); Interview 10, industry player (IP2); Interview 11, industry player (IP3); Interview 12, industry player (IP4); Interview 13, industry player (IP5); Interview 14, non-for-profit organisation (NP1); Interview 15, policymaker (PM1). This case study is part of a wider EU Horizon 2020-funded project, TRANSrisk (Transition Pathways and Risk Analysis for Climate Change Policies); see http://transrisk-project.eu/.

2 Previous and Current Studies Integrating Indigenous Knowledge and Climate Change

There has been a wider acknowledgement on the role of Indigenous knowledge and addressing climate mitigation and adaptation respectfully. The dialogue between Western scientific studies and traditional ecological knowledge (TEK) is still relatively new. TEK is a branch of Indigenous knowledge, practices and beliefs that is passed down over generations on the relationship between living beings and the environment and provides valuable knowledge on addressing climate change (see Lertzman 2010, Berkes 1999, Huntington 1992, Albert 2001, Fox 2003, Brewster 2004, Eloka 2010, qtd. in Alexander et al. 2011). TEK is still a largely untapped resource and can significantly contribute to addressing climate change issues (Alexander et al. 2011).

One starting point for creating links can be the use of narratives in a collaborative relationship, sharing TEK and Western scientific findings on climate change. Examples include the international Indigenous Conference Snowchange organised in 2005. All members of circumpolar nations and Indigenous peoples met in the conference to discuss and share their observations (through narratives) and presented recommendation and actions forward to address climate change. The collaboration was built on sharing knowledge and gaining consensus on the issues related to the northern regions (Mustonen 2005).

Another example of collaboration between TEK and Western science is seen in the Fort McKay Berry Focus Group. This group, formed from members of the Fort McKay First Nation and scientists from the Wood Buffalo Environmental Association (WBEA), has developed a community-based programme for monitoring traditionally consumed berries in the area. The key elements on this project have been the development of a scientific methodology sensitive to the culture and tradition of the community members (Wood Buffalo Environmental Association 2017). Since 2010, this group has been successful in monitoring the health and quality of the berries, as well as important factors that impact these indicators. Beyond the results, the group has allowed the community stakeholders to more accurately inform environmental decisions (Wood Buffalo Environmental Association 2010).

The Arctic Climate Impact Assessment (ACIA) is another study of the Arctic regions carried over the span of 5 years with contributions from over 300 scientists and other experts and members of Indigenous communities. The study was the first comprehensive, integrated assessment of climate change and UV radiation over the Arctic regions that applied several climate models and was complemented with Indigenous knowledge (Symon et al. 2005).

Another example at global study that integrates Indigenous knowledge in addressing climate changes is seen in a panel session at the 2008 conference on Sustaining Cultural and Biological Diversity in a Rapidly Changing World: Lessons for Global Policy (AMNH 2008). The panel, comprising of Indigenous leaders and scholars in the area, compiled 57 Indigenous narratives from all over the world describing climate change and its impacts. The narratives provided valuable information over an extended period of time that was not captured in scientific records.

While studies about the inclusion of TEK are growing and there are more empirical studies on monitoring climate change, few studies have provided methods or tools that place Indigenous knowledge and interests as a starting point in developing climate change mitigation and adaptation action. In fact, the Western perspective of sustainability tends to be the dominating narrative both in a global context and in the Canadian context (Dearden and Mitchell 1998; Draper 1998 qtd. in McGregor and Debroah 2004). Authoritative climate change knowledge is institutionalised through the Intergovernmental Panel on Climate Change (IPCC) in such a way that promotes the "privileging of positivist science and technocratic perspectives, and the marginalisation of other ways of knowing (for example, local, traditional and Indigenous knowledge) . . ." (Ford et al. 2016, p. 349). In other words, Indigenous knowledge would typically contribute to the dominating Western perspectives and play a secondary role in contributing to understanding climate change rather than respecting differing perspective of understanding and addressing climate change issues.

The Western narratives can also take on a colonising tone and can especially dominate in the area of resource management projects, which have significant environmental, cultural and socioeconomic impacts. Some studies have explored the ethical approaches for resource extraction and Indigenous peoples (Lertzman and Vredenburg 2015), but much more needs to be done to better reflect the voices and interests of Indigenous peoples. This imbalanced perspective calls for the implementation of a flexible methodology for consultation in issues that impact Indigenous peoples. A new collaboration is needed that considers the wide range of interests of Indigenous peoples while ensuring integration and free, prior and informed consent (FPIC) of Indigenous communities to develop low-carbon pathways capable of meeting the needs of the local, national and global interests.

3 Inclusion of Indigenous Interests: Free, Prior and Informed Consent

Before entering in the discussion about the consultation process in resource development in the Canadian context, we consider it relevant to clarify the concepts of consultation and free, prior and informed consent (FPIC) used for this work.

According to the Oxford dictionary, consultation can be defined as:

> "The action or process of formally consulting or discussing" and can include "a meeting with an expert [. . .] in order to seek advice".

The discussion implies an exchange of information from an expert group. Indigenous communities are the experts in identifying the changes and impacts of activities performed in their lands, since they hold observation records for generations. Seeking advice requires a flow from knowledge from the expert group to the group who receives the knowledge. This knowledge flow in the consultation process, if not carefully implemented, can reinforce negative power structures if information is not respected and reciprocated by non-Indigenous parties.

The term free, prior and informed consent (FPIC) is a well-established concept in international human rights law affirmed in the United Nations Declaration on the Rights of Indigenous Peoples (UNDRIP). The term consent is defined in the Oxford dictionary as: "Permission for something to happen or agreement to do something". In order for this process of agreement to incorporate the FPI elements, the UN Office of the High Commissioner for Human Rights has established the following conditions (Pillay 2013):

> *Free*, implies that there is no coercion, intimidation or manipulation; *Prior*, implies that consent is to be sought sufficiently in advance of any authorisation or commencement of activities and respect is shown to time requirements of Indigenous consultation/consensus processes; and *Informed*, implies that information is provided that covers a range of aspects, including the nature, size, pace, reversibility and scope of any proposed project or activity; the purpose of the project as well as its duration; locality and areas affected; a preliminary assessment of the likely economic, social, cultural and environmental impact, including potential risks [...]. (Pillay 2013, p. 2)

Obtaining FPIC can reduce the risks associated with large development projects. However, obtaining consent in itself is not sufficient and needs to be accompanied with a mutual understanding of what constitutes consent (see Baker 2013, p. 30).

4 Indigenous Legal Rights and Consultation Process in Canada

Government—Indigenous relations in Canada were initially formalised in the eighteenth century in the form of accords between the Crown and Indigenous communities. These accords, known as treaties, outline the terms for land distribution and were used to create peace between Indigenous people and settlers. Nearly 50% of the Canadian land mass is covered under these treaties, which include 59% of the total Indigenous communities in the country (Indigenous and Northern Affairs 2008). Indigenous communities in the Athabasca region are included on Treaty Eight, which was signed in 1899 and has been submitted to include more communities. The terms of this treaty established reserves; annuities; hunting, trapping and fishing rights; agricultural aids; and other benefits for the Indigenous communities in exchange for the surrender of the land to the Crown (federal government) (Daniel et al. 1980). The implementation of the terms and conditions of Treaty Eight has been historically linked to misinterpretation and disagreements, since its creation responded mainly to economic considerations from the government instead of humanitarian concerns for the Indigenous communities. Today, the Athabasca region is rich in natural resources, and the increasing industrial activity in the area has raised pressure on the federal government to reassess the Treaty Eight terms and clarify the treaty by considering contemporary legal and political realities (Indigenous and Northern Affairs 2009).

As established in Section 35 of the Constitution Act (1982), the Government of Canada is required to guarantee proper consultation of Aboriginal peoples and to accommodate Indigenous interests when the Crown contemplates action(s) that may negatively impact potential or established Aboriginal or Treaty rights (Government of Canada 1982). The element of accommodation has been reinforced in a number of important Supreme Court decisions over the past decades (1973–2014) (see Joseph 2017 for overview). The court cases have not only upheld Indigenous land rights and interests but emphasised the importance of the duty of the government to consult and accommodate Indigenous interests through negotiation rather than through litigation. These court cases present potential risk for the oil sand industry and also called for further government obligation to properly consult not only to satisfy legal obligations but to reach consensus.

Previous government consultation and accommodation processes have also been outlined in the Aboriginal Consultation and Accommodation, Updated Guidelines for Federal Officials to Fulfil the Duty to Consult (Indigenous and Northern Affairs 2011). The guidelines emphasise the need to consult and accommodate by providing "policy-based guidance to assist officials in their efforts to effectively incorporate consultations and, where appropriate, accommodation into government activities and processes" (Government of Alberta 2013, 2016a). The Guiding Principles and Consultation Directives were developed from lessons learnt through court cases and other best practice and engagement experiences. However, accommodation does not imply "consent". In fact, the term "consent" only appears once in the updated guidance as a reference to the United Nations Declaration on the Rights of Indigenous Peoples.

Although Canada has issued a Statement of Support to endorse the UN Declaration, the government has expressed concern with the interpretation of free, prior and informed consent (FPIC), which can imply the power of veto (United Nations 2008; Indigenous and Northern Affairs 2011). In 2014, the UN Special Rapporteur noted the importance of developing a policy framework for implementing the duty to consult with a genuine opportunity for Indigenous peoples to provide input and involvement through FPIC in early stages of project developments. Moreover, the Rapporteur emphasised that conducting consultation in "good faith" should not be regarded as granting "veto power". Instead, performing consultation in the aim of achieving consent implies meaningful and informed dialogue and accommodation (TORYS 2016). Echoing these concerns, the Truth and Reconciliation Commission of Canada in 2015 called on the corporate sector in Canada to adopt the United Nations Declaration on the Rights of Indigenous Peoples and commit to meaningful consultation, building respectful relationships and obtaining the free, prior and informed consent of Indigenous peoples before proceeding with economic development projects (Truth and Reconciliation Commission 2015).

With the (inter)national context as their backdrop, in February 2018, the Government of Canada proposed legal requirements in Bill C69 that requires full impact assessments of economic development projects. Certain listed projects are obliged to carry out impact assessments that consider impacts on the environment, health, society and the economy. Previously, social and economic impacts were broadly considered under environmental impact assessments. The bill "prohibits proponents" or companies "from carrying out a designated project if the designated project is likely to cause certain environmental, health, social or economic effects" (Minister of Environment and Climate Change, p. ii). However, the Government can still override the decision if the project is in the public interest but must "tak[e] into account the impacts on the rights of the Indigenous peoples of Canada, [and] all effects that may be caused by the carrying out of the project". The new bill also established the Impact Assessment Agency, who is obliged to form research and advisory committees that include "the interests and concerns of Indigenous peoples of Canada, and appoint as a member of any such bodies" (Minister of Environment and Climate Change, p. 78).

5 Risks Identified in the Current Consultation Process Within the Canadian Context

Considering the 1982 Constitution, Supreme Court cases and Federal Guidance and the Impact Assessment Act that recognises Indigenous rights, interests and the legal obligation to consult and accommodate, we argue for there is an urgent need to focus on both consultation and inclusion in the engagement process. Free, prior and informed consent can be gained through consultation and inclusion of Indigenous knowledge and interests in engagement processes, leading to decisions that accommodate Indigenous needs and rights. Protecting Indigenous rights and respecting their world views will also help Canada achieve the aims of the Paris Agreement.

In this section, we will identify current issues in the consultation process that leads to risk for the government, Indigenous community and industry.

5.1 Government Aspect

The Government of Canada has a legal responsibility to ensure that Indigenous rights (both constitutionally enshrined and those protected in treaty agreements) and interests are upheld. Government is responsible for establishing the consultation system and setting policies and legislative requirements, and importantly, they hold treaty agreements with many Indigenous communities. Within the context of Alberta, the initial energy resource lease is granted by the Alberta government (Alberta Energy) within a particular jurisdiction without needing to consult Indigenous communities, thus setting up an immediate likelihood for conflict in many situations, especially if those areas are culturally significant for an Indigenous community. Energy leases are granted to the proponent, and this then creates a legal expectation by the proponent that they will have the right to develop the resource, provided their project meets environmental impact assessment and other regulatory standards. Although the government formally advocates for consultation (through pressures from the legal system), the government bypasses direct consultation with communities. However, there are regulatory requirements for proponents to carry out consultation about their project(s). Thus, when the consultation process is officially initiated, proponents are often met with resistance. Proponents have effectively become the direct interface to address Alberta's desire to develop the oil sand resource and its associated decision-making and policies, which may conflict with Indigenous community interests. It is the current system itself that leads to increasing risks within the consultation process (Interview, E1, 2017).

The government has been criticised by both Indigenous communities and industry for playing a role that can interfere with meaningful consultation (Interviews, IP1-5, 2016–2017). They do not take part in the consultation process, but are responsible for assessing if there has been sufficient "communication" between the Indigenous community and the proponent. Yet the government does not ensure that the issues raised by Indigenous communities are accommodated or addressed. The Alberta government appears to oversee the process by reviewing consultation records in what seems to be an exercise to ensure legal risk has been minimised when project approvals are granted. It should be noted that First Nation communities in Alberta do not support the province's consultation policy (Interview, E1, 2017). At the project approval stage, governments and their regulatory agencies (e.g. Aboriginal Consultation Office of Alberta, ACO) do not necessarily apply conditions to the project that protect the communities' interests and address the concerns raised during consultation. Rather the consultation processes are assessed as a process rather than as a contribution to the final outcome.

The government can in fact reduce their own risk of litigation by not being directly involved in the consultation process. But there is a trade-off associated with the government's risk-averse behaviour. While the government reduces its own risk, the risk increases for other parties. For the proponents, the risk may involve significant delays and exploratory access to the land revoked due to litigation. For the Indigenous community, if the project is approved without meaningfully addressing the risky negative impacts to the environment, culture and community, they can take the proponent to court, leading to project delays or halting the project altogether. This requires resources from the community, and there may also be risks that the court decisions may not favour the community interests.

An example of litigation resulting from a lack of inclusion and consensus is seen in the case of an in situ oil sand development near the Fort McKay First Nation's Moose Lake reserves. The community of Fort McKay filed a lawsuit in 2016 against the province of Alberta to challenge the decisions passed by the province's Aboriginal Consultation Office (ACO), who granted consultation adequacy on the proposed development in Fort McKay's treaty land. The ACO's decision was a procedural clearance that allowed the Alberta Energy Regulator (AER) to approve the project. The community stated that the cumulative impacts (among other facts) of the development violated Fort McKay's Treaty rights. The ACO claimed that they were only permitted to consider site-specific impacts related to the development's direct impacts (Henton 2016; Weber 2016). The lack of meaningful consideration of the community's concerns and stated resultant impacts in the consultation process not only put the proposed project at risk but also eroded trust in the consultation process. Consultation may be a good starting point for discussion, but on its own does not imply an equal opportunity to exchange knowledge and interests if all parties concerned are not included as part of the decision-making process.

5.2 Indigenous Aspect

Currently Indigenous communities are over-consulted due to traditional land use research and the legal obligations for consultation (i.e. guidelines discussed above) introduced by provincial and federal governments (Interviews, CM1-3, 2016–2017). Particular members of the community including elders are frequently invited to consultation meetings, but their inputs are often not included to influence outcomes (Baker and Westman 2018). As a result, elders experience fatigue and are weary of consultation as the project often goes ahead regardless of their inputs and subsequently choose not to participate, thereby perpetuating the misconception of assimilation (King 2012, Simpson 2007 qtd. in Baker and Westman 2018). The needs and concerns of the community can sometimes be in conflict with the development, which can threaten a community's way of life and negatively impact the environment and their culture. Therefore, when Indigenous communities' views are not appropriately addressed and accommodated, there is a higher risk that the project will not effectively consider the long-term environmental and social impact (Baker 2013).

5.3 Industry Aspect

Consultation becomes a bureaucratic process for projects that require traditional land use assessment and where results assert that projects have no substantial impacts on the land. From the business perspective, the consultation obligation is viewed as a cumbersome bureaucratic process that prolongs business decisions and that do not necessarily benefit participants, including Indigenous right holders (Calgary Chamber 2015). Increasing approval times creates costs and financial risks, and the consultation process itself may lead to a negative outcome that places the entire project at risk. In order to satisfy the legal requirement, proponents must demonstrate that they have communicated with the Indigenous community. Proponents secure participation in the consultation process by paying an honorarium to community members when they attend meetings and share their traditional knowledge with a project proponent (Interviews, IP1-5 and E1, 2016–2017). However, consultation alone does not guarantee that the outcomes truly accommodate Indigenous interests and needs. In fact, consultation without inclusion of interests and needs reduces the ability of Indigenous right holders to influence outcomes in the consultation process, thus creating false hope and increasing the likelihood for litigation.

5.4 Who Bears the Responsibility?

Even if Indigenous interests are included in the decision-making process, without consensus during this process, the accommodation of interest does not guarantee that Indigenous concerns are sufficiently addressed. In fact, the "First Nations Consultation & Accommodation Handbook" states that "The Crown is ultimately responsible for accommodation, but project proponents may have a role in accommodating First Nations" (Laidlaw and Passelac-Ross 2014, p. 70). Therefore, the accommodation actions that directly or indirectly affect the socioeconomic welfare of the community rely on external parties that base their decision on a different set of values than those of the affected community. Without a common consensus of views, the accommodation process (if it exists at all) is likely to overlook the main needs of the community.

This takes us to the concept of "inclusion", where Indigenous rights, interests and knowledge are included in every step of the engagement and decision-making process in order to develop consensus. We define inclusion within the context of the engagement process as consciously incorporating Indigenous views, interests, rights and knowledge as a part of the outcomes in a decision-making process. The outcome is therefore built on consensus and reflects free, prior and informed consent.

6 Understanding Indigenous Ways of Knowing and World Views as Essential Step Towards Inclusion

Effective inclusion throughout the consultation process can help to build consensus and requires a number of considerations including learning and respecting Indigenous knowledge systems embedded in world views, engaging in effective listening, creating respectful spaces for conversation and reciprocating.

Acknowledging that Indigenous knowledge differs from Western knowledge can provide a starting point to effective consultation and inclusion. This involves understanding the fundamental principles of Indigenous knowledge, which stems from relational knowing based on inner metaphysical knowledge and the outer physical environment (Ermine 2000; Kovach 2010). In other words, Indigenous knowledge can be explained as "the peoples' cognitive and wise legacy as a result of their interaction with nature in a common territory" (Maurial 1999 p. 62, qtd. in Hart 2010).

Simpson (2000) identifies seven main Indigenous world views. The first recognises that "knowledge is holistic, cyclic, and dependent upon relationships and connections to living and non-living beings and entities" (Simpson 2000, p. 62). The second accepts that there are many truths that vary according to individual experiences. The third acknowledges that everything is alive, while the fourth emphasises that all things are equal. The fifth views the land as sacred, and the sixth recognises the importance of the relationship between people and the spiritual world. The last considers humans as the least important in this world (Simpson 2000 qtd. in Hart 2010).

A number of the Indigenous world views emphasise a strong relationship with the environment and are, in fact, more aligned to climate justice, sustainability and a low-carbon future compared to modern economic Western paradigms that promote economic growth as a core element of societal development.

The inclusion of Indigenous knowledge and interests (e.g. in building consensus for addressing resource developments or addressing sustainability) requires putting aside the colonial perspective that Western knowledge is the basis of understanding. Rather, the starting point of inclusion can begin by what Kirkness and Barnhardt (1991) describe as the 4Rs of respect, relevance, reciprocity and responsibility. The 4Rs emerged as a means of promoting a more effective "two-way exchange" between universities and Indigenous students who must enter into post-secondary institutions and adapt to a very different context and notion of hierarchy.

In the context of Canada, Indigenous communities have been making significant efforts to learn and adapt to the Western institutions and values in order to effectively communicate with the government and decision-making bodies; yet non-Indigenous Canadians have not learnt to reciprocate in conversing with Indigenous communities as whole (Harrington 1991 qtd. in Kirkness and Barnhardt 1991). Therefore, considering the 4Rs when engaging with Indigenous communities may help to bridge the communication and learning gap.

6.1 Respect

The first step for inclusion is "respecting" Indigenous culture. The meaning and value of Indigenous knowledge are embedded in its cultural context and present in everyday activities and life experiences (Kirkness and Barnhardt 1991). There should also be a recognition on how Western education imposed literacy, language, values and perspectives on generations of Indigenous peoples (Scollon and Scollon 1981), and how Indigenous communities survived colonisation. On the other hand, non-Indigenous approaches often take a less respectful position when addressing Indigenous communities. Indigenous peoples are approached and frequently asked to comment on existing agenda topics as opposed to being asked to share their own story (Archibald 2008). One means of avoiding a knowledge colonisation is to respectfully approach Indigenous communities without preconceived notions and to listen to their individual perspectives.

6.2 Relevant

Another starting point is to address issues that are "relevant" to Indigenous perspectives and experience as a legitimate source of knowledge. The Western perspective of "objectivity" and evidence-based facts "make oral histories suspect and unreliable" in the Canadian legal system, which is based on Eurocentric perspective that provides little space for diverging cultural world view (Pryce 1992, 35 qtd. in Archibald 2008, p. 106). The Western Canadian justice systems attempt to control divergent behaviour through punishment, requiring conformation to social norms. Indigenous views of justice however work at restoring peace and equilibrium in the community by reconciling the wrongdoer to his/her own conscience with the individual or family that has been wronged (Archibald 2008, p. 102). Thus, understanding the relevance for Indigenous perspectives then requires "a re-valuation of forms of knowledge that are not derived from books" (Goody 1982, p. 201, qtd. in Kirkness and Barnhardt 1991) or Western institutionalised knowledge.

Additionally, the concept of value can differ between Indigenous and Western communities. Indigenous peoples are deeply impacted by nature, relationships, tradition and "valuables" that are passed down from elders in the form of stories, language and traditions. Stories are the basis for teaching and learning, and "life is in them for those who know how to ask and learn" (Cajete 1994, 41 qtd. in Archibald 2008, p. 17). Value in modern Western society is institutionalised and primarily measured by economic or monetary terms through, e.g. gross domestic product or purchasing power parity and knowledge passed from education institutions. These perspectives are not necessarily "better" or "worse", but recognising the differences may help to bridge gaps between perspectives.

6.3 Reciprocating

Appreciating differences will help lead to the next inclusion value of reciprocating relationships (although different value systems can result in goals that are not easily reconciled with each other). This step can include accommodation on outcomes that meet the needs and interests of Indigenous peoples and not merely meeting legal or other or organisational obligations. One-sided consultation (often referred to by Fort McKay as "drive-by consultation") can reinforce unequal power dynamics if the consultation favours the powerful institution or organisation who enters into Indigenous communities with the objective of extracting information without listening or engaging in conversation or reciprocity. Kovach (2010) notes that ethical misconduct can occur when non-Indigenous researchers enter into Indigenous communities without cultural knowledge. She further also notes there are mutual benefits in developing a reciprocal relationship: "Because of the interconnection between all entities, seeking this information ought not to be extractive but reciprocal, to ensure an ecological and cosmological balance. Much insight comes to an individual inwardly and intuitively" (Kovach 2000). Kovach encourages a reflective and thoughtful approach to maintain a balance of give and take, a value that should be generally intuitive in our present day.

6.4 *Responsibility*

Finally, "responsibility" is encouraged through participation for both Indigenous peoples and proponents. Responsibility is a form of accountability where proponents make efforts to accommodate Indigenous interests, for example, by creating more inclusive spaces for dialogue. When agreeing to be part of an engagement process, Indigenous peoples agree to participate and have their own responsibility to uphold their values. One important means of participation is through the oral tradition of storytelling. Kovach (2010) states that "The act of sharing through personal narrative, teaching story, and general conversation are methods by which each generation is accountable to the next in transmitting knowledge". When non-Indigenous parties are invited to listen to stories, there is an "implication that the listener is or becomes a member of the community" (Archibald 2008, p. 26). Inclusion also requires responsible use of the stories told, that is, to accurately reflect the knowledge shared through stories.

From the Western perspective, engagement and consultation are a means of creating a dialogue between two parties usually occurring within the context of conversations in meetings, workshops and interviews. Storytelling, therefore, can be a medium that unifies Indigenous world views and Western methods of gathering knowledge for research (Thomas 2005; Wilson 2001).

7 Framework

As indicated earlier in the chapter, consultation is a prior requirement when developing resources on treaty land, but Indigenous inputs from the consultation do not typically influence the outcomes, and this often leads to litigation. Consultation on its own is not sufficient, and the inclusion of Indigenous values is needed to gain consensus. Consensus should be achieved at each stage of the engagement process before moving on to the next stage. A criticism can be that consensus building prolongs the engagement processes. But ignoring or bypassing Indigenous right holders can potentially lead to even longer delays if the litigation route is taken. This can potentially jeopardise the entire projects (as discussed earlier).

In Fig. 1, we present a generic engagement process that focuses on step-by-step consensus building through consultation and inclusion based on the 4Rs. The consensus-building engagement process (CBEP) is broadly based on existing consultation processes indicated in the Alberta government duty to consult (Government of Alberta 2016b) and Fort McKay Sustainability Department consultation process (Fort McKay Sustainability Department 2012). The CBEP is shown as a cyclical process with five stages intended to support decision-making processes that impact Indigenous communities, particularly in resource development projects. For instance, the framing can be applied in oil sand development projects to support the consultation process when exploring the impacts of the project on Indigenous communities.

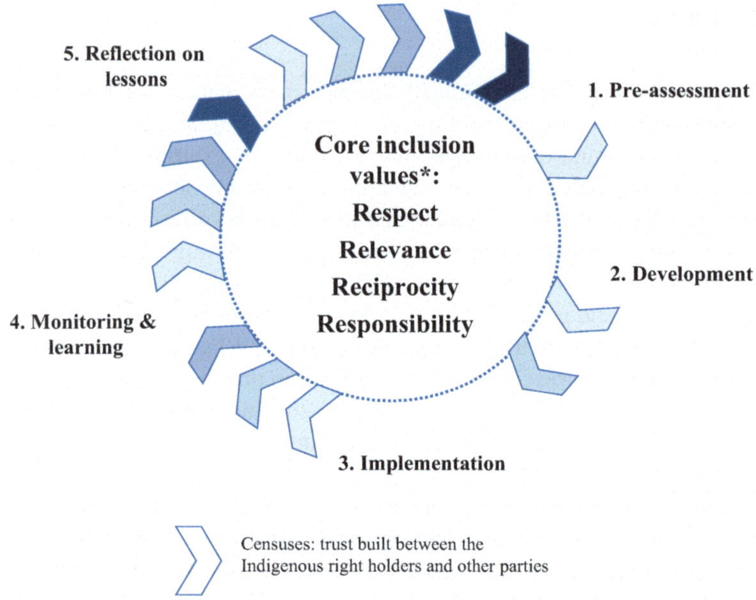

Fig. 1 Consensus-building engagement process. Source: Authors; 4Rs from Kirkness and Barnhardt (2001)

In between the CBEP stages, arrows represent steps needed to build up trust among parties during the consultation process. Effectively, the Indigenous community has veto power in each stage, as consensus is required before moving to the next step. Although consensus may prolong the entire engagement process, we argue that addressing the issues of concern within a respectful space defined by the Indigenous community and other parties is more constructive for building long-term relationships. Concerns can also be addressed through court but as stated earlier is costly for all sides involved. Consensus building will also help build trust in the long run between all parties involved and can further support future decision-making processes.

In describing the framework, we will define non-Indigenous peoples who wish to engage in a dialogue with the Indigenous community as "interested parties". Interested parties can include businesses (proponents), government, researchers or any other non-Indigenous community members. The Indigenous community is the community that will be affected by the interested party's actions, whether it is to meet the legal requirement of consultation or for research purposes.

The circle in the framework represents Indigenous ways of knowledge, cultural traditions and context of the community and reflects the holism or interrelatedness of the intellectual, spiritual, emotional and physical aspects of life. The circle is the core element of the engagement process and represents balance, inclusion, continuity and equality. Western processes and ways of knowledge are introduced and revolve around the circle, which signifies the Indigenous ways of knowledge. When seeking consultation, interested parties enter into the Indigenous community. This explicitly recognises that non-Indigenous perspectives need to respect Indigenous viewpoints. This potentially shifts the power dynamics away from interested parties, who are dependent on Indigenous peoples' consent and goodwill to achieve their desired outcomes. Consensus is gained step by step, with mutual trust as the building blocks of the engagement process. With time, greater trust is gained, and towards the end of the interaction, all participants can gain from the relationship established.

We acknowledge the differences across the many groups of Indigenous peoples in Canada and worldwide and recognise that there are a wide range of cultures, traditions, values and interests. We aim to present a framing with a central position of consulting Indigenous peoples to understand and include their interests and knowledge in shaping future pathways. We also acknowledge the efforts from the Canadian government and industries to involve consultation as part of the decision-making process but argue that much more, meaningful efforts need to be made to actively include Indigenous values, interests and rights through consensus building in order to truly accommodate the needs and interests of Indigenous communities and meet the spirit of UNDRIP.

The consensus-building engagement process (CBEP) framework is not intended to be an all-encompassing prescriptive engagement guideline for parties interested in initiating engagement with an Indigenous community. Rather, the framework intends to encourage a reflective engagement process that is codeveloped by the Indigenous right holders and interested parties in order to come to a consensus on decisions such as developing resources impacting Indigenous communities. The outcomes of the reflective engagement process should be mutually beneficial for the Indigenous community and the interested parties. We recognise each engagement process will differ based on the unique context. The CBEP framing broadly highlights key stages that can be generically considered in engagement processes that include (1) pre-assessment, (2) development, (3) implementation, (4) monitoring and learning and (5) reflection. The phases may not necessarily occur in a linear fashion, and there will likely be overlaps between phases or some phases may begin at the similar timeframes (e.g. implementation and monitoring/learning). The main idea is to recognise that each step is built from consensus gained through trust and the inclusions of the 4Rs.

7.1 Pre-assessment

In the initial stage before the project is underway, the issue at hand is raised by the party wishing to initiate the engagement process. A balance is needed between presenting the issues at hand (e.g. identifying low-carbon pathways) and setting the agenda (e.g. presenting set pathways and asking for validation). More time should be spent at this stage to gather resources by learning about Indigenous values, the community context, protocols and communication methods (Government of Canada 2015). Tools including stakeholder mapping methods (see Nikas et al.

2017; Karakosta et al. 2011) can help to identify a broader stakeholder group through participatory approaches that may not be typically included in conventional consultation processes, for example, single parents, young members of the community and other groups that in number are not the dominant voice in the area. This technique can ensure a heterogenous representation can participate on the CBEP. The mapping method can also help to identify the interactions between stakeholders and make more explicate the important relationships that need greater attention.

Aside from drawing on existing literature which may provide general context, informal contacts can be made with Indigenous associations or independent Indigenous consultants in order to understand the specific protocols required for engagement and the (in)formal decision-making structure in the community. The informal consultation may help to identify problems early on that may be unknown to the parties and potentially surface later on. As a result of the consultation, the issue definition may need to be reassessed and the issues redefined to better reflect Indigenous viewpoints. Language and content of the issues may need to be adapted to consider Indigenous perspectives and use of words.

7.2 Development: Listening and Conversations

The development phase begins when the engagement is formally initiated and primarily consists of planning for the implementation phase where the full engagement takes place with the community. The community should be acknowledged and formally thanked for their participation and goodwill.

This planning phase provides a space for the Indigenous community and interested parties to discuss mutual interest before rolling out the full engagement process. The objective, motivations and outputs of the consultation need to be transparently discussed and the engagement timeframe agreed for the implementation phase. Not all community members may be involved in this phase depending on where the decision-making structures in the community may be focused. For instance, this phase might primarily include the project champions and other decision-makers in the community who may help to decide the form and nature of the engagement process.

Interested parties need to engage in respectful listening and respond responsibly to the insight gained in the planning process with the Indigenous communities. Listening will allow interested parties to gain valuable insights into understanding the communities' priorities and their knowledge base, which together motivate decisions. This sharing of views and interests can help to develop an implementation plan through consensus that is relevant to the community and also helps to meet the objectives of interested parties.

In terms of setting project meetings or deliverable dates, an important note is to allow for flexibility in delivery timeframes and formats. Western project management protocols often require rigid reporting templates and fixed timelines with limited flexibility (and often do not begin consultation early enough). But when engaging with Indigenous communities, the concept of reporting and time needs to be re-evaluated and expectations on both sides communicated clearly and respectfully. It may be preferable to initiate the engagement process earlier and adjustments made accordingly, if possible, to accommodate any reporting alternative and additional time needed. Once a consensus is agreed, the next step involves implementing the engagement plan. Additionally, the timing of the consultation needs to respect the timeframe of the communities by working around cultural practices, ceremonies and other activities. The location of the consultation is also important in terms of inclusion and fair representation. If the location is not thoughtfully arranged and considers the needs of Indigenous communities, it can lead to marginalisation at the spatial level or "social-spatial exclusion" (Trudeau and McMorran 2011; Kühn 2015). For instance, if members of the Indigenous communities are expected to travel far distances to participate in the consultation process, they will be isolated from their community support and may be more vulnerable to external pressures.

7.3 Implementation: Inclusion and Accommodation

The engagement process takes place during the implementation phase. This phase may contain multiple sub-steps and include a number of engagement rounds. Based on the first two phases, interested parties should have a relatively good idea about the context, protocols required and the interests of the Indigenous community. If interested parties are still uncertain about aspects of the implementation process, these concerns should be honestly and openly discussed with the Indigenous representative or community members to help with the learning process and in building trust.

7.4 Monitoring and Learning: Responsibility and Accommodation

The monitoring and learning process occurs when implementation phase is carried out (during and/or after). This phase is a continuous process where the engagement outcomes are being realised. It is essential during this phase that engagement outcomes are responsibly communicated to accurately reflect consultation outcomes. Changes may be needed in this phrase to correct a misunderstanding of results or to more accurately convey outcomes. This helps to ensure that reciprocity is maintained by following up on engagement results. Community members can be consulted on how frequent they would like the communication of results to occur and in what form. This step ensures accountability and responsibility when conveying results from the implementation stage and helps to assess if the results are relevant to Indigenous interests.

7.5 *Reflection: Lessons*

This is the final phase of the engagement process, which can consist of presenting outcomes of the engagement process to the Indigenous community members and other relevant audiences (together or separately as mutually agreed). Project outputs codeveloped with the community should also be made fully available to the community, as they are also co-owned by the community. The explicit acknowledgement of the co-ownership of results shows respectful consideration of participation from all parties and acknowledges that results are an outcome of consensus building between Indigenous communities and interested parties.

8 Conclusions

The development and implementation of effective policies for climate change mitigation action require a true inclusion of the concerns, cultural values and interests of communities affected by these policies. To achieve true inclusion, an inclusive communication system can be followed throughout the engagement life cycle, considering the fundamental principles of the community as the core elements for consultation, participation and implementation. This approach is especially needed when the communities at risk of the negative impacts associated with resource development are conformed by Indigenous peoples. The fundamental principles of Indigenous knowledge not only play a role in communication and cultural exchange but also provide generation-long information of the local land and valuable perspectives on sustainability and climate change.

Indigenous world views emphasise the unity of the individual with the environment and all living things, and these values are aligned with sustainability and low-carbon futures. Meanwhile, modern economic Western paradigms perceive societal development primarily from an economic growth perspective. These different world views can increase the risks for climate change mitigation action at the government, industry and community level. Acknowledging the existence of different world views requires that more effective and collaborative methods of communication are needed to develop a true inclusive action plan. As an approach to achieving inclusion and reducing risks of negatives outcomes, a consensus-building engagement process (CBEP) was proposed with Indigenous values and interests as core elements.

In the CBEP, the Indigenous community essentially has veto power in the decisions made in each stage for the decision-making process (i.e. developing oil sand resources). Reaching consensus is necessary before moving forward to the subsequent steps as a means to ensure mutual understanding among the entities involved before a final decision is taken. Although it is advised to consider Indigenous knowledge and interests as a starting point, we acknowledge a total shift in the thought process of the parties is not yet possible since individuals (non-Indigenous) tend to rely on their own set of values in decision-making. However, we consider that by having the intention to place Indigenous knowledge and interests first, we can provide a framing to allow all parties to make more balanced analyses of both perspectives.

On the other hand, considering the resources and influence of government institutions and companies, it is not realistic (yet) to assume there will be an equal level of power for parties participating in an engagement process. Therefore, a way to mitigate the unequal power dynamics can be through inviting a neutral third party to oversee, facilitate and champion the process. The facilitator can be an expert that has knowledge and experience working with Indigenous communities. The third-party facilitator may improve communication and act as a mediator when needed. The government also needs to rethink its risk-averse approach by participating directly in the consultation in a meaningful way. This will allow for better communication between the government and Indigenous peoples, a relationship that has historically been difficult. However, cooperation is unavoidable if Canada wishes to seriously address climate change and sustainability and the impact and risks of resource development on Indigenous communities.

In addition to the consensus-building engagement process that considers local communities' interests and values, the equal inclusion of Indigenous communities in the decision-making process (e.g. including veto power) is essential to the development of a sustainable and legitimate low-carbon pathway. Additional studies are also needed to develop best corporate and government practices for inclusion to reduce the risks associated with resource development in communities and the environment. More Indigenous voices and perspectives are needed in the scientific climate change community, including IPCC reports, journal articles and other media sources, to disseminate values and concepts that balance socioeconomic needs and promote environmental stewardship relevant for the climate change mitigation needs of today and the future.

Canada, along with many high-emitting countries, can no longer continue to develop its high-carbon energy resources without facing risks of litigation and backlash not only from the Indigenous communities but from international pressures. The Paris Agreement and its objectives have changed the playing field, and countries must address climate change as well as consider the rights and interests of local communities, particularly Indigenous people who are impacted by risk related to climate change and mitigation action. Canada can lead in this area by setting a precedence for engaging in fair and effective consultation with Indigenous communities to address the complex problems of climate change.

References

Alexander, C., Bynum, N., Johnson, E., King, U., Mustonen, T., Neofotis, P., Oettlé, N., Rosenzweig, C., Sakakibara, C., Shadrin, V., Vicarelli, M., Waterhouse, J., & Weeks, B. (2011). Linking indigenous and scientific knowledge of climate change. *BioScience, 61*(6), 477–484.

AMNH, American Museum of Natural History. (2008). Sustaining cultural and biological diversity in a rapidly changing world. In *Sustaining cultural and biological diversity in a rapidly changing world: Lessons for global policy*. New York City. Accessed from http://www.wennergren.org/history/sustaining-cultural-and-biological-diversity-rapidly-chaning-world-lessons-global-policy.

Archibald, J. A. (2008). *Indigenous storywork: Educating the heart, mind, body, and spirit*. Vancouver: UBC Press.

Baker, S. H. (2013). *Why the IFC's free, prior, and informed consent policy doesn't matter (yet) to indigenous communities affected by development projects*.

Baker, M. J., & Westman, C. (2018). Extracting knowledge: Social science, environmental impact assessment, and Indigenous consultation in the oil sands of Alberta, Canada. *The Extractive Industries and Society, 5*(1), 144–153.

Boothe, P., & Boudreault, F.-A. (2016). *By the numbers: Canadian GHG emissions*. Lawrence National Centre for Policy and Management. Accessed from http://www.ivey.uwo.ca/news/news-ivey/2016/1/by-the-numbers-canadian-ghg-emissions/.

Cajete, G. (1994). *Look to the mountain: An ecology of Indigenous education*. Skyland, NC: Kivaki Press.

Calgary Chamber. (2015). *The consultation conundrum: Examining aboriginal consultation in Alberta*.

Daniel, R., Wright, T., & Limited, D. (1980). Treaties of the Northwest, 1871–1930. In *A history of native claims processes in Canada, 1867–1979* (p. 9). Ottawa: Department of Indian and Northern Affairs.

Dearden, P., & Mitchell, B. (1998). *Environmental change and challenge: A Canadian perspective*. Toronto, ON: Oxford University Press.

Draper, D. (1998). *Our environment: A Canadian perspective*. Scarborough, ON: International Thomson.

Energy Resources Conservation Board, A. (2010). *ST98-2010: Alberta's energy reserves 2009 and supply/demand outlook 2010–2019*. Accessed from http://www.aer.ca/documents/sts/ST98/st98_2010.pdf.

Environment Canada. (2017). *Canadian economic sector provinces territories 1990–2015*. Accessed from http://donnees.ec.gc.ca/data/substances/monitor/national-and-provincial-territorial-greenhouse-gas-emission-tables/D-Tables-Canadian-Economic-Sector-Provinces-Territories/?lang=en.

Ermine, W. (2000). Aboriginal epistemology. In M. Battiste & J. Barman (Eds.), *First nation education in Canada: The circle unfolds* (pp. 101–111). Vancouver, BC: UBC Press.

Ford, J. D., et al. (2016). Including Indigenous knowledge and experience in IPCC assessment reports. *Nature Climate Change, 6*(4), 349–353. https://doi.org/10.1038/nclimate2954.

Fort McKay Sustainability Department. (2012). *Consultation process – Proponent guide*.

Government of Alberta. (2013). *The Government of Alberta's policy on consultation with first nations on land and natural resource management, 2013*.

Government of Alberta. (2016a). *The Government of Alberta's guidelines on consultation with Metis settlements on land and natural resource management 2016*.

Government of Alberta. (2016b). *Oil Sands Emissions Limit Act, Statutes of Alberta, 2016* (p. Chapter O-7.5). Accessed from http://www.qp.alberta.ca/1266.cfm?page=O07p5.cfm&leg_type=Acts&isbncln=9780779794768.

Government of Canada. (1982). Rights of the aboriginal peoples of Canada. In *Part II. Constitution Act, 1982*. Accessed from http://laws-lois.justice.gc.ca/eng/Const/page-16.html.

Government of Canada. (2015). *First nations communications toolkit*.

Hart, M. A. (2010). Indigenous worldviews, knowledge, and research: The development of an indigenous research paradigm. *Journal of Indigenous Voices in Social Work, 1*(1), 1–16.

Henton, D. (2016, April 26). Fort McKay First Nation sues Alberta over energy development. *Calgary Herald*. Accessed from http://calgaryherald.com/business/energy/fort-mckay-first-nation-sues-alberta-over-energy-development.

Indigenous and Northern Affairs, C. (2008). *Treaty-making in Canada*. Accessed from http://www.aadnc-aandc.gc.ca/eng/1100100028574/1100100028578.

Indigenous and Northern Affairs, C. (2009). *Treaty research report – Treaty Eight (1899)*. Accessed from http://www.aadnc-aandc.gc.ca/eng/1100100028809/1100100028811.

Indigenous and Northern Affairs, C. (2011). *Aboriginal consultation and accommodation – Updated guidelines for Federal officials to fulfil the duty to consult*.

Joseph, B. (2017). *Aboriginal rights, title, and the duty to consult*. Port Coquitlam.

Karakosta, C., Doukas, H., & Psarras, J. (2011). CDM sustainable technology transfer grounded in participatory in-country processes in Israel. *International Journal of Sustainable Society, 3*(3), 225–242.

Kirkness, V. J., & Barnhardt, R. (1991). First nations and higher education. *Journal of American Indian Education, 30*(3), 1–18.

Kirkness, V. J., & Barnhardt, R. (2001). First nations and higher education: The four R’s – respect, relevance, reciprocity, responsibility. In R. Hayoe & J. Pan (Eds.), *Knowledge across cultures: A contribution to dialogue among civilizations*. Hong Kong: Comparative Education Research Centre, The University of Hong Kong.

Kovach, M. E. (2000). *Indigenous methodologies: Characteristics, conversations, and contexts*. Toronto: University of Toronto Press.

Kovach, M. (2010). Conversational method in indigenous research. *First Peoples Child & Family Review, 5*(1), 40–48.

Kühn, M. (2015). Peripheralization: Theoretical concepts explaining socio-spatial inequalities. *European Planning Studies, 23*(2), 367–378.

Laidlaw, D., & Passelac-Ross, M. (2014, March). *Alberta First Nations consultation & accommodation handbook* (p. 122). Accessed from http://prism.ucalgary.ca/bitstream/1880/50216/1/ConsultationHandbookOP44.pdf.

Leach, A., et al. (2015). *Executive summary. CLIMATE LEADERSHIP*. Report to Minister. Alberta Minister of Environment and Parks. Accessed from http://www.alberta.ca/climate-leadership-plan.aspx.

Lertzman, D., & Vredenburg, H. (2015). Indigenous peoples, resource extraction and sustainable development: An ethical approach. *Journal of Business Ethics, 56*(239), 239–254.

McGregor and Debroah. (2004). Traditional ecological knowledge and sustainable development: Towards coexistence. In M. Blaser, H. A. Feit, & G. McRae (Eds.), *In the way of development: Indigenous peoples, life projects, and globalization* (pp. 72–91).

Mustonen, T. (2005). Stories of the Raven – Snowchange 2005 Conference Report Anchorage, Alaska, USA.

Natural Resources Canada, N. R. C. (2016). *Energy fact book*. Accessed from https://www.nrcan.gc.ca/sites/www.nrcan.gc.ca/files/energy/pdf/EnergyFactBook_2016_17_En.pdf.

Nikas, A., et al. (2017). Managing stakeholder knowledge for the evaluation of innovation systems in the face of climate change. *Journal of Knowledge Management, 21*(5), 1013–1034.

Pillay, N. (2013). *Free, prior and informed consent of indigenous peoples*. Accessed from http://www.ohchr.org/Documents/Issues/IPeoples/FreePriorandInformedConsent.pdf.

Pryce, P. (1992). The manipulation of culture and history: A critique of two expert witnesses. *Native Studies Review, 8*(1), 35–46.

Scollon, R., & Scollon, S. (1981). *Narrative literacy and face in interethnic communication*. Norwood, NJ: Ablex Publishing Corporation.

Simpson, L. (2000). Anishinaabe ways of knowing. In J. Oakes, R. Riew, S. Koolage, L. Simpson, & N. Schuster (Eds.), *Aboriginal health, identity and resources* (pp. 165–185). Winnipeg, MB: Native Studies Press.

Symon, C., Arris, L., & Heal, B. (Eds.). (2005). *Arctic climate impact assessment*. Cambridge University Press. Accessed from http://caid.ca/ACIA.html.

The Constitution Act. (1982). *Schedule B to the Canada Act 1982 (UK), c 11*.

Thomas, R. (2005). Honoring the oral traditions of my ancestors through story-telling. In S. Strega & L. Brown (Eds.), *Research as resistance* (pp. 237–254). Toronto: Canadian Scholars’ Press.

TORYS. (2016). *Free prior and informed consent in Canada. Towards a new relationship with indigenous peoples.*
Trudeau, D., & McMorran, C. (2011). The geographies of marginalization. In *A companion to social geography* (pp. 437–453). Wiley-Blackwell.
Truth and Reconciliation Commission. (2015). *Truth and reconciliation commission: Calls to action.*
UNFCCC, S. (2015). *Canada intended nationally determined contributions (INDCs).* United Nations Framework Convention on Climate Change.
UNFCCC, S. (2017). *The Paris Agreement – Main page.* Accessed October 5, 2017, from http://unfccc.int/paris_agreement/items/9485.php.
United Nations. (2008). *United Nations declaration on the rights of indigenous peoples.*
University of Alberta Faculty of Native Studies. (2015). *'Indigenous Canada: Looking Forward/Looking Back: Worldview'. Module 1.* Edmonton: University of Alberta.
Weber, B. (2016, April 27). Alberta to repeal Prentice's First Nations consultation law Bill. *CBC News.* Accessed from http://www.cbc.ca/news/canada/edmonton/alberta-to-repeal-prentice-s-first-nations-consultation-law-1.3555973.
Wilson, S. (2001). What is Indigenous research methodology? *Canadian Journal of Native Education, 25*(2), 175–179.
Wood Buffalo Environmental Association. (2010). *Traditional ecological knowledge/Berry Monitoring Project Factsheet.*
Wood Buffalo Environmental Association. (2017). *Wood Buffalo Environmental Association | Traditional knowledge.* Accessed from http://www.wbea.org/traditional-knowledge.
World Resources Institute, W. R. I. (2016). *CAIT climate data explorer – Historical emissions.* Accessed from http://cait.wri.org/historical/, Environment Canada and World Bank Population data?indicator[]=Total GHG Emissions Excluding Land-Use Change and Forestry&indicator[]=Total GHG Emissions Including Land-Use Change and Forestry&year[]=2012&sortIdx=NaN&chartT.

An Application of Calibration and Uncertainty Quantification Techniques for Agent-Based Models

Sotiris Papadelis and Alexandros Flamos

Abstract In this chapter, a step-by-step application of calibrating an agent-based model is presented. In particular, an agent-based model for small-scale PV adoption was calibrated on the historical data for the small-scale solar PV capacity additions that took place in Greece from January 2010 to February 2013. The process of the model calibration allowed to (a) quantify and take into consideration uncertainties that are related to the characteristics and the decision-making criteria of the agents (i.e. independent PV power producers), in contrast to the more obvious uncertainties, such as technology costs, and (b) use the calibration results to explore the plausible—given the historical data—behaviour of the potential PV adopters in Greece under the new net-metering scheme (in effect as of mid-2015).

Keywords Agent-based · Calibration · Stochastic emulators · History matching · Uncertainty quantification

1 Introduction

Models try to narrow the differences between decision-makers' thinking, reasoning, representation and computing (Doukas 2013). Anytime we decide to develop a model, we have two choices. The first one is to rely on physical laws that we know they hold true or mathematical relationships that are valid as long as the underlying theory is valid too. The second choice is to come up with a plausible model that captures relationships and dynamics that we expect to be true and acknowledge the fact that many parts of the model will inevitably be arbitrary.

Agent-based models (ABMs) fall into the second choice (Flamos 2016). This means that developing a new ABM is an interesting endeavour, but unless we

S. Papadelis · A. Flamos (✉)
Technoeconomics of Energy Systems Laboratory (TEESLab), Department of Industrial Management & Technology, University of Piraeus (UNIPI), Piraeus, Greece
e-mail: aflamos@unipi.gr

H. Doukas et al. (eds.), *Understanding Risks and Uncertainties in Energy and Climate Policy*, https://doi.org/10.1007/978-3-030-03152-7_3

identify the uncertainty that governs the applicability of the model, it is just a modelling exercise. While ABMs give us increased flexibility in modelling complex systems driven by agent actions and interactions, they give rise to the question of when an ABM represents reality well enough to extrapolate into the future.

A common approach to validating the extent at which an ABM is realistic is through calibration using historical data. If done right, this process is an opportunity to deal with model uncertainty, which is the specific type of uncertainty that stems from the fact that there exist many variations of an ABM that are all plausible under the same set of historical data.

Agreement between model and data does not imply that the modelling assumptions accurately describe the processes producing the observed behaviour; it merely indicates that the model is one (of maybe several) that is plausible. As a result, dealing with model uncertainty during calibration allows us to address the trade-off between a very flexible model that can easily fit on the historical data and model uncertainty.

In this chapter, we present a method for quantifying model uncertainty. The ABM used aims at replicating the dynamics of small-scale PV adoption in Greece. Its outputs are determined by the values of its ten free parameters. As a result, before the model can be used for forward-looking simulations, these parameters must be constrained into a subspace of plausible values.

Accordingly, the model was calibrated on the data for the period from 2010 to 2013. This period was the period with the largest additions in small-scale PV capacity (Anagnostopoulos et al. 2017; Papadelis et al. 2016). The calibration data corresponds to:

a) Past prices and market conditions (prices for small-scale PV systems and the relevant Greek feed-in tariff (FIT) prices)
b) The requests for grid connections, the records for which are available at the individual request level[1]

The chapter is structured as follows: Sect. 2 presents a brief description of the employed ABM, Sect. 3 presents the concept of Gaussian process emulators that play a central role in the calibration method that was applied, Sect. 4 presents the details of fitting a Gaussian process emulator on the ABM results and Sect. 5 presents the method and results of the model calibration. The chapter concludes with a discussion.

2 The ABM for the Diffusion of Small-Scale Solar PV

The parameters that govern the model's behaviour are as follows:

1. **Agent initial beliefs**. Each agent has a private initial belief for the expected annual cash inflows from investing in 300 Wp of solar PV. This belief is

[1] http://www.rae.gr/site/categories_new/regirsty/licences.csp

expressed as a Gaussian distribution with a mean value μ^{CF} and a precision ρ^{CF}. Low values of ρ^{CF} reflect flexibility in beliefs (i.e. little evidence is enough to shift the agent's beliefs). Low values of μ^{CF} reflect pessimism about the payoffs of the investment. The initial beliefs of each agent are randomly drawn from two global probability distributions—one for the mean value μ^{CF} and one for the precision ρ^{CF} of each agent—while the parameters of both distributions are regarded as parameters that need to be set through calibration.

2. **Social learning**. To capture the effects of social learning, each agent receives information from the agents in its social circle that have already invested in PV. This information concerns the actual profitability of their investments so far, and it is used to update[2] the agent's beliefs. Accordingly, each agent has a social circle. This condition is modelled as a "small-word" network (Watts and Strogatz 1998).
3. **Resistance towards PV investments**. Each agent is characterized by their resistance towards investing in solar PV. Resistance is defined as a weighted sum of two parameters:

   (a) The profitability of the investment expressed by its payback period, so that the larger the profitability, the shorter the payback period, and, thus, the lower the resistance.

   We assume that agents are able to use their beliefs regarding the expected cash inflows to estimate the profitability of investing. Given the fact that the agents' beliefs are expressed probabilistically, we can calculate the following z_i value, where $i = 1,2,\ldots$ represents the years from the simulation's "now":

$$z_i = \frac{\mu^{\mathrm{CF}} \cdot \sum_{t=1}^{i} \left(\frac{1}{(1+d)^t}\right) - \mathsf{capex}}{i/\rho^{\mathrm{CF}}} \sim N(0,1)$$

where capex is the capital expenditure and d the discount rate.

   The term z_i gives us the probability that the (discounted) payback period of the investment is equal to i. Making the assumption that all agents evaluate an investment based on when it will have paid itself back with probability 90%; this formula gives us the respective payback period i.

   (b) The difference between the total number of agents in the simulation and the number of them that has already invested in PV—so that the larger the installed base, the smaller the resistance.

   Baranzini et al. (2017) provide evidence that both learning and imitation are important components of social contagion for the adoption of solar PV technology. By making attitude towards PV a function of the installed base, we aim to capture the imitation (social influence) aspect.

[2]This is equivalent to updating a Gaussian prior given a new observation.

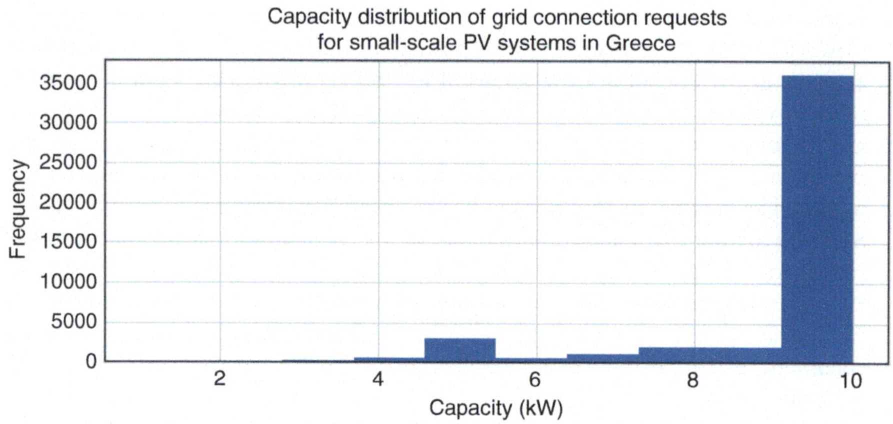

Fig. 1 Size distribution of grid connection requests for small-scale PV systems in Greece

The weights of this sum are derived from a global probability distribution, the parameters of which are regarded as parameters that need to be set through calibration.

4. **Propensity to invest**. Each agent has a threshold value for their resistance parameter. When the latter falls under the threshold value, action could—but not necessarily will—be induced. For the calibration phase, we assumed that when agents decide to invest in a PV system, its size is given by the empirical probability distribution that was derived from the available historical data (Fig. 1).

Finally, Young (2009) notes that inertia is the simplest reason why innovations take time to diffuse: people delay in acting on new information. Accordingly, we have included inertia in the model by defining a global parameter (i.e. it has the same value for all agents in the model) that represents the probability that any agent would actually invest if their resistance threshold is crossed.

3 The Concept of Emulators

An emulator is a probabilistic approximation of the ABM. The probabilistic nature of the emulators makes them ideal for the quantification of the uncertainty regarding their estimations, as well as of the way the uncertainty of the ABM parameters gets propagated into its results. In addition, emulators can be evaluated significantly faster in comparison to the actual model, which allows us to employ Monte Carlo sampling methods that would be otherwise prohibitively expensive in terms of computation resources.

In the application presented in this chapter, Gaussian processes are used as emulators. Accordingly, a brief description of Gaussian processes is provided.

3.1 Gaussian Processes for Regression

Suppose that we have some function $f(x)$ and we want to estimate the value of this function at the input points x_1, x_2 and x_3. As long as we believe that a suitable prior of each variable $f(x)$ could follow a Gaussian distribution, the marginalization property of the Gaussian allows us to write

$$\boldsymbol{f} = \begin{bmatrix} f(x_1) \\ f(x_2) \\ f(x_3) \end{bmatrix} \sim N\left(\begin{bmatrix} f(x_1) \\ f(x_2) \\ f(x_3) \end{bmatrix} \mid \begin{bmatrix} m_f(x_1) \\ m_f(x_2) \\ m_f(x_3) \end{bmatrix}, \begin{bmatrix} \sigma_f^2(x_1) & 0 & 0 \\ 0 & \sigma_f^2(x_2) & 0 \\ 0 & 0 & \sigma_f^2(x_3) \end{bmatrix} \right) \tag{1}$$

where $m_f(x_i)$ is the mean value of $f(x_i)$ and $\sigma_f^2(x_i)$ its variance.

If we also believe that the values of this function are related to each other (i.e. knowing one can tell us something about the others), we can assume that the function values follow a joint Gaussian distribution that imposes covariance between these values. Covariance is introduced by defining a covariance function $k_f(\boldsymbol{x},\boldsymbol{x}^*)$ that models the covariance between the function values at different inputs $\boldsymbol{x}$ and $\boldsymbol{x}^*$:

$$k_f(\boldsymbol{x},\boldsymbol{x}^*) = \mathrm{Cov}[f(\boldsymbol{x}),f(\boldsymbol{x}^*)] \tag{2}$$

For the case of the three-dimensional Gaussian distribution of this example, the prior distribution becomes now of the form:

$$f(\boldsymbol{x}) = \begin{bmatrix} f(x_1) \\ f(x_2) \\ f(x_3) \end{bmatrix} \sim N\left(\begin{bmatrix} f(x_1) \\ f(x_2) \\ f(x_3) \end{bmatrix} \mid \begin{bmatrix} m_f(x_1) \\ m_f(x_2) \\ m_f(x_3) \end{bmatrix}, \begin{bmatrix} k_f(x_1,x_1) & k_f(x_1,x_2) & k_f(x_1,x_3) \\ k_f(x_2,x_1) & k_f(x_2,x_2) & k_f(x_2,x_3) \\ k_f(x_3,x_1) & k_f(x_3,x_2) & k_f(x_3,x_3) \end{bmatrix} \right) \tag{3}$$

If we write $\boldsymbol{x} = [x_1, x_2, x_3]^{\mathrm{T}}$, the distribution in Eq. (3) can be written in a more compact form as:

$$f(\boldsymbol{x}) = \begin{bmatrix} f(x_1) \\ f(x_2) \\ f(x_3) \end{bmatrix} \sim N(\boldsymbol{f} \mid m_f(\boldsymbol{x}), K_f(\boldsymbol{x},\boldsymbol{x})) \tag{4}$$

where the mean function $m_f(\boldsymbol{x})$ reflects the expected function value at input $\boldsymbol{x}$:

$$m_f(\boldsymbol{x}) = E[f(\boldsymbol{x})] \tag{5}$$

A *Gaussian process* (GP) generalizes the aforementioned multivariate normal to infinite dimensions. It is defined as a collection of random variables f, any finite number of which has a joint Gaussian distribution (Rasmussen and Williams 2006). If we assume a GP prior over a set of function values f, we actually assume that:

1. **Their marginal distribution is Gaussian**. For any $\boldsymbol{x} = \boldsymbol{x}_i$, the prior distribution for $f(\boldsymbol{x}_i)$ is

$$f(\boldsymbol{x}_i) \sim N\big(m_f(\boldsymbol{x}_i), k_f(\boldsymbol{x}_i, \boldsymbol{x}_i)\big) \tag{6}$$

The covariance function k_f is called the kernel of the GP.

2. **Their joint distribution is also Gaussian**. For a sample $\mathbf{X} = \{\boldsymbol{x}_i,\ \ i = 1, \ldots, n\}$ consisting of n pairs of inputs, the probability density function of the f's values is

$$p(f|\mathbf{X}) = N(f \mid m_f(\mathbf{X}), K_{ff}(\mathbf{X})) \tag{7}$$

where $K_{ff}(\mathbf{X})$ is the covariance matrix obtained by evaluating the kernel k_f on all data points in $\mathbf{X}$:

$$K_{ff}(\mathbf{X}) = \begin{bmatrix} k_f(\boldsymbol{x}_1, \boldsymbol{x}_1) & k_f(\boldsymbol{x}_1, \boldsymbol{x}_2) & \cdots & k_f(\boldsymbol{x}_1, \boldsymbol{x}_n) \\ k_f(\boldsymbol{x}_2, \boldsymbol{x}_1) & k_f(\boldsymbol{x}_2, \boldsymbol{x}_2) & \cdots & k_f(\boldsymbol{x}_2, \boldsymbol{x}_n) \\ \vdots & \vdots & \vdots & \vdots \\ k_f(\boldsymbol{x}_n, \boldsymbol{x}_1) & k_f(\boldsymbol{x}_n, \boldsymbol{x}_2) & \cdots & k_f(\boldsymbol{x}_n, \boldsymbol{x}_n) \end{bmatrix} \tag{8}$$

Once a mean function and a kernel have been chosen, we have actually decided on a prior for the function f. This is a distribution over possible functions. Since it is not conditioned on any observed data, it represents our prior assumption of the function space from which the data may have been generated.

When we do observe some actual data $\boldsymbol{D} = \{(\boldsymbol{x}_{i,:}, y_i),\ \ i = 1, \ldots, n\}$, the GP prior is conditioned on $\boldsymbol{D}$ to obtain a posterior GP process that represents all functions that are consistent with both the observed data and the prior. This posterior GP process $p(f|\boldsymbol{D})$ is used as a proxy for the original ABM.

The main advantage from using a GP is that it becomes analytically tractable to estimate the values of the approximated function on inputs that are not part of the observed data. In particular, we can make predictions for new inputs $\boldsymbol{X}_*$ by drawing f_* from the posterior $p(f|\boldsymbol{D})$. This posterior predictive distribution can be written as[3]:

$$p(f_*|\boldsymbol{D}) = N\Big(f_* \mid m_f(\boldsymbol{X}_*) + K_{*f}K_{ff}^{-1}\big(y - m_f(\boldsymbol{X})\big),\ \ K_{**} - K_{*f}K_{ff}^{-1}K_{f*}\Big) \tag{9}$$

[3]Assuming that there are no observation errors.

This means that we can fit a GP on the results from a small number of parameter combinations for the ABM and then use the GP's posterior predictive distribution to estimate the model's results for other parameter combinations without a need to re-run the ABM.

3.2 Benefits of Using Gaussian Processes as Emulators

An important implication is that we have a formula that calculates the predictive uncertainty at each test input X_*:

$$\mathrm{Var}(f|X_*) = K_{**} - K_{*f}K_{ff}^{-1}K_{f*} \tag{10}$$

This allows us to employ a sequential sampling scheme:

1. The GP emulator is fitted on the results from a small number of parameter combinations at which the ABM is run.
2. Additional parameter combinations for the ABM can be chosen corresponding to regions where the emulator's predictive uncertainty is high.
3. The GP emulator is updated to include the new ABM results.

The function $\mathrm{Var}(f|X_*)$ can accept matrices as inputs. This allows us to evaluate it on a large set of possible inputs and then either select only the inputs (i.e. matrix rows) that correspond to a predictive uncertainty that is higher than a predetermined threshold or select a predetermined number of inputs that correspond to the highest predictive uncertainty.

4 The Design and Validation of the GP Emulator

4.1 Options for the Emulator Form

Following Haylock and O'Hagan (1996), the general form for the emulator to fit on the model output j is

$$g_j(\boldsymbol{x}) = \sum_{i=1}^{q} h_i(\boldsymbol{x})\beta_i + \boldsymbol{u}(\boldsymbol{x}) \tag{11}$$

The first term is a regression term, where $h_i(\cdot)$ are deterministic functions of the model parameters and β_i the regression coefficients. The second term is a GP with a zero mean function and an appropriately chosen kernel. One way to interpret this formulation is that the ABM is approximated by a regression term, which is then improved by a GP. Alternatively, this formulation can be interpreted as an emulator that follows a GP that is given by:

$$p\big(g_j(\boldsymbol{x})\big) = N\left(\boldsymbol{x} \mid \sum_{i=1}^{q} h_i(\boldsymbol{x})\beta_i, k_f(\boldsymbol{x}, \boldsymbol{x} \mid \boldsymbol{\theta})\right) \tag{12}$$

where $\boldsymbol{\theta}$ denotes the parameters of the kernel.

The most common approach for estimating the parameters of the GP emulator (i.e. the coefficients β_i and the parameters $\boldsymbol{\theta}$ of the kernel) is the "plug-in" approach of Kennedy and O'Hagan (2001), where the parameters are estimated using maximum likelihood and, then, treated as fixed during the calibration of the model.

As far as the GP kernels are concerned, the most popular choice is the squared exponential kernel, which implies that the function to approximate has infinitely many derivatives. If the smoothness assumption of the squared exponential kernel is unrealistic for a specific modelling case, the Matérn kernel class can be used as an alternative (Genton 2002).

Both the squared exponential kernel and the kernels of the Matérn class include the following term:

$$k\big(\boldsymbol{x}_i, \boldsymbol{x}_j\big) \propto \frac{\left\|\boldsymbol{x}_i - \boldsymbol{x}_j\right\|}{\ell} \tag{13}$$

The length scale ℓ determines how close two data points x_i, x_j have to be to influence each other significantly. If we allow for different values of ℓ per input dimension, then we achieve automatic relevance determination, so named because estimating the length scale parameters $\ell_1, \ell_2, \ldots, \ell_q$ implicitly determines the "relevance" of each dimension. Input dimensions with relatively large length scales imply relatively little variation/impact on the function being approximated.

4.2 *Fitting the GP Emulator*

In the application presented in this chapter, a GP emulator—the STatistical approximation-based modEl EMulator (STEEM) of the Technoeconomics of Energy Systems Laboratory (TEESlab)—was fitted on the results from a number of parameter combinations at which the ABM was run. This means that the parameter combinations were the input data and the ABM results were the output data for the GP emulator. The mean function of the emulator was set to zero, while the Matérn 3/2 kernel was used as the covariance function.

To generate the input data, preliminary ranges for the value of each model parameter were derived, and then the ABM was run for 150 different parameter combinations. These initial combinations were chosen using a maximin Latin

Table 1 ABM parameter description and initial ranges

	Description	Min	Max
1.	The shape parameter of the global distribution that assigns μ^{CF} to each agent in the model	100	250
2.	The scale parameter of the global distribution that assigns μ^{CF} to each agent in the model	10	50
3.	The shape parameter of the global distribution that assigns ρ^{CF} to each agent in the model	10	50
4.	The scale parameter of the global distribution that assigns ρ^{CF} to each agent in the model	5	20
5.	The shape parameter of the global distribution that assigns the weight of the profitability to each agent's resistance	0.5	5
6.	The scale parameter of the global distribution that assigns the weight of the profitability to each agent's resistance	0.1	1
7.	The shape parameter of the global distribution that assigns the weight of the installed base to each agent's resistance	0.5	5
8.	The scale parameter of the global distribution that assigns the weight of the installed base to each agent's resistance	0.1	1
9.	The shape parameter of the global distribution that assigns each agent's threshold value for their resistance parameter	10	30
10.	The scale parameter of the global distribution that assigns each agent's threshold value for their resistance parameter	5	10

hypercube design, to fill the entire input space by maximizing the minimum distance between the points generated.

The initial ranges for the ABM parameters were defined as presented in Table 1.

There are two possible caveats when selecting the initial ranges for the model parameters to be calibrated:

1. The ranges are wider than necessary. Sensitivity analysis will be carried out to test if and for which parameter this is true.
2. The ranges are too narrow. At the end of the calibration, one should revisit these ranges.

The corresponding PV capacity pathways were scaled to the [0, 1] range. This allowed the comparison of the simulated pathways with the—also scaled—pathway of the actual PV investments during the calibration period.

The ABM calibration was conducted with the assumption that the model represents reality well if it can replicate the historical growth rates for PV installations. In other words, calibration looked for similar shapes of the cumulative capacity of PV installations, while the scale is inevitably different. At the same time, calibrating based only on the shape of the simulated pathways creates a problem of non-identifiability; we need a final target to use as a point of reference. To this end, the median of the capacities achieved at the end of the simulated period was mapped to the value of 1 during the scaling. In addition, the input data was normalized.

The ABM results were split into four periods (mid-2010, end-2010, end-2011, end of simulation period), and four GP emulators (four implementations of the STEEM) were fitted on the respective results. The hyperparameters of each emulator were estimated using likelihood maximization, and the Python package GPy[4] was employed for this purpose.

4.3 Diagnostics

Bastos and O'Hagan (2009) present different validation methods for GP emulators. In this chapter, the generalization capability of the fitted emulator was judged by performing k-fold cross-validation on the emulator that corresponds to the results at the end of the simulated period. K-fold cross-validation was selected to make sure that there is no region in the input space that the emulator has been poorly fitted. As a metric for the validation, the coefficient of Nash-Sutcliffe efficiency was used, which is given by:

$$E = 1 - \frac{\sum_{i=1}^{m} \left(y_i - y_{i,\mathrm{GP}}\right)^2}{\sum_{i=1}^{m} \left(y_i - \bar{y}\right)^2} \tag{14}$$

where m represents the number of data in each fold, y_i are the respective outputs of the ABM, $\bar{y}$ is the mean of these outputs and $y_{i,\mathrm{GP}}$ are the estimations of the GP emulator for the same parameter inputs. This coefficient has an optimal value of 1; a value of 0 indicates that a forecast using the mean of the ABM outputs will have the same utility as the result of the GP emulator.

4.4 Sensitivity Analysis

The preliminary ranges for the value of each model parameter were chosen arbitrarily with the goal of fitting a GP emulator on a reasonably large input space. However, after the emulator is fitted, it is useful to revisit the ranges by performing a sensitivity analysis that is facilitated by the emulator. Sensitivity analysis explains which of the d features in the input dataset $\mathbf{X} = \{\boldsymbol{x}_j,\ j = 1,\ldots,d\}$ are most responsible for the uncertainty in the model's result. It makes sense to restrict the range of the parameters that have only a small impact on the uncertainty of the model's outputs.

For each feature $j : 1,2,\ldots d$, we split the dataset $\mathbf{X}$ into two parts, one including the selected feature and the other the remaining ones:

[4] https://github.com/SheffieldML/GPy/

$$EX = \left(\boldsymbol{x}_j, \mathbf{X}_{-j}\right) \tag{15}$$

Then, we can assess the sensitivity of the model's output to the uncertainty regarding $\boldsymbol{x}_j$ through the expected reduction in the output's variance if the true value of $\boldsymbol{x}_j$ is learnt. Saltelli et al. (2010) provide the following formula for the calculation of the expected reduction in variance:

$$\mathrm{Var}_{x_j}\left(E\left(y \mid x_j\right)\right) = \frac{1}{N} \sum_{i=1}^{N} \left\{ f(\boldsymbol{B})_i \left(f\left(\boldsymbol{A}_{\boldsymbol{B}}^{(j)}\right)_i - f(\boldsymbol{A})_i \right) \right\} \tag{16}$$

$\boldsymbol{A}$ and $\boldsymbol{B}$ are two independent $N \times d$ matrices that contain samples of the model's inputs. Following Saltelli et al. (2010), a Sobol sequence was used to derive these matrices. Sobol sequences were designed to cover the unit hypercube with lower discrepancy than completely random sampling. The index j runs from 1 to d, where d is the number of parameters. The index i runs from 1 to N, where N is the number of input samples. The term $f(\boldsymbol{A})_i$ represents the ith element of the vector that is the output of the GP emulator when evaluated at $\boldsymbol{X}_* = \boldsymbol{A}$. The term $\boldsymbol{A}_{\boldsymbol{B}}^{(j)}$ represents a matrix where column j comes from matrix $\boldsymbol{B}$ and all other columns come from matrix $\boldsymbol{A}$. The matrices $\boldsymbol{A}$ and $\boldsymbol{B}$ can be generated from a Sobol sequence of size $N \times 2d$: $\boldsymbol{A}$ is the left half of the sequence and $\boldsymbol{B}$ is the right part of it.

Given $\mathrm{Var}_{x_j}\left(E\left(y \mid x_j\right)\right)$, we can compute the first-order sensitivity coefficient S_j that captures the main effect of x_j on y as:

$$S_j = \frac{\mathrm{Var}_{x_j}\left(E\left(y \mid x_j\right)\right)}{\mathrm{Var}(y)} \in [0, 1] \tag{17}$$

As it turns out, the ABM's output variance is mainly driven by:

1. The shape parameter of the global distribution that assigns each agent's threshold value for their resistance parameter
2. The shape parameter of the global distribution that assigns the weight of the profitability to each agent's resistance
3. The shape parameter of the global distribution that assigns μ^{CF} to each agent in the model

Accordingly, all other parameters were fixed to the mean value of their initial ranges.

5 Model Calibration

5.1 The History Matching Method

The calibration of the model was based on the history matching method (Craig et al. 1997). History matching works by excluding subsets of the model's parameter space, which are unlikely to provide a good match between the model outputs and the observed reality. As the plausible space becomes smaller and smaller, emulators become smoother and more accurate, allowing us to zoom in the parameter space we explore. As a result, the implausible parameter values are excluded in iterations, known as waves, where new emulators are built after each wave. History matching has been successfully applied across a wide range of scientific domains, including the case of calibrating ABM (Andrianakis et al. 2015).

A central concept of the history matching method is the quantification of the major uncertainties that have an impact on the calibration process (Kennedy and O'Hagan 2001):

1. **Ensemble variability** (V_{es}). This uncertainty represents the stochastic nature of the ABM; different runs with the same parameters will give different outcomes. To calculate the ensemble variability, one combination for the model's parameters was selected—it was assumed that the variability is independent from the inputs, so one combination will suffice—and run the model $K = 25$ times to capture the variability of the results. Accordingly, the ensemble variability was calculated to be V_{es}=0.002.
2. **Emulation uncertainty** (V_{em}). This type of uncertainty originates in the possible discrepancy between the ABM and the GP-based STEEM. To calculate the emulation uncertainty, the model runs used for the calculation of the ensemble uncertainty were utilized. The emulation uncertainty was given by:

$$V_{\text{em}} = \frac{1}{K-1} \sum_{k=1}^{K} \left(f_{j,k}(x) - g_j(x)\right)^2 \tag{18}$$

where $f_{j,k}(x)$ is the output of the ABM for the period j at the kth run and $g_j(x)$ is the emulator's result. Accordingly, the emulation uncertainty was calculated to be $V_{em} = 0.002$.

Having specified the aforementioned uncertainties, the implausibility of a specific parameter combination $\boldsymbol{x}$ is given by (Vernon et al. 2010):

$$I_j(\boldsymbol{x}) = \frac{\left|z_j - g_j(\boldsymbol{x})\right|}{\sqrt{V_{\text{es}} + V_{\text{em}}}} \tag{19}$$

where z_j is the observed data for the period j.

A natural cut-off value for implausibility is 3, i.e. any parameter combination $\boldsymbol{x}$ with $I(\boldsymbol{x}) > 3$ should be considered implausible.

For the calculation of the implausibility function, the following steps took place:

1. $z_j \rightarrow$ The actual data was scaled to the [0, 1] range.
2. $\boldsymbol{x} \rightarrow$ A new and much larger (20,000) set of combinations was derived through Latin hypercube design, this time taking random points within the sampling intervals.

3. $g_j(\boldsymbol{x}) \rightarrow$ We predicted the cumulative capacity additions at the end of the period j for each combination.
4. $I_j(\boldsymbol{x}) \rightarrow$ The implausibility function was used as an indicator function to split the samples into plausible and implausible ones through the Patient Rule Induction Method (Friedman and Fisher 1999).

5.2 *The Patient Rule Induction Method*

The Patient Rule Induction Method (PRIM) is a heuristic algorithm that aims to identify rectangular partitions (called boxes) of the input space where the average response is much larger than the average response across the whole input space. When the responses correspond to classification classes, the goal is to identify partitions with high homogeneity or, equivalently, low impurity as measured by the Gini index:

$$\text{GI} = 1 - \sum_i p_i^2 \tag{20}$$

where p_i is the fraction of responses labelled with class i in the partition.

For binary responses (as is our case), both aforementioned goals are equivalent if the responses of no interest (i.e. with implausibility function values greater than 3) are assigned the zero value.

The PRIM algorithm proceeds as follows:

1. Identify a rectangle box that includes all input data, B_1.

   Identify a dimension j that when removing the slice of data below the a quantile or above the $1 - a$ quantile of the corresponding variable's distribution, we achieve the greatest decrease in the impurity of the remaining data. The parameter a is user-defined.

   The objective function that we used divides the gain in purity by the loss in mass:

$$\max_j \frac{\text{GI}_{B1} - \text{GI}_{B1-j}}{\beta_{B1} - \beta_{B1-j}} \tag{21}$$

where GI_{B1} is the Gini index of the initial box B_1 and GI_{B1-j} the Gini index of the box resulting from removing the amount of β_{B1-j} data from dimension j.

2. Repeat step 2 until a minimum number β of data remains in the rectangle. This process is called "peeling". The parameter β is also user-defined.
3. Reverse the peeling process by expanding the rectangle in the dimension where adding a slice of data leads to the greatest decrease in the impurity of the expanded partition.

4. Repeat step 4 until no decrease in impurity is possible. This process is called "pasting" and aims to refine the box boundaries.

Based on the aforementioned, PRIM was applied so as to find boxes that contain plausible parameter combinations (i.e. with $I(\boldsymbol{x}) \leq 3$). The quality of the derived partition was judged by measuring its density. Density is the ratio of the total number of plausible combinations inside the box to the number of all combinations inside this box.

In addition, the partition boxes were evaluated according to their coverage. Coverage is the ratio of the total number of plausible combinations inside the box to the total number of plausible combinations found across the whole dataset. If the coverage is low, it is possible that either the data has high levels of noise or there are additional partitions with high density as well.

To explore the latter possibility, the method of Guivarch et al. (2016) can be used, according to which the data in the identified box are marked as uninteresting (i.e. they are assigned the zero value), and the PRIM algorithm is re-run so as to identify a new box if possible.

5.3 Calibration and Extrapolation Results

After considering all periods, the PRIM constrained the parameters 1, 5 and 9 of Table 1 as presented in Table 2.

Subsequently, the calibrated model was used to explore the expected effectiveness of the Greek net-metering scheme in driving investments in small-scale PV during the period 2018–2025. To this end, a set of scenarios for different plausible values of the model parameters was run, assuming that the retail prices remained unaffected and taking into consideration the provisions of the Greek net-metering scheme.

The ABM that was used for the forward-looking simulations differs from the one used for calibration in the following ways:

1. **Available options**. When the agents decide to invest in solar PV, they can choose one option from a limited set of different ones, all of which are available to all

Table 2 Plausible ranges for the ABM parameters after calibration

		After calibration	
	Description	Min	Max
1.	The shape parameter of the global distribution that assigns μ^{CF} to each agent in the model	181	192
5.	The shape parameter of the global distribution that assigns the weight of the profitability to each agent's resistance	1.9	3
9.	The shape parameter of the global distribution that assigns each agent's threshold value for their resistance parameter	11.6	19.6

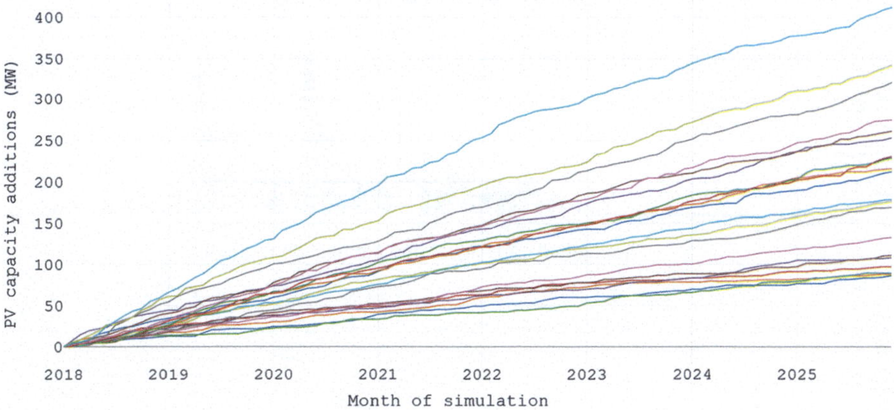

Fig. 2 New PV capacity additions due to the net-metering scheme

agents. These options concern the size of the PV system installed: 2.4 kWp, 4.8 kWp or 9.6 kWp.

2. **The option selection rule**. If more than one option is evaluated favourably by an agent, the selection is based on the softmax rule; the probability of investing in option j is related to the agent's resistance to it (r_j) according to:

$$P(j) = \frac{e^{-r_j}}{\sum_k e^{-r_k}}$$

The results are presented in Fig. 2.

6 Discussion

The ambition of this chapter was to highlight the fact that when calibrating ABMs to historical data, we should aim to explicitly quantify model uncertainty. For the model under study, this uncertainty can be regarded as uncertainty that is related to the characteristics and the decision-making criteria of the agents (i.e. independent PV power producers), and it is evident by the range of the different plausible parameters of the model and the range in the results that they produce.

The main takeaway is that model uncertainty exists whether we choose to quantify it or not. However, by quantifying it, we are able to understand the utility and the limits of our model. In this case, we can compare the future scenarios with the historical data (Fig. 3). This reveals that the expected results from the net-metering scheme in Greece are positive but not as strong as the results of the FIT period; even with the most favourable of the plausible parameter combinations, it will take 7 years to achieve the same PV capacity additions that the FIT scheme achieved from 2010 up to the end of 2012.

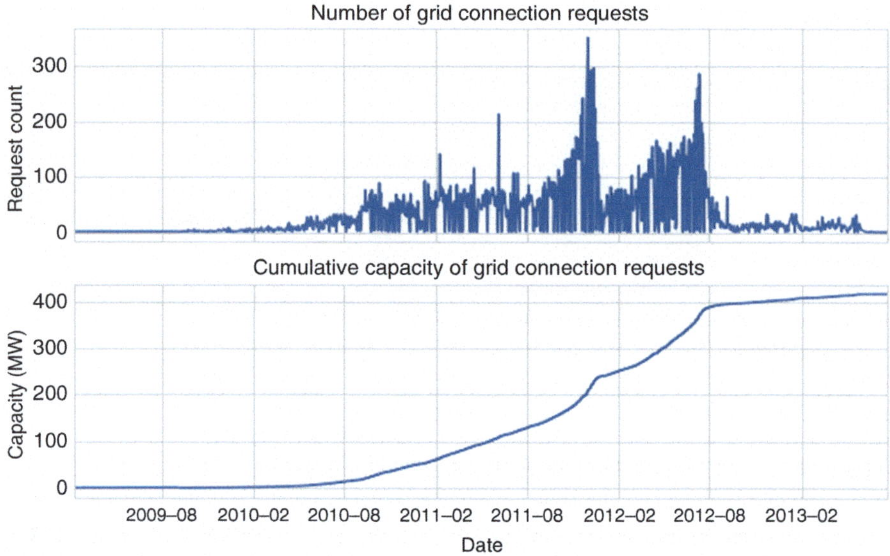

Fig. 3 Grid connection requests for small-scale PV systems in Greece

In addition, highlighting model uncertainty makes it sensible that a policy design process that utilizes ABMs is structured around the concept of adaptability; as new data on the actual decisions of the relevant actors is accumulated, the initial policy should adapt, the same way as it should adapt to changes in its environment, such as technology costs.

Finally, ABMs can help policies succeed by focusing on directly affecting the agents' characteristics. As an example, dissemination of trusted information can help align agents' beliefs, while marketing campaigns and appropriately devised narratives can help increase the degree of influence by social learning and imitation.

Acknowledgements This research was funded by the EC Horizon 2020 Project titled "Transitions pathways and risk analysis for climate change mitigation and adaption strategies" (TRANSrisk) with Grant Agreement No. 642260. The authors would like to acknowledge the support from the EC. The content of the paper is the sole responsibility of its authors and does not necessary reflect the views of the EC.

References

Anagnostopoulos, P., Spyridaki, N.-A., & Flamos, A. (2017). A "New-Deal" for the development of photovoltaic investments in Greece? A parametric techno-economic assessment. *Energies, 10* (8), 1173.

Andrianakis, I., Vernon, I. R., McCreesh, N., McKinley, T. J., Oakley, J. E., et al. (2015). Bayesian history matching of complex infectious disease models using emulation: A tutorial and a case study on HIV in Uganda. *PLOS Computational Biology, 11*(1), e1003968.

Baranzini, A., Carattini, S., & Peclat, M. (2017). *What drives social contagion in the adoption of solar photovoltaic technology*. GRI Working Papers 270, Grantham Research Institute on Climate Change and the Environment.
Bastos, L. S., & O'Hagan, A. (2009). Diagnostics for Gaussian process emulators. *Technometrics, 51*, 425–438.
Craig, P. S., Goldstein, M., Seheult, A. H., & Smith, J. A. (1997). Pressure matching for hydrocarbon reservoirs: A case study in the use of Bayes linear strategies for large computer experiments (with discussion). In C. Gastonis et al. (Eds.), *Case studies in Bayesian statistics* (Vol. III, pp. 37–93). New York: Springer.
Doukas, H. (2013). Linguistic multicriteria decision-making for energy systems: Building the "RE_2S" framework. *Wiley Interdisciplinary Reviews Energy and Environment (WIREs), 2*(5), 571–585.
Flamos, A. (2016). A sectoral micro-economic approach to scenario selection and development: The case of the Greek power sector. *Energies, 9*, 77.
Friedman, J. H., & Fisher, N. I. (1999). Bump hunting in high-dimensional data. *Statistics and Computing, 9*(2), 123–143.
Genton, M. G. (2002). Classes of kernels for machine learning: A statistics perspective. *Journal of Machine Learning Research, 2*, 299–312.
Guivarch, C., Rozenberg, J., & Schweizer, V. (2016). The diversity of socio-economic pathways and CO2 emissions scenarios: Insights from the investigation of a scenarios database. *Environmental Modelling & Software, 80*, 336–353.
Haylock, R. G., & O'Hagan, A. (1996). On inference for outputs of computationally expensive algorithms with uncertainty on the inputs. *Bayesian Statistics, 5*, 629–637.
Kennedy, M. C., & O'Hagan, A. (2001). Bayesian calibration of computer models. *Journal of the Royal Statistical Society: Series B (Statistical Methodology), 63*, 425–464.
Papadelis, S., Stavrakas, V., & Flamos, A. (2016). What do capacity deployment rates tell us about the efficiency of RES-E support measures in Greece? *Energies, 9*(1), 38.
Rasmussen, C. E., & Williams, C. K. I. (2006). *Gaussian processes for machine learning*. Cambridge, MA: MIT Press.
Saltelli, A., Annoni, P., Azzini, I., Campolongo, F., Ratto, M., & Tarantola, S. (2010). Variance based sensitivity analysis of model output. Design and estimator for the total sensitivity index. *Computer Physics Communications, 181*(2), 259–270.
Vernon, I., Goldstein, M., & Bower, R. G. (2010). Galaxy formation: A Bayesian uncertainty analysis. *Bayesian Analysis, 5*(4), 619–670.
Watts, D. J., & Strogatz, S. H. (1998). Collective dynamics of 'small-world' networks. *Nature, 393*, 440–442.
Young, H. (2009). Innovation diffusion in heterogeneous populations: Contagion, social influence, and social learning. *American Economic Review, 99*(5), 1899–1924.

Investments in the EU Power System: A Stress Test Analysis on the Effectiveness of Decarbonisation Policies

Pedro Crespo del Granado, Christian Skar, Haris Doukas, and Georgios P. Trachanas

Abstract Ambitious emission reduction targets are challenging the status quo on designing effective strategies for electricity generation portfolios. In this chapter, we consider the role of low-carbon technologies and determine the cost-benefits of policy strategies to mitigate greenhouse gas emissions in the EU. In particular, we look into how long-term scenarios for transmission expansion and decarbonisation policies influence the evolution of the EU power system infrastructure. We use an EU electricity investment model to determine the optimal portfolio of electricity generation technologies and compute their respective costs and emissions achieved towards 2050. Based on the investment model's results (strategies and suggested portfolios), we investigate how these portfolios perform under divergent policy or geopolitical developments. For this purpose, we apply a robust optimisation tool based on the min-max and the min-max regret criteria, which selects ideal portfolios by stress testing a particular scenario or policy choice under uncertainty of input parameters.

Results show that pursuing a strong transmission expansion strategy under the EU PRIMES reference case leads to the maximum regret, while relying on EU scenarios with strong prospects for decarbonisation, either with possibilities or with limitation on transmission expansion, leads to portfolios that exhibit the least variance. However, applying regret analysis on investment costs and total emissions indicates a limited transmission investment case as the more robust one, also noting that a high carbon price will accelerate the energy transition.

Keywords Investments · Decarbonisation · Robustness · Optimisation · Strategies · Energy transition

---

P. Crespo del Granado (✉) · C. Skar
Department of Industrial Economics and Technology Management, Norwegian University of Science and Technology (NTNU), Trondheim, Norway
e-mail: pedro@ntnu.no; christian.skar@ntnu.no

H. Doukas · G. P. Trachanas
Decision Support Systems Laboratory, School of Electrical and Computer Engineering, National Technical University of Athens, Athens, Greece
e-mail: h_doukas@epu.ntua.gr; gtrachanas@epu.ntua.gr

H. Doukas et al. (eds.), *Understanding Risks and Uncertainties in Energy and Climate Policy*, https://doi.org/10.1007/978-3-030-03152-7_4

1 Introduction

Population growth and economic development constitute the primary factors of the increase in global carbon emissions (Raupach et al. 2007) in the last decades. Anthropogenic greenhouse gas (GHG) emissions have contributed to the increase in annual average global temperature. According to the Intergovernmental Panel on Climate Change, the range of climate change projections for global warming from 1900 to 2100 is from 1.4 °C to 5.8 °C (IPCC 2014). In order to limit GHG emissions, the electricity sector is facing the challenge to undertake a major transformational phase towards a low-carbon system by substituting the existing carbon-intensive technologies with climate-friendly ones. In this regard, energy analysts and policymakers face a future that is technologically, institutionally and politically complex and uncertain (Nikas et al. 2017); furthermore, assessments of the energy transition must take into account different perspectives to reflect the interests of numerous stakeholder groups (Papapostolou et al. 2017). On this matter, portfolio-based approaches are one of the established methods to evaluate national energy strategies and climate policies. They provide an analytical basis to devise generation mixes that take into account security of supply, climate targets, technological progression and costs.

Standard portfolio-based techniques usually provide an optimisation-based model to determine the mix of power generation technologies. To name but a few, McLoughlin and Bazilian (2006) apply a mean-variance portfolio (MVP) optimisation to analyse the Irish electricity generation mix. A similar approach is developed in White (2007) for California's electric utility resource planning, recommending an optimal generating portfolio for the inclusion of greater shares of renewable technologies. Awerbuch and Berger (2003) apply MVP analysis for the European Union by reflecting the risk of fuel, operation and maintenance, and construction period costs. By applying portfolio theory, Zhu and Fan (2010) evaluate China's 2020 midterm plans for generating technologies, while reflecting the risk of relevant generating cost streams and including CO_2 emission scenarios. Also, Fuss et al. (2012) derive portfolios across various socio-economic scenarios for a range of stabilisation targets by including alternative risk measures.

All in all, uncertainty in portfolio optimisation techniques might be considered in the form of risk. However, another form of uncertainty in the model are its inputs and assumptions, which typically induces the formulation of different cases and scenario analyses. These scenarios are strongly related to different socio-economic, supply and/or stabilisation targets and assumptions. To cope with this kind of uncertainty, it is imperative to stress test the obtained solutions across different scenarios. In this direction, various robust decision support tools have been developed to provide solutions that perform well, independently of any scenario's realisation. The present study applies the min-max and the min-max regret criteria,[1] both lying in the core of

[1]The robust decision under min-max criterion is that for which the lowest (highest) level of benefit (cost) taken across all possible input scenarios is as high (low) as possible. Regret is defined as the difference between the resulting benefit (cost) and the benefit (cost) from the decisions that would

the so-called robustness analysis, to examine the performance of optimal technology portfolios by considering energy transition scenarios of the power system. The main goal is to identify a solution that performs well against the worst and/or the best case performance. Pure min-max criterion is appropriate for conservative decision makers, since it is associated with the worst-case scenario. Since hedging against uncertainty within energy planning is complex and fraught with multiple forms of uncertainties, robustness approaches have received increasing attention over the last years. For instance, in power systems planning, the uncertainties across the processes of transmission, conversion or distribution are treated with the adoption of interval programming in conjunction with regret analysis (see Dong et al. 2011 and the references therein). Furthermore, we refer to van der Weijde and Hobbs (2012) and Munoz et al. (2014, 2017) concerning transmission planning, and to Fan et al. (2010) and Morris et al. (2018) regarding investment decisions under policy uncertainty, and risk aversion and CO_2 regulatory uncertainty in generation investments, respectively.

In this chapter, to analyse the decarbonisation of the EU power system, we apply a stochastic power investment model to determine the optimal technology portfolios under certain scenarios. Namely, we use the EMPIRE (*E*uropean *M*odel for *P*ower System *I*nvestment with *R*enewable *E*nergy) model, developed by Skar et al. (2016a, b). EMPIRE, formulated as a multi-horizon stochastic programme, incorporates long- and short-term system dynamics while optimising investments under operational uncertainty. It is well known that the consideration of renewable technology in the generation mix, in particular wind and solar power, impacts the supply and demand balance, due to the intermittent and uncontrollable nature of these technologies. EMPIRE is designed to handle these challenges. In contrast to other power sector models, the major contribution of EMPIRE is that it simultaneously incorporates short- and long-term dynamics, in conjunction with short-term uncertainty. Dynamics refer to multiple investment periods coexisting with multiple sequential operational decision periods, while uncertainty is enhanced through multiple input scenarios that describe operating conditions. That is, EMPIRE is a capacity and transmission expansion model, designed to determine optimal capacity investments and system operation over long-term planning horizons, extended in a 40–50-year basis. A central planner's perspective is adopted, minimising a system's cost while serving a price inelastic demand. Regarding the effect of short-term uncertainty on investment decisions, the methodology used is based on the principles of multi-horizon stochastic programming, as proposed by Kaut et al. (2014). Related expansion models in the literature are, for example, the DIMENSION (Richter 2011) model used by Jägemann et al. (2013), who analyse the costs for the decarbonisation of the European power sector. Similarly, another dynamic investment model, the LIMES-EU+, was adopted by Haller et al. (2012), where carbon capture and storage

have been taken knowing prior to the decision time which particular input scenario would occur. Applying the min-max criterion to the regret values, we obtain the robust min-max regret decisions (Kouvelis and Yu 1997).

(CCS) and nuclear power are excluded. Both optimisation models are deterministic, in contrast to EMPIRE's stochastic features. In the multi-stage stochastic model E2M2 (Swider and Weber 2007), short-term wind uncertainty is analysed for the German power system. Another similar model is the two-stage stochastic model TIMES (Seljom and Tomasgard 2015), which also includes short-term uncertainty.

In our stress testing framework assessment, EMPIRE determines the optimal portfolio of electricity generation technologies and calculates their respective costs and emissions achieved for the 2015–2050 period. Each long-term strategy obtained by EMPIRE assumes the realisation (model inputs) of a future technological progression or geopolitical scenario assumptions. For instance, technology evolution differentiation represents the variability of the scenarios. To hedge against the possible realisation of certain technological or geopolitical scenarios, we apply robust optimisation through the min-max and min-max regret criteria. In the first case, a safe performance is guaranteed, independently of any selection within the input parameter dataset. In the regret analysis case, the stress test of input data means that the decision maker measures the deviation from optimality as long as input varies within the scenario set, and then the min-max criterion is applied to guarantee a safe "distance" from that optimality.

We chose to implement robust optimisation, since we refer to risk-averse decision makers, either concerning the best or the worst case. The literature with applications of min-max regret analysis in the energy sector is quite extensive. For example, in Dong et al. (2011), min-max regret analysis is incorporated in combination with interval linear programming for the study of power management systems under multiple supply and demand scenarios. In Li et al. (2016), electrical power generation planning is studied while considering discrete scenarios of possible climate change outcomes. In Kazakci et al. (2007), energy crop supply is measured through a linear mathematical programme. In the framework of climate change mitigation policy, we refer to Loulou and Kanudia (1999), where a min-max regret formulation is proposed to determine strategies for GHG emission reduction. In Li et al. (2011), an interval model is developed to support planning of GHG mitigation within an uncertain energy system. In addition, these approaches are also applied to solid waste management by combining tools for interval and robust optimisation (Li and Huang 2006; Chang and Davila 2007). However, we should mention that minimising maximum cost or regret is a rather extreme form of risk aversion and that there exist other alternatives to model it, such as using utility functions (preferred by economists) or Conditional Value at Risk (preferred by engineers). To summarise, the objective of this study is to (1) analyse the robustness of different pathways for the energy transition of the power sector by stress testing their outcomes across individual scenarios and (2) propose a multidisciplinary method to complement a power system capacity expansion model with a robustness-based approach (min-max regret analysis).

The next section presents the main features of the EMPIRE model. This is followed by Sect. 3, which describes the implementation of four distinct cases and their respective results. Then, Sect. 4 presents the two robustness tools that apply the stress test analysis on optimal investment portfolios across the different cases.

2 Modelling Investments in Electricity Generation and Transmission

2.1 Short-Term vs. Long-Term Considerations for Optimal Portfolio

There is a multitude of energy models looking at different dimensions of the power system. Main difference among models is the level of technical detail, representation of uncertainty (e.g. RES or demand variations), temporal considerations, spatial aggregation and planning horizons (for a review, see Crespo del Granado et al. 2018). Typically, the fundamental problem is to decide the short-term scheduling of power plants (e.g. coal, gas, hydro or nuclear) based on generation costs, plants' operational limitations and RES-load interactions. This model, in the literature, is known as the "unit commitment problem", the "optimal economic dispatch" or simply the generation dispatch. To decide the power dispatching of a determined number of coal or gas power plants, nuclear reactors or oil generators, the model determines an optimal supply portfolio based on functions representing economic decisions and energy generation physics. While this kind of model represents the supply-demand balance in detail with a high time resolution (hourly decisions), it usually does not consider long-term decisions for capacity planning. In contrast, models for long-term capacity planning leave behind detail engineering aspects of operations. Raising a causality dilemma since the short-term decisions require an adequate generation capacity to satisfy demand, while long-term investment decisions are a consequence on how much adequacy is required in the short term. As both planning horizons are nonmutually exclusive (see Fig. 1), investment models have come up with different assumptions to represent operational decisions and features of the power system. For the EMPIRE model, these assumptions are as follows:

- Operations and investment horizon: As hourly operations in 1 year compromise 8760 periods, repeating them for multiple years increases the dimensionality of the problem. Long-term planning problems typically analysed years to decades ahead and simulating short-term problems for large time spans might create an intractable problem. A common approach is to sample representative weeks. In EMPIRE, we sample typical weeks per season along with 2 weeks representing extreme cases (e.g. high peak demand and low RES availability). These weeks' parameters are updated for incoming investment periods. EMPIRE considers a planning horizon from 2015 to 2050. Investment windows are every 5 years in which the representative operational weeks are scaled up to resemble the 5 years in operation. That is, all investment periods are in a single optimisation along with operation snapshots (see similar approach in Haller et al. 2012).
- Spatial aggregations: EMPIRE model covers the European Economic Area countries (see Fig. 2, 31 European countries along with 55 interconnectors). EMPIRE models each country as a single node together with existing capacity

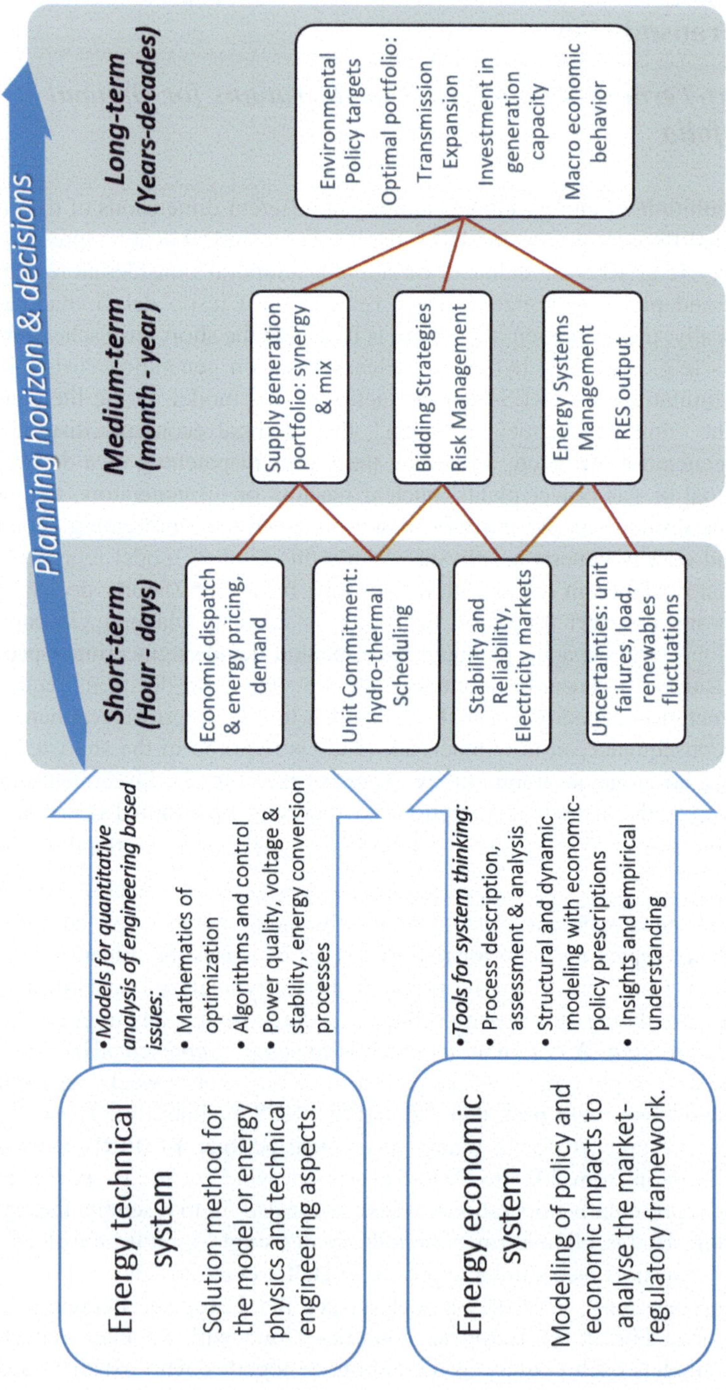

Fig. 1 Assessment of power systems portfolios by considering technical and economic perspectives based on the joint representation of short-term to long-term decisions

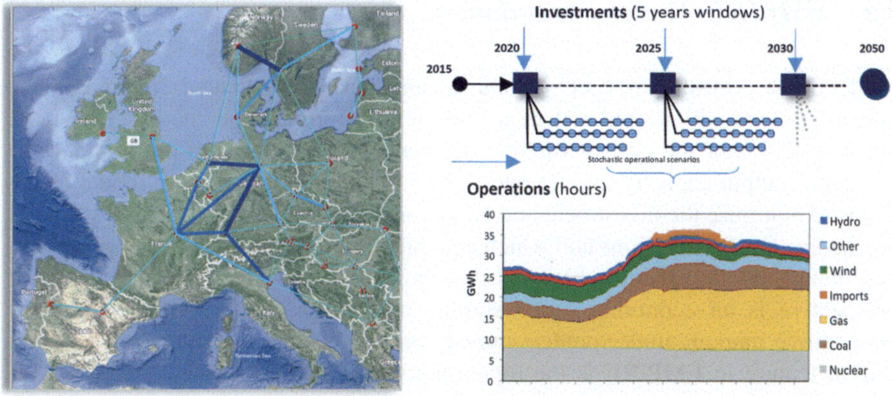

Fig. 2 EMPIRE model EU spatial coverage combining short (operations)- and long-term (investments) decisions to determine optimal portfolios for the power system

and estimated demand. The net transmission capacity defines the country connection capacity to neighbouring regions. That is, the interconnection among nodes in the model resembles a transportation problem.

- Technical engineering aspects: EMPIRE short-term operational aspects represent hourly decisions of the generators and transmission exchange between countries. Generators are aggregated by type, for example, nuclear generators have a single variable per country. As an energy system model, EMPIRE does not consider power flows and voltage relationships. As the main interest is the long-term expansion of the system, the model also assumes linear production cost profiles for all generators.
- Policy and economic perspectives: EMPIRE can set RES targets and calculates CO_2 emissions per country. Main model outputs are the usage factor of power plants, load shedding, RES curtailments, transmission infrastructure investments and cooperation among countries to meet policy objectives. Hence various policies can be tested by changing the model parameters (technology costs, demand or RES targets), including the prospects of development of technologies (e.g. CCS) and considering assumptions on transmission expansion possibilities.

In summary, EMPIRE intends to combine as much as it can from both worlds: short-term (operational) and long-term (strategic) decisions. In addition, the model provides enough rich technological details to provide rational insights for the long-term planning of the power system. Figure 1 epitomises this discussion in which the modelling approach might take a technical engineering perspective or an economic viewpoint mainly focuses on the empirical understandings of impacts in policymaking. The EMPIRE model intends to encompass a compromise of this vision since its formulation convenes as much as it can from different perspectives. For a further discussion on tools for integrated assessments, combining models and strength-weaknesses of different modelling approaches, refer to Crespo del Granado et al. (2018).

2.2 EMPIRE Model Formulation

Modelling a portfolio of energy units, as an hourly coordinated supply-demand system, is typically formulated as a multiperiod optimisation model. This takes into consideration the cost of running the generation units, ramp-up time and the unit input-output capacity (power rates). As a result, the optimisation model objective is to minimise the investment and operational costs. The main decision variables are the investment decisions and which generation units to use and when: peak time, base load, smoothing renewables and the usage of transmission capacity. EMPIRE perspective is an economic social surplus maximisation that assumes perfectly competitive markets under predetermined consumer decisions. As noted earlier, a central feature in EMPIRE is the representation of two timescales, the long term (strategic) and the short term (operational). Operational decisions are associated with a strategic stage in order to co-optimise long-term investment decisions and the short-term operational decisions. That is, strategic decisions face supply-demand balancing decisions under uncertainty. For example, in Fig. 2, observe the structure of the 5-year investment windows subjected to operational uncertainty. Investment variables (generation and transmission expansion) in 2020 have specific hourly load profiles per country and must determine the operations of the units and hence decide the optimal portfolio from 2025 onwards. The model includes a discount rate to calculate the net present value (NPV) of investments. This structure follows a multi-horizon stochastic framework in which the operational uncertainty is in the load profiles and wind and solar power generation. For a more comprehensive discussion on setting up the EMPIRE model, refer to Skar et al. (2016a, b).

EMPIRE Objective Function

As aforementioned, to take into account long-term and short-term decisions, EMPIRE minimises the NPV of investments based on operational decisions. On one hand, the objective function contains the investment decisions for generation (x_{gi}^{gen}, capacity investment in generator g at year i) and transmission (x_{lj}^{tran}, investing in transmission line l at year i) under costs c_{gi}^{gen} and c_{li}^{tran}, respectively. On the other hand, the presentation of the operation decisions comes from the production of generator g and $y_{ghi\omega}^{\text{gen}}$ for operational hour h in year i under stochastic scenario ω. In addition, the operations consider load shedding ($y_{nhi\omega}^{\text{LL}}$) at node n (country), under operational hour h in year i and stochastic scenario ω. Both $y_{ghi\omega}^{\text{gen}}$ and $y_{nhi\omega}^{\text{LL}}$ face costs of producing electricity (q_{gi}^{gen}) and the cost of using load shedding (q_{ni}^{VoLL}). In short, EMPIRE objective function is as follows:

$$\begin{aligned}\min_{\mathbf{x},\mathbf{y}} z = \sum_{i\in\mathcal{I}} \delta_i \times \Bigg\{ & \sum_{g\in\mathcal{G}} c_{gi}^{\text{gen}} x_{gi}^{\text{gen}} + \sum_{l\in\mathcal{L}} c_{li}^{\text{tran}} x_{li}^{\text{tran}} + \sum_{\omega\in\Omega} p_\omega \times \sum_{h\in\mathcal{H}} \alpha_h \\ & \times \sum_{n\in\mathcal{N}} \left(\sum_{g\in\mathcal{G}_n} \left[q_{gi}^{\text{gen}} y_{ghi\omega}^{\text{gen}} \right] + q_{ni}^{\text{VoLL}} y_{nhi\omega}^{\text{LL}} \right) \Bigg\} \end{aligned} \quad (1)$$

To compute the NPV in Eq. (1), we set a discount factor δ_i per year i. For the operations, we apply a factor α_h to scale up and match the total number of hours of the investment period. Recall that the time span of the investments is 5 years for which we use representative weeks of the year and hence the scale up factor accounts for the remaining hours. Note that since we consider stochastic scenarios for renewables and demand, we assign the respective probabilities p_ω for operations of that scenario.

EMPIRE Constraints

Since it would be unrealistic to invest in certain generators for certain countries (e.g. wind offshore for Switzerland) or allow sudden large investments on each period, Eq. (1) is subjected to investment constraints (period-wise and cumulative), that is, a restriction by an upper bound ($\bar{x}_{tn*}^{\text{gen}}$) on investments in new capacity for generator g along with the cumulative installed generation over the planning horizon. This also considers the retirement of power plants based on the retired share (ρ_{gi}) of generator g's initial capacity by year i. These constraints are

$$\begin{aligned}
&\sum_{g \in \mathcal{G}_{nt}} x_{gj}^{\text{gen}} \leq \bar{x}_{nti}^{\text{gen, Period}}, \quad n \in \mathcal{N}, t \in \mathcal{T}, i \in \mathcal{I}. \\
&\sum_{j=1}^{i} \sum_{g \in \mathcal{G}_{nt}} x_{gj}^{\text{gen}} \leq \bar{x}_{nt}^{\text{gen, Cumulative}} - (1 - \rho_{gi}) \bar{x}_{g0}^{\text{gen}}, \quad n \in \mathcal{N}, t \in \mathcal{T}, i \in \mathcal{I}.
\end{aligned} \tag{2}$$

Likewise, investment constraints for transmission (exchange) capacity are also set up:

$$x_{li}^{\text{tran}} \leq \bar{x}_{li}^{\text{tran, Period}}, \quad l \in \mathcal{L}, i \in \mathcal{I}. \tag{3}$$

As for the equations to represent the operations, these are based on the usage of generators g at node n, the interaction (flow) with neighbouring nodes, the capacity and usage characteristics of the generators, hydro or alike storage technologies and the emission standards per country. The details of these operational constraints are as follows:

1. Supply-demand balance (production + losses*import−exports−pumping + load shedding = load)

$$\sum_{g \in \mathcal{G}_n} y_{ghi\omega}^{\text{gen}} + \sum_{a \in \mathcal{A}_n^{\text{in}}} (1 - \eta_a^{\text{line}}) y_{ahi\omega}^{\text{flow}} - \sum_{a \in \mathcal{A}_n^{\text{out}}} y_{ahi\omega}^{\text{flow}} - y_{nhi\omega}^{\text{pump}} + y_{nhi\omega}^{\text{LL}} = \xi_{nhi\omega}^{\text{load}}, n \in \mathcal{N}, h \in \mathcal{H}, \omega \in \Omega, i \in \mathcal{I}. \tag{4}$$

2. Generation capacity constraint based on existing a prior capacity plus the invested capacity for year i under

$$y^{\text{gen}}_{ghi\omega} \leq \xi^{\text{gen}}_{ghi\omega} \times \left((1-\rho_{gi})\bar{x}^{\text{gen}}_{g0} + \sum_{j=1}^{i} x^{\text{gen}}_{gj} \right), \quad g \in \mathcal{G}, h \in \mathcal{H}, i \in \mathcal{I}, \omega \in \Omega. \quad (5)$$

3. Upward ramping constraints to simulate appropriate load following of the generators:

$$y^{\text{gen}}_{ghi\omega} - y^{\text{gen}}_{g(h-1)i\omega} \leq \gamma^{\text{gen}}_{g} \times \left((1-\rho_{gi})\bar{x}^{\text{gen}}_{g0} + \sum_{j=1}^{i} x^{\text{gen}}_{gj} \right), \quad g \in \mathcal{G}^{\text{Thermal}}, \quad (6)$$
$$s \in \mathcal{S}, h \in \mathcal{H}^{-}_{s}, i \in \mathcal{I}, \omega \in \Omega,$$

4. Flow constraint—limit flow on arcs (directional arcs and lines are symmetric). It also considers a prior line capacity plus the capacity expansion investment for year i:

$$y^{\text{flow}}_{ahi\omega} \leq \bar{x}^{\text{tran}}_{l0} + \sum_{j=1}^{i} x^{\text{tran}}_{lj}, \quad l \in \mathcal{L}_n, a \in \mathcal{A}_l, h \in \mathcal{H}, i \in \mathcal{I}, \omega \in \Omega. \quad (7)$$

5. Hydro energy constraint—limit total hydropower production within a season (due to water availability)

$$\sum_{h \in \mathcal{H}_s} y^{\text{gen}}_{ghi\omega} \leq \xi^{\text{RegHydroLim}}_{gsi\omega}, \quad g \in \mathcal{G}^{\text{RegHydro}}, s \in \mathcal{S}, i \in \mathcal{I}, \omega \in \Omega. \quad (8)$$

6. Pump-storage upper reservoir capacity and inter-temporal balance for storage:

$$\begin{aligned} w^{\text{upper}}_{n(h-1)i\omega} + \eta^{pump}_{n} y^{\text{pump}}_{nhi\omega} - y^{\text{gen, pump}}_{nhi\omega} &= w^{\text{upper}}_{nhi\omega} \\ w^{\text{upper}}_{nhi\omega} &\leq \bar{w}^{\text{upper}}_{n} \end{aligned}, n \in \mathcal{N}, h \in \mathcal{H}_s, i \in \mathcal{I}, \omega \in \Omega. \quad (9)$$

Appendix 1 notes the nomenclature description of the sets, parameters and variables used in the above formulation. Note that EMPIRE has been used in other studies which present a more comprehensive model formulation, data sources and other details (see, for example, Skar et al. (2014, 2016a, b) and ZEP (2013, 2014)). Note that exogenous drivers of investments in the EMPIRE model are changes in demand, retirement of the existing generation fleet, changes in fuel prices and a price of carbon. Technological advancement, such as investment cost reductions and efficiency improvements for thermal generation, plays an important role in the design of the optimal generation portfolio and is included in the model (see Appendix 2 for data inputs). EMPIRE computational dimension includes approximately

15 million variables and 22 million constraints, and it takes from an hour to 5 hours to solve (depending on scenario, technology choice and solver configurations).

3 Energy Transition: Cases and EMPIRE Model Results

3.1 *Defining Cases*

The EU Energy Roadmap 2050 and various stakeholders' discussions with the European commission outlined four main decarbonisation routes for the energy sector; these are energy efficiency, RES, nuclear and CCS. These decarbonisation options have the premise that European integration will be one of the driving forces in its success. Cooperation among countries and political determination will generate measures for the integration of European electricity markets, viability of infrastructure projects of common interest (PCI), common climate targets (e.g. the Paris agreement) and joint policies to accommodate higher shares of RES in the system. The cooperation among different EU actors and countries towards 2050 is assumed to be one of the cornerstones of the energy transition. As a result, some scenario-building studies (Bauer et al. 2017) have discussed the degree of cooperation achievable in the long term as a measure to weigh-in and to formulate different scenarios towards 2050.

In this spirit, to understand the consequences of assuming different evolutions of the generation portfolio of the electricity system, we define four cases[2] that are in line with the scenarios developed by the PRIMES model (National Technical University of Athens 2010) for the EU Low Carbon Roadmap 2050 (European Commission 2011). That is, we use the following two EU scenarios:

- The "PRIMES Reference scenario" projects energy trends to 2050 based on policies already adopted by March 2010. It includes policies agreed in the EU climate and energy package of 2009.
- The "PRIMES Decarbonisation scenario". We use the EUCO 27 variant, which assumes international agreement on an effective global action plan complemented with policies for carbon pricing across all sectors. It positions the adoption of major low-carbon technologies in the energy sector, e.g. energy efficiency and RES, CCS, nuclear and electrification of transport.

Since the energy transition is affected by technological development, climate change commitments, energy security and international agreements, both PRIMES scenarios might evolve differently under different contexts. For example, if cooperation among nations does not occur as expected, this might limit the integration of EU electricity markets and the creation of PCI. We can reflect this hypothetical

---

[2]Note that we refer to cases as the instances in which we perform an analysis and model implementation, while scenarios are the input assumptions to the modelling exercise.

situation by limiting the expansion of transmission capacity in EMPIRE. Limiting transmission expansion could reflect the view that EU nations find difficult to develop stronger cooperation. In short, this argument defines four cases:

- PRIMES reference case, transmission options available or optimised (**OTR**)
- PRIMES decarbonisation case, transmission options available or optimised (**OTD**)
- PRIMES reference case, transmission expansion limited (**LTR**)
- PRIMES decarbonisation case, transmission expansion limited (**LTD**)

The overall results illustrated in Fig. 3 show the developments of the energy mix for these four cases. For the years 2010 and 2015, historic data are used for all generation and transmission capacities; however, the economic dispatch comes from the model. The year 2020 is the first investment period for generation capacities. As for the transmission investment cases (LTR and LTP), this is set to the reference capacities in ENTSO-E's 10-year network development plan (TYNDP) 2016 (see ENTSO-E (2015) for more information). The major differences between the decarbonisation and the reference cases are as follows: The demand for electricity is higher, and the price of carbon is higher in the decarbonisation cases. The main effect on EMPIRE is that the need for investments is higher and that the generation mix is forced to be cleaner.

3.2 Results for 2020–2030 Period: All Cases

As expected, there are similarities in all scenarios in 2020 and 2030—regardless of input data used (reference or decarbonisation scenario) and whether transmission investments are restricted beyond 2020. This is because all the cases share similar features for these periods since transmission expansion is the same and difference between reference and decarbonisation is not significant for demand projections. The first effect that stands out is the expansion of the share of coal generation in 2020 for all scenarios. On one hand, this is because of the retirement of ageing nuclear power capacity with zero emission that is not replaced by new capacity. On the other hand, due to moderate carbon prices (15 EUR/tCO_2) and a high price ratio of gas to coal (at about 3.4), new coal generation is the lowest cost option to replace the retired capacity. Hence, this new coal capacity takes over the share of the existing natural gas installed in the system, leading to a small increase in power sector emissions from 2015 which is not necessarily in line with the European Union's climate goals set for 2020 (a 20% emission reduction compared to 1990s level). In reality the power sector emissions, subject to the EU Emissions Trading System, would not be allowed to increase, and the economic advantage of coal compared to lower emission technologies such as renewables, natural gas and nuclear would be counteracted by an increase in the ETS price. However, as the ETS price increases in the period beyond 2020, this becomes less of an issue as the emissions reduction from EMPIRE approach levels is in line with EU's climate policy.

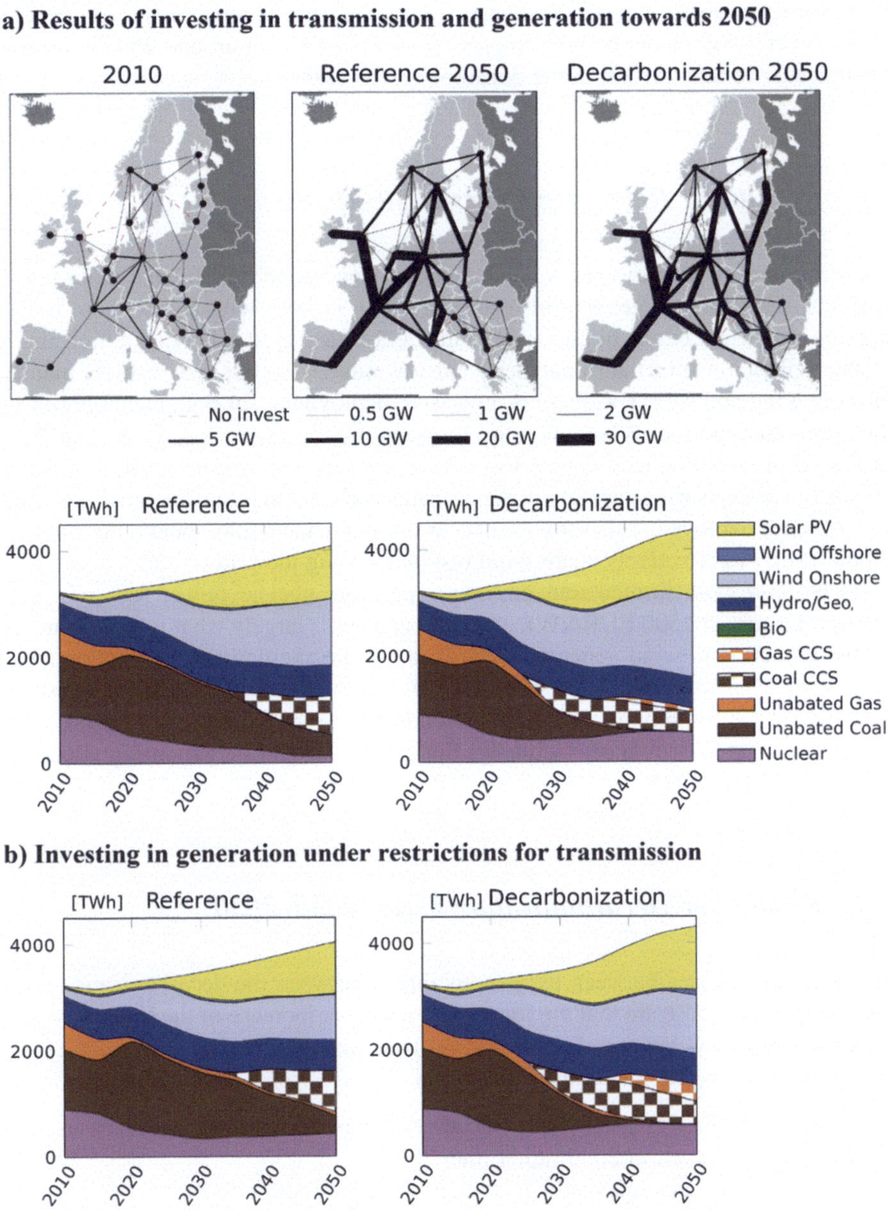

Fig. 3 Reference and decarbonisation cases under transmission capacity available (**a**) or restricted (**b**)

As for renewables, most of the investments in the period until 2030 are in onshore wind power. Only moderate investments are in solar PV—of around 25 GW for the reference cases and 75 GW in the decarbonisation cases.

3.3 Results for Reference Cases 2030–2050

For the reference scenario cases, unabated coal generation remains the main fossil fuel technology in the generation mix until 2040, in both with and without transmission expansion cases. In 2040 coal with CCS is deployed achieving a high share of the generation mix than unabated coal towards 2050. For renewables, mainly onshore wind and solar PV, there is a massive deployment for both technologies in the 2030–2050 period. When allowing for transmission expansion, more than 50% of the electricity produced comes from these sources. One of the most significant effects of not expanding capacity in the transmission system is that less onshore wind generation is deployed. This shows that cost-optimal wind deployment relies heavily on the ability to effectively share wind resources using the grid.

In the case with limited transmission expansion, nuclear power sees reinvestments (at a cost of 4500 EUR/kW), and this capacity is largely what is used to make up the reduction in wind generation compared to the alternative transmission case. As can be seen in the EU grid maps of Fig. 3a, there is a significant increase of capacity on the interconnectors between countries. For example, the cross-border connections between France, Germany, Spain, the UK and neighbouring countries are reinforced heavily.

3.4 Results for Decarbonisation Cases 2030–2050

The main differences between exogenous inputs between the decarbonisation case and the reference case are that the former has a steeper increase of the ETS price and a higher electricity demand. In comparison with the results of the reference cases, there are effects worth mentioning:

- Deployment of CCS occurs a full decade earlier in the decarbonisation cases than the reference cases, in 2030 rather than 2040.
- Nuclear power sees reinvestment in both decarbonisation cases.
- Solar and onshore wind expansion is more aggressive in these cases than the reference case, reaching a level of around 65% of the generation mix in the transmission expansion cases. By 2050 all unabated fossil generation has been completely phased out.
- As in the reference case, we observe that less wind onshore is installed when there are no investments in the transmission capacity beyond 2020. This is made up by fossil fuel with CCS, most notably gas CCS. But a small amount of offshore wind is deployed in the restricted transmission case.

For the cases with investments in transmission capacity, the overall picture is alike to the reference case, although with a few distinctions. There are significant investments in Central Europe, but some connections see different investments in the decarbonisation case than the reference. For instance, the interconnector between France and Germany has less capacity in the decarbonisation case, whereas the links from France to Switzerland and Switzerland to Germany are reinforced. Another notable difference is the increased investments from the Baltic countries in the north to the Balkans in the south in the decarbonisation case. This is a result of the increased penetration of variable renewable resources in the decarbonisation case, which has a strong effect on the optimal design of the transmission system for balancing supply and demand throughout the continent.

4 Robustness Tool and Stress Testing the Optimal Portfolios

In this section, we present the mathematical formulation and the structure of the min-max and min-max regret criteria. Uncertainty is represented deterministically through the concept of scenario (see also Kouvelis and Yu 1997). Potential future realisation of the model is represented through a particular scenario, which occurs with a positive but unknown probability.

Our aim is to identify robust strategies corresponding to a plausible objective performance, along all scenarios of our decision model. First, we apply the min-max criterion, pointing out that the robust decision is that having the best worst-case performance across all future scenarios. Let us explain this in the discrete scenario case.

Consider the following optimisation problem:

$$\min_{x \in X} f(c, x), \tag{10}$$

where $x = (x_1, \ldots, x_N) \in X \subseteq \mathbb{R}^N$ and $c = (c_1, \ldots, c_N) \in \mathbb{R}^N$. The feasible set X contains the admissible decision variables $(x_1, \ldots, x_N)$ satisfying some prespecified constraints of the model. The input parameters $(c_1, \ldots, c_N)$, inserted exogenously, define the uncertainty of the model in the following sense: if $S = \{s_1, \ldots, s_p\}$ denotes the finite set of all potentially realisable input data scenarios, then realisation of a certain scenario $s \in S$ means that

$$c = c^s = \left(c_1^s, \ldots, c_N^s\right).$$

While S is not identically a singleton, uncertainty is then inherent. Let X and S be the feasible set and the discrete scenario set, respectively, for the minimisation problem (10). The corresponding min-max decision is exactly

$$x_{minimax} = arg \min_{x \epsilon X} \max_{s \epsilon S} f(c^s, x).$$

Subsequently, consider the optimal solution for scenario's $s \in S$ realisation, which is

$$x_s^{opt} = arg\,\min_{x \epsilon X} f(c^s, x),$$

and the corresponding optimal performance:

$$f_s^{opt} := f(c^s, x_s^{opt}).$$

Given an admissible decision $x \in X$, its *regret* $R(x, s)$, under scenario's $s \in S$ realisation, is defined as its deviation from scenario's $s \in S$ optimal performance, that is,

$$R(x, s) := \left| f(c^s, x) - f_s^{opt} \right|.$$

For any decision, we are interested in identifying the worst deviation from optimality across the whole range of uncertainty. Therefore, it is reasonable to obtain information on the worst regret. For any decision $x \in X$, its *maximum regret* is defined as

$$R_{max} := \max_{s \epsilon S} R(x, s).$$

Then, the min-max regret criterion aims at identifying the solution presenting the best worst-case deviation from optimality, independently of the input data realisation. The corresponding min-max regret decision is exactly

$$x_{regret} = arg\,\min_{x \in X} \max_{s \in S} R_{max} = arg\,\min_{x \in X} \max_{s \in S} \left| f(c^s, x) - f_s^{opt} \right|. \quad (11)$$

Subsequently, we carry out a stress testing for the investments in electricity generation and transmission with respect to their performance across different scenarios. More precisely, we measure the variation of the cumulative 2010–2050 investment costs and cumulative 2010–2050 emissions across different scenarios on transmission and decarbonisation. Regarding scenarios, we apply the four cases previously described: the two PRIMES scenarios (reference and decarbonisation) with assumptions on transmission expansion (either optimised investments or limited to just the 10-year development plan by ENTSO-E).

First, we consider the objective referring to the cumulative 2010–2050 costs in billion € (2010) as they are presented in Table 1. In the first column, we consider the optimal portfolio investments per scenario, and we examine their performance across the remaining investment scenarios. Then, we carry out min-max and min-max regret analysis to obtain the most robust state. According to the min-max criterion, the best worst performance across all scenarios is located at the LTD case. In particular, the cumulative capacity costs do not exceed 2224.7 billion €, for any scenario's realisation. This is the risk-averse case decision-making.

Table 1 Cumulative 2010–2050 investment capacity costs of the optimal portfolios across different scenarios (in billion € (2010))

Portfolios/scenarios	Scenario OTR	Scenario OTD	Scenario LTR	Scenario LTD
OTR-optimal	1995.1	2507.3	9248.1	13,079.3
OTD-optimal	2065.0	2157.2	6678.4	7890.4
LTR-optimal	2027.0	2448.6	2039.1	2641.8
LTD-optimal	2097.3	2206.5	2112.9	2224.7

Table 2 Regret values for the investment costs

Portfolios/scenarios	Scenario OTR	Scenario OTD	Scenario LTR	Scenario LTD
OTR-optimal	0.0	350.1	7209.0	10,854.6
OTD-optimal	69.9	0.0	4639.3	5665.7
LTR-optimal	31.9	291.4	0.0	417.1
LTD-optimal	102.2	49.3	73.8	0.0

Table 3 Cumulative 2010–2050 emissions (in GtCO2)

Portfolios/scenarios	Scenario OTR	Scenario OTD	Scenario LTR	Scenario LTD
OTR-optimal	41.0	42.5	43.1	44.3
OTD-optimal	28.2	28.2	31.0	31.4
LTR-optimal	42.3	42.6	42.5	43.2
LTD-optimal	30.6	29.7	30.4	29.9

Table 4 Emissions regret values

Portfolios/scenarios	Scenario OTR	Scenario OTD	Scenario LTR	Scenario LTD
OTR-optimal	12.8	14.3	12.7	14.4
OTD-optimal	0.0	0.0	0.6	1.5
LTR-optimal	14.1	14.4	12.1	13.3
LTD-optimal	2.4	1.5	0.0	0.0

To apply the min-max regret criterion, we need the regret values for the costs across all scenarios (Table 2). In this case, the two approaches coincide, i.e. the more robust strategy is still located at the LTD-optimal portfolio.

Regarding the second objective, that is, the cumulative emissions for the 2010–2050 period, these are presented in Table 3. The corresponding regret values in terms of emissions are presented in Table 4. The min-max strategy corresponds to the LTD case, bounding the total emissions at the level of 30.6 GtCO2, independently of any scenario's realisation. On the contrary, min-max regret analysis points to the OTD strategy. This is the classical case between the risk-averse and the risk-seeking decision maker. More precisely, the OTD portfolio contains better performances (28.2 GtCO2) from the LTD portfolio. On the other hand, based on the

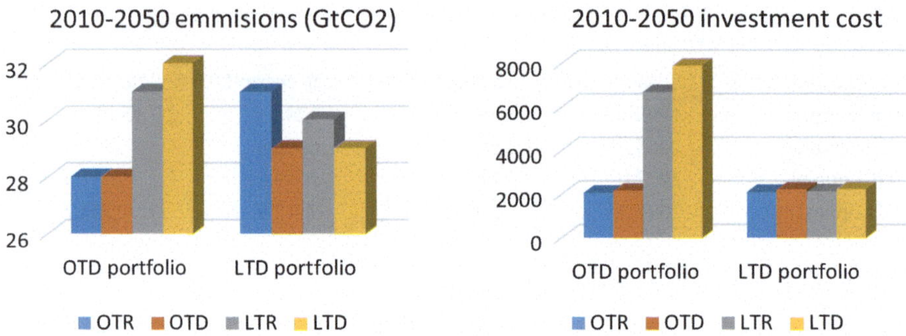

Fig. 4 Cumulative emissions and investment capacity costs of the optimal portfolios across different scenarios

Table 5 Relative joint objectives

Portfolios/scenarios	Scenario OTR	Scenario OTD	Scenario LTR	Scenario LTD
OTR-optimal	(1.00,1.45)	(1.16,1.49)	(3.53,1.41)	(5.87,1.48)
OTD-optimal	(1.03,1.00)	(1.00,1.00)	(2.27,1.01)	(3.54,1.05)
LTR-optimal	(1.01,1.50)	(1.13,1.51)	(1.00,1.39)	(1.18,1.44)
LTD-optimal	(1.05,1.08)	(1.02,1.05)	(1.03,1.00)	(1.00,1.00)

min-max regret criterion, the LTD scenario results in higher emissions (29.9 GtCO2). A visual representation reflecting the robustness structure of the two different objectives (investment costs and emissions) is presented in Fig. 4.

If we consider simultaneously the two objective functions (costs and emissions), we notice that a conflict occurs. Following the min-max regret analysis, the cost-oriented robust strategy is located in LTD optimality, while in the emissions case, regret robustness corresponds to the OTD-optimal portfolio. Coping with this conflict, we apply the min-max regret criterion for the joint objective $E^s = \left(E_1^s, E_2^s\right)$, where E_1^s and E_2^s represent the cumulative costs and emissions, respectively, across scenarios $s \in \{\text{OTR, OTD, LTR, LTD}\}$. First, we identify the optimal state per scenario, represented by $\left(E_{1*}^s, E_{2*}^s\right)$. Next, to avoid the curse of non-homogeneity, we consider the relative performance (RP) per scenario, with respect to the corresponding optimal situation, that is,

$$RP(s) := \left(\frac{E_1^s}{E_{1*}^s}, \frac{E_2^s}{E_{2*}^s}\right), s \in \{\text{OTR, OTD, LTR, LTD}\}.$$

Then, we obtain the relative joint performances in Table 5. Note that, in terms of relative performances, optimal situations correspond to the unit values. This means that, for each scenario, the optimal situation corresponds to (1,1). To identify the more robust joint strategy, we will apply the min-max regret criterion for the relative joint objectives. In this case, since the elements belong to $\mathbb{R}_+^2$, we measure the deviation from optimality in terms of the Euclidean norm, that is,

Table 6 Relative regret values

Portfolios/scenarios	Scenario OTR	Scenario OTD	Scenario LTR	Scenario LTD
OTR-optimal	0.44	0.52	0.4	4.89
OTD-optimal	0.03	0.00	1.27	2.54
LTR-optimal	0.50	0.52	0.38	0.47
LTD-optimal	0.09	0.05	0.03	0.00

$$\text{dist}(x, y) = \left((x_1 - x_2)^2 + (y_1 - y_2)^2\right)^{1/2},$$

for any $(x_1, y_1), (x_2, y_2) \in \mathbb{R}_+^2$. Then, the min-max regret criterion for the relative case is expressed as follows:

$$\text{MMR} = \text{arg min}_P \text{max}_s \left(\text{dist}(RP(s), (1, 1))\right),$$

where P refers to the set of the four examined optimal investment portfolios. The regret values, in terms of the relative objective functions, are presented in Table 6. Applying the min-max regret criterion to the relative regret values, we conclude that the joint robust strategy is LTD, being in accordance with the distinguished application of the pure min-max criterion.

Based on the above analysis, we highlight the following observations:

- In the min-max setting, the LTD strategy is the optimal, expressing the desire of the conservative decision maker in any of the examined future development.
- With the min-max regret approach, the LTD strategy for cost and the OTD strategy for emissions create the least regret for the decision maker. If, however, joint relative regret analysis is applied to the future optimal development, then LTD strategy is established to be the robust one.

To provide some policy recommendations emerging from the above analysis, it is observed that:

- Portfolios for decarbonisation scenarios are more robust. This is due to the fact that nuclear remains in the mix until 2050, while the introduction of CCS technologies in the mix comes much earlier than in the reference scenario.
- In the optimised transmission portfolios, even though an important priority is considered, if the new generation built is based on this scenario and not realised in terms of infrastructure and investments, then the resulting regret could be enormous. From a technical point of view, this situation is realistic, considering, for example, that balancing issues will be far more challenging.

5 Conclusions

In this study, different emissions reduction technologies are considered for the identification of optimal policy mixes towards mitigating GHG emissions. Optimal electricity generation portfolios are determined through an EU electricity investment model, by considering different scenarios on transmission expansion and decarbonisation. Then, the performance of portfolios' long-term strategies is stress tested on the basis of investment costs and cumulative emissions. In this regard, we notice that technologies such as nuclear and CCS are found to be crucial, in the sense that forming portfolios with optimised transmission expansion proves to be inconsistent with the stress test analysis results. To this end, robustness tools such as the min-max regret criterion contributed to explore a posteriori more insight into the scenarios' assumptions, showing that combining modelling approaches provide a new dimension on assessing pathways of the energy transition.

Concluding, we found that decarbonisation scenarios are the preferred strategy for all cases. However, the limited decarbonisation strategy is the more robust, resulting in the least regret.

In this study, the limitations primarily involve the selection of the four scenarios; furthermore, future research may be directed to the investigation of the impact of additional objective functions to apply the min-max and the min-max regret criteria. Also, future research should consider analysing the evolution of CCS technologies and the respective risks of investing in CCS. The technology is still not commercially viable and perhaps does not become a reality. It is thus important to calculate the costs (regret), if pathways consider this development.

Acknowledgement The current paper was primarily based on the research conducted within the framework of the Horizon 2020 European Commission projects "Navigating the Roadmap for Clean, Secure and Efficient Energy Innovation (SET-Nav)" grant agreement No 691843 and "Transitions pathways and risk analysis for climate change mitigation and adaptation strategies (TRANSrisk)" grant agreement No. 642260. The content of the paper is the sole responsibility of its authors and does not necessary reflect the views of the European Commission.

Appendices

Appendix 1 Nomenclature Used in the EMPIRE Model Formulation

Sets	
$\mathcal{N}$	Nodes (one per country)
$\mathcal{G}$	Generators. The set $\mathcal{G}_n$ is the set of all generators at node n
$\mathcal{L}$	Transmission lines (exchange corridors) between neighbouring nodes in the transmission system
$\mathcal{A}_n^{\mathrm{in/out}}$	Arcs to/from neighbouring nodes in the transmission system. Note that for every line connecting two nodes in the transmission system, there exist two arcs. These are used to represent directional flow
$\mathcal{H}$	Operational hours. The set $\mathcal{H}_s$ is the set of all operational hours in season s. The $\mathcal{H}_s^-$ is the set of all operational hours except the first hour in season s
$\mathcal{S}$	Seasons: 1 representative week per season and 2 weeks with extreme load and RES situations
Ω	Stochastic scenarios $\omega \in \Omega$
$\mathcal{T}$	Aggregate generation technologies (e.g. coal, gas, wind, solar, etc.)

(continued)

Decision variables (all non-negative)	
x_{gi}^{gen}	Investment in capacity for generator g in year i
x_{lj}^{tran}	Investment in capacity for transmission line l in year i
$y_{ghi\omega}^{\text{gen}}$	Production on generator g, operational hour h, year i, stochastic scenario ω
$y_{ahi\omega}^{\text{flow}}$	Flow on arc a, operational hour h, year i, stochastic scenario ω
$y_{nhi\omega}^{\text{pump}}$	Energy used for pumping on pump p, operational hour h, year i, stochastic scenario ω
$y_{nhi\omega}^{\text{LL}}$	Load shedding at node n, operational hour h, year i, stochastic scenario ω
$w_{nhi\omega}^{\text{upper}}$	Water-level upper reservoir for pump storage in node n, op. hour h, year i, scenario ω
Parameters	
δ_i	Discount factor year i (at rate interest rate r, this is $\delta_i = (1 + r)^{-5i}$)
α_h	Operational hour scale factor. This factor represents the total number of hours in a year represented by the operational hour h. Summing a variable/parameter scaled by α_h for all $h \in \mathcal{H}$ yields a yearly total, e.g. $\sum_{h\in\mathcal{H}} \alpha_h \xi_{nhi\omega}^{\text{load}}$ is the total electric energy consumption for node n in year i, scenario ω
p_ω	Probability of scenario ω for the stochastic parameters
c_{gi}^{gen}	Total cost (fixed and capital costs) incurred by investing in 1 MW new capacity for generator g
c_{li}^{tran}	Total cost (fixed and capital costs) incurred by investing in 1 MW new exchange capacity for line l
q_{gi}^{gen}	Variable costs (fuel + emission + O&M) incurred by producing 1 MWh of electric energy on generator g in year i
q_{ni}^{VoLL}	Cost of using load-shedding variable $y_{nhi\omega}^{\text{LL}}$
$\xi_{nhi\omega}^{\text{load}}$	Load at node n in operational hour h, year i and stochastic scenario ω
$\xi_{ghi\omega}^{\text{gen}}$	Available share of generation capacity for generator g in operational hour h, year i, stochastic scenario ω. Note that for thermal generation technologies and regulated hydropower the availability parameters are constant across all $\omega \in \Omega$
$\xi_{gsi\omega}^{\text{RegHydroLim}}$	Total energy available for production in season s
ρ_{gi}	Retired share of generator g's initial capacity by year i
γ_g^{gen}	Limit on total upward ramping as a fraction of total installed capacity for generator g
$\bar{x}_{g0}^{\text{gen}}$	Initial installed capacity generator g
$\bar{x}_{l0}^{\text{tran}}$	Initial exchange capacity line l
$\bar{x}_{tn*}^{\text{gen}}$	Upper bound on (period-wise/cumulative) investments in new capacity for generator g
$\bar{x}_{l*}^{\text{tran}}$	Upper bound on (period-wise) investments in new exchange capacity line l
η_a^{line}	Exchange losses on arc a (given as a share of the total flow)
η_n^{pump}	Pump efficiency for pump storage in node n
hr_{gi}	Heat rate generator g, year i
e_f	Carbon content fuel f

Appendix 2 Technological Assumptions for EMPIRE Implementation

Table 7 Investment costs of generation technologies in EMPIRE

Technology	2020	2025	2030	2035	2040	2045	2050	Unit
Lignite	1600	1600	1600	1600	1600	1600	1600	€$_{2010}$/kW
Lignite CCS adv		2600	2530	2470	2400	2330	2250	€$_{2010}$/kW
Coal	1500	1500	1500	1500	1500	1500	1500	€$_{2010}$/kW
Coal CCS adv		2500	2430	2370	2300	2230	2150	€$_{2010}$/kW
Gas OCGT	400	400	400	400	400	400	400	€$_{2010}$/kW
Gas CCGT	800	800	800	800	800	800	800	€$_{2010}$/kW
Gas CCS adv		1350	1330	1310	1290	1270	1250	€$_{2010}$/kW
Bio	2250	2250	2250	2250	2250	2250	2250	€$_{2010}$/kW
Nuclear	4500	4500	4500	4500	4500	4500	4500	€$_{2010}$/kW
Hydro regulated	3000	3000	3000	3000	3000	3000	3000	€$_{2010}$/kW
Hydro RoR	4000	4000	4000	4000	4000	4000	4000	€$_{2010}$/kW
Wind onshore	1033	1002	972	942	912	881	851	€$_{2010}$/kW
Wind offshore	3205	2770	2510	2375	2290	2222	2172	€$_{2010}$/kW
Solar	826	653	481	463	445	427	409	€$_{2010}$/kW

Note: Data for fossil fuel technologies (incl. advanced CCS) come from ZEP (2013). Source of wind onshore and offshore (Gerbaulet and Lorenz 2017). Solar PV costs are based on the medium scenario in Fraunhofer ISE (2015)

Table 8 Efficiency of thermal power plants in EMPIRE

Technology	2015	2020	2025	2030	2035	2040	2045	2050	Unit
Lignite exist	35	36	36	36	36	36	37	37	%
Lignite	44	45	45	46	47	48	48	49	%
Lignite CCS adv			37	39	40	41	42	43	%
Coal exist	38	38	38	38	38	39	39	39	%
Coal	46	46	47	47	48	48	49	49	%
Coal CCS adv			39	40	41	41	42	43	%
Gas exist	49	50	51	52	52	53	54	55	%
Gas OCGT	40	41	41	41	41	42	42	42	%
Gas CCGT	60	60	60	61	63	64	65	66	%
Gas CCS adv			52	54	56	57	58	60	%
Oil exist	38	38	38	38	38	38	38	38	%
Bio exist	35	35	35	35	35	35	35	35	%
Bio	36	36	37	38	38	39	39	40	%
Nuclear	36	36	36	37	37	37	37	37	%

Source: ZEP (2013)

Table 9 Fixed operation and maintenance costs in EMPIRE

	2020	2025	2030	2035	2040	2045	2050	Unit
Lignite	32.4	32.4	32.4	32.4	32.4	32.4	32.4	$€_{2010}$/kW/an
Lignite CCS		51.4	50.0	48.7	47.4	46.1	44.7	$€_{2010}$/kW/an
Coal	31.1	31.1	31.1	31.1	31.1	31.1	31.1	$€_{2010}$/kW/an
Coal CCS		47.0	45.9	44.7	43.6	42.5	41.4	$€_{2010}$/kW/an
Gas OCGT	19.5	19.5	19.5	19.5	19.5	19.5	19.5	$€_{2010}$/kW/an
Gas CCGT	30.4	30.4	30.4	30.4	30.4	30.4	30.4	$€_{2010}$/kW/an
Gas CCS		46.9	46.9	46.9	46.9	46.9	46.9	$€_{2010}$/kW/an
Nuclear	127.0	123.3	119.5	115.8	112.1	108.3	104.6	$€_{2010}$/kW/an
Wave	153.8	153.8	153.8	153.8	153.8	153.8	153.8	$€_{2010}$/kW/an
Geo	92.3	92.3	92.3	92.3	92.3	92.3	92.3	$€_{2010}$/kW/an
Hydro regulated	125.0	125.0	125.0	125.0	125.0	125.0	125.0	$€_{2010}$/kW/an
Hydro RoR	125.0	125.0	125.0	125.0	125.0	125.0	125.0	$€_{2010}$/kW/an
Bio	46.3	45.3	44.3	43.3	42.3	41.3	40.3	$€_{2010}$/kW/an
Wind onshore	52.6	51.7	50.9	50.0	49.1	48.2	47.3	$€_{2010}$/kW/an
Wind offshore	127.6	122.4	117.2	112.0	106.8	101.6	96.4	$€_{2010}$/kW/an
Solar	18.6	17.1	15.7	14.3	12.9	11.4	10.0	$€_{2010}$/kW/an

Source: ZEP (2013). Solar PV costs from Fraunhofer ISE (2015)

Table 10 Variable operation and maintenance costs in EMPIRE

	2020	2025	2030	2035	2040	2045	2050	Unit
Lignite	0.5	0.5	0.5	0.5	0.5	0.5	0.5	€2010/MWh
Lignite CCS	0.0	3.3	3.3	3.3	3.3	3.3	3.3	€2010/MWh
Coal	0.5	0.5	0.5	0.5	0.5	0.5	0.5	€2010/MWh
Coal CCS	0.0	2.5	2.5	2.5	2.5	2.5	2.5	€2010/MWh
Gas OCGT	0.5	0.5	0.5	0.5	0.5	0.5	0.5	€2010/MWh
Gas CCGT	0.5	0.5	0.5	0.5	0.5	0.5	0.5	€2010/MWh
Gas CCS	0.0	1.9	1.9	1.9	1.9	1.9	1.9	€2010/MWh
Nuclear	1.7	1.7	1.6	1.6	1.5	1.5	1.4	€2010/MWh

Source: ZEP (2013). Variable costs and operation and maintenance costs of other technologies assumed to be included in the fixed operation and maintenance costs. For CCS technologies there is an additional cost component of 20 $€_{2010}/tCO_2$ stored to account for transport and storage costs (this cost is flat for all years)

Table 11 Investment costs of storage technologies in EMPIRE

Technology	2020	2025	2030	2035	2040	2045	2050	Unit
Pump storage (power)	1000	1000	1000	1000	1000	1000	1000	$€_{2010}$/kW
Pump storage (energy)	100	100	100	100	100	100	100	$€_{2010}$/kWh
Li-ion utility battery	246	198	198	198	198	198	198	$€_{2010}$/kWh

Source: Pump-storage costs are based on own assumption. Lithium-ion battery costs are based on an adapted version of the medium cost scenario in Cole et al. (2016). For pump storage the power and energy capacity investments are decoupled. Li-ion batteries are assumed to be 0.5 C (i.e. capable to discharge from full to empty in 2 h)

Table 12 Investment costs of interconnectors in EMPIRE

	2020	2025	2030	2035	2040	2045	2050	Unit
HV lines	662	662	604	604	604	604	604	€2010/MW/km
HV cables	2769	2769	2160	2160	1551	1551	1551	€2010/MW/km

Table 13 Total European-installed power generation capacity in 2015 in EMPIRE by technology

Lignite	Coal	CCGT	Oil	Bio	Nuclear	Hydro	Wind	Solar PV	Other	Unit
60	108	232	26	25	125	158	142	95	1	GWp

Sources: Solar potentials from Gils et al. (2017). Wind potentials from IEA (2016). Nuclear maximum capacities based on visions 1 and 2 presented in ENTSO-E (2015). Hydro capacities are own assumptions

References

Awerbuch, S., & Berger, M. (2003). *Applying portfolio theory to EU electricity planning and policy-making*, IEA/EET Working Paper, Report Number EET/2003/03, Paris.

Bauer, N., Calvin, K., Emmerling, J., Fricko, O., Fujimori, S., Hilaire, J., & de Boer, H. S. (2017). Shared socio-economic pathways of the energy sector–quantifying the narratives. *Global Environmental Change, 42*, 316–330.

Chang, N. B., & Davila, E. (2007). *Minimax regret optimization analysis for a regional solid waste management system. Waste Management, 27*, 820–832.

Cole, W. J., Marcy, C., Krishnan, V. K., & Margolis, R. (2016). *Utility-scale lithium-ion storage cost projections for use in capacity expansion models*. 2016 North American Power Symposium (NAPS), pp. 1–6.

Crespo del Granado, P., van Nieuwkoop, R. H., Kardakos, E. G., & Schaffner, C. (2018). *Modelling the energy transition: A nexus of energy system and economic models. Energy Strategy Reviews, 20*, 229–235.

Dong, C., Huang, G. H., Cai, Y. P., & Xu, Y. (2011). *An interval-parameter minimax regret programming approach for power management systems planning under uncertainty. Applied Energy, 88*, 2835–2845.

ENTSO-E. (2015). *TYNDP 2016 Scenario Development Report – Final after public consultation*, Brussels. https://www.entsoe.eu/Documents/TYNDP documents/TYNDP 2016/rgips/TYNDP2016 market modelling data.xlsx

European Commission. (2011). Energy Roadmap 2050, COM(2011) 885/2. https://ec.europa.eu/energy/en/topics/energy-strategy-and-energy-union/2050-energy-strategy

Fan, L., Hobbs, B. F., & Norman, C. S. (2010). Risk aversion and CO_2 regulatory uncertainty in power generation investment: Policy and modeling implications. *Journal of Environmental Economics and Management, 60*(3), 193–208.

Fraunhofer ISE. (2015). Current and future cost of photovoltaics. Long-term scenarios for market development, system prices and LCOE of utility-scale PV systems. Freiburg: Study on behalf of Agora Energiewende

Fuss, S., Szolgayová, J., Khabarov, N., & Obersteiner, M. (2012). Renewables and climate change mitigation: Irreversible energy investment under uncertainty and portfolio effects. *Energy Policy, 40*, 59–68.

Gerbaulet, C., & Lorenz, C. (2017). *dynELMOD: A dynamic investment and dispatch model for the future European electricity market*. DIW Data Documentation.

Gils, H. C., Scholz, Y., Pregger, T., Tena, D. L., & Heide, D. (2017). Integrated modelling of variable renewable energy-based power supply in Europe. *Energy, 127*, 173–188.

Haller, M., Ludig, S., & Bauer, N. (2012). Decarbonization scenarios for the EU and MENA power system: Considering spatial distribution and short term dynamics of renewable generation. *Energy Policy, 47*, 282–290.

IEA. (2016). *Nordic energy technology perspectives 2016*. Paris: OECD.

IPCC. (2014). *Climate Change 2014: Synthesis Report*. In Core Writing Team, R. K. Pachauri & L. A. Meyer (Eds.), Contribution of Working Groups I, II and III to the Fifth Assessment Report of the Intergovernmental Panel on Climate Change (151p.). Geneva, Switzerland: IPCC.

Jägemann, C., Fürsch, M., Hagspiel, S., & Nagl, S. (2013). Decarbonizing Europe's power sector by 2050 – Analyzing the economic implications of alternative decarbonization pathways. *Energy Economics, 40*, 622–636.

Kaut, M., Midthun, K. T., Werner, A. S., Tomasgard, A., Hellemo, L., & Fodstad, M. (2014). Multi-horizon stochastic programming. *Computational Management Science, 11*(1–2), 179–193.

Kazakci, A. O., Rozakis, S., & Vanderpooten, D. (2007). Energy supply in France: A min-max regret approach. *Journal of the Operational Research Society, 106*, 1470–1479.

Kouvelis, P., & Yu, G. (1997). *Robust discrete optimization and its applications, Nonconcex optimization and its applications* (Vol. 14). Dordrecht: Kluwer Academic.

Li, Y. P., & Huang, G. H. (2006). Minimax regret analysis for municipal solid waste management: An interval-stochastic programming approach. *Journal of the Air & Waste Management Association, 56*(7), 931–944.

Li, Y. P., Huang, G. H., & Chen, X. (2011). An interval-valued minimax-regret analysis approach for the identification of optimal greenhouse-gas abatement strategies under uncertainty. *Energy Policy, 39*, 4313–4324.

Li, S., Coit, D. W., & Felder, F. (2016). Stochastic optimization for electric power generation expansion planning with discrete climate change scenarios. *Electric Power Systems Research, 140*, 401–412.

Loulou, R., & Kanudia, A. (1999). Minimax regret strategies for greenhouse gas abatement: Methodology and application. *Operations Research Letters, 25*, 219–230.

McLoughlin, E., & Bazilian M. (2006). *Application of portfolio analysis to the Irish electricity generating mix in 2020*, Sustainable Energy Ireland Working Paper.

Morris, J., Srikrishnan, V., Webster, M., & Reilly, J. (2018). Hedging uncertainties: Electricity investment decisions under policy uncertainty. *The Energy Journal, 39*(1), 101–122.

Munoz, F. D., Hobbs, B. F., Ho, J. L., & Kasina, S. (2014). An engineering-economic approach to transmission planning under market and regulatory uncertainties: WECC case study. *IEEE Transactions on Power Systems, 29*(1), 307–317.

Munoz, F. D., van der Weijde, A. H., Hobbs, B. F., & Watson, J. P. (2017). Does risk aversion affect transmission and generation planning? A Western North America case study. *Energy Economics, 64*, 213–225.

National Technical University of Athens. (2010). *The PRIMES energy systems model*. Technical report. http://www.e3mlab.ntua.gr/manuals/PRIMsd.pdfS

Nikas, A., Doukas, H., Lieu, J., Alvarez Tinoco, R., Charisopoulos, V., Charisopoulos, V., & van der Gaast, W. (2017). Managing stakeholder knowledge for the evaluation of innovation systems in the face of climate change. *Journal of Knowledge Management, 21*(5), 1013–1034.

Papapostolou, A., Karakosta, C., Nikas, A., & Psarras, J. (2017). Exploring opportunities and risks for RES-E deployment under cooperation mechanisms between EU and Western Balkans: A multi-criteria assessment. *Renewable and Sustainable Energy Reviews, 80*, 519–530.

Raupach, M. R., Marland, G., Ciais, P., Quéré, C. L., Canadell, J. G., Klepper, G., et al. (2007). Global and regional drivers of accelerating CO_2 emissions. *Proceedings of the National Academy of Sciences Of the USA, 104*, 10288–10293.

Richter, J. (2011). *DIMENSION – A dispatch and investment model for European electricity markets*. EWI Working Paper No. 11/03.

Seljom, P., & Tomasgard, A. (2015). Short-term uncertainty in long-term energy system models – A case study of wind power in Denmark. *Energy Economics, 49*, 157–167.

Skar, C., Doorman, G. L., & Tomasgard, A. (2014). *The future European power system under a climate policy regime*. In IEEE International Energy Conference, EnergyCon (pp. 337–344).
Skar, C., Doorman, G., Pérez-Valdés, G. A., Tomasgard, A. (2016a). *A multi-horizon stochastic programming model for the European power system*, CenSES Working Paper 2/2016.
Skar, C., Egging, R., & Tomasgard, A. (2016b). The role of transmission and energy storage for integrating large shares of renewables in Europe. *IAEE Energy Forum, Q1*, 35–36.
Swider, D. J., & Weber, C. (2007). The costs of wind's intermittency in Germany: Application of a stochastic electricity market model. *European Transactions on Electrical Power, 17*(2), 151–172.
Van der Weijde, A. H., & Hobbs, B. F. (2012). The economics of planning electricity transmission to accommodate renewables: Using two-stage optimization to evaluate flexibility and the cost of disregarding uncertainty. *Energy Economics, 34*(6), 2089–2101.
White, B. (2007). *A mean-variance portfolio optimization of California's generation mix to 2020*, Report no. CEC-300-2007-009-D, California Energy Commission.
ZEP. (2013). *CO_2 capture and storage (CCS): Recommendations for transitional measures to drive deployment in Europe*. Report by the European Technology Platform for Zero Emission Fossil Fuel Power Plants, Brussels.
ZEP. (2014). *CCS and the electricity market: Modelling the lowest-cost route to decarbonising European power*. Report by the European Technology Platform for Zero Emission Fossil Fuel Power Plants, Brussels.
Zhu, L., & Fan, Y. (2010). Optimization of China's generating portfolio and policy implications based on portfolio theory. *Energy, 35*, 1391–1402.

Impact Assessment of Climate and Energy Policy Scenarios: A Multi-criteria Approach

Hera Neofytou, Charikleia Karakosta, and Natalia Caldés Gómez

Abstract Over the last 20 years, numerous studies have shown significant changes in the global climate which negatively affect life in many aspects. The perpetual problems due to climate change impacts have created the urgent need to find efficient ways to tackle them. In this frame, European countries are moving towards the creation of energy and climate policies in order to achieve specific targets and mitigate the consequences of greenhouse gas emissions. They have defined a number of scenarios comprised of different targets' combinations and ambition levels. The targets to be achieved are defined as to the CO_2 emission reduction, the improvement in energy efficiency and the increase of the share of renewable energy sources until 2030. Thus, the aim is to lead EU to counteract the increasing energy demand and its negative effects on the environment as well as to abate the fossil fuel dependency. In this context, the scope of the particular paper is to examine which of the defined scenarios could respond adequately to the European region's profile and which could affect positively living conditions. Subsequently, the research focuses on the assessment of each alternative climate and energy policy scenario and its socioeconomic, environmental and energy impacts with the application of multi-criteria decision analysis. The method used in the herein analysis is the PROMETHEE II, which ranks the proposed scenarios based on the decision-makers' preferences. In order to ensure the robustness of the results, a sensitivity analysis is also performed.

H. Neofytou (✉)
Decision Support Systems Laboratory (EPU-NTUA), School of Electrical & Computer Engineering, National Technical University of Athens, Athens, Greece
e-mail: hneofytou@epu.ntua.gr

C. Karakosta
Decision Support Systems Laboratory (EPU-NTUA), School of Electrical & Computer Engineering, National Technical University of Athens, Athens, Greece

Department of Industrial Management and Technology, University of Piraeus, Piraeus, Greece

N. C. Gómez
Department of Energy, Energy System Analysis Unit, CIEMAT, Madrid, Spain

H. Doukas et al. (eds.), *Understanding Risks and Uncertainties in Energy and Climate Policy*, https://doi.org/10.1007/978-3-030-03152-7_5

Keywords EU 2030 framework · Climate policy · Energy policy · Socioeconomic impacts · Environmental impacts · Energy impacts · Multi-criteria decision analysis · PROMETHEE

1 Introduction

Nowadays one of the most important and challenging problems is the climate change (EC 2017a). Health, environment and economy are affected directly by the climate change, and it is vital to seek solutions so as to counteract the rising dangers. At the same time, countries' growth and human activities increase the energy demand and the fossil fuel dependency, thus contributing to the increase of greenhouse gas (GHG) emissions. Several initiatives and agreements between countries have emerged to address the repercussions of climate change and carbon use, and some of them have already accomplished significant targets.

In industrialised countries, it is a great challenge to restructure consumption and production patterns into a low-carbon system. In 2014 the European Commission (EC) agreed upon a 2030 framework (EC 2014) for climate and energy policies including targets for 40% greenhouse gas reduction, 27% increase in the share of renewable energy sources (RES) and 27% increase in energy efficiency (EE) achievements, all compared to the 1990 levels. During the process of new measures' formulation, emphasis should be given in the methods employed, the impacts covered, the coverage of scenarios and the comparison of costs and benefits (De Bruyn and Warringa 2014) in order to ensure a safe, secure, sustainable and affordable energy for EU citizens (EC 2012).

International climate and energy policymaking is currently surrounded by many uncertainties. At the level of member state and EU-level policymakers, there is uncertainty about whether the EU climate and energy policy ambitions will eventually be matched by other countries' ambitions and whether the policy instruments and knowledge could be transferred to others. A process of decarbonising the EU economy will result in new economic opportunities for EU businesses, but it also requires that energy will increasingly be provided within the EU using renewable and biomass-based energy sources. This requires investments in secure energy systems, as well as social acceptance of different ways to supply and use energy (Flamos 2016; Papadelis et al. 2013).

In addition to the above, an impact assessment is important in order to ensure a sustainable development and since EU is obliged to conduct impact assessments of any new legislation before it is approved (EC 2009, 2017b). The impacts that would arise as a result of the implementation of a scenario are critical guidelines during the acceptance of policies between policymakers and the public (POLIMP 2012). For example, an energy policy scenario, apart from the direct impacts, such as energy savings, affects indirectly other sectors as well. It is possible that while some measures have positive consequences from a point of view, they also have negative, or less positive, results from another. In this sense it is important to strike a balance between the different levels and types of impacts (Streatfield and Markless 2009).

In this paper emphasis will be given to the socioeconomic, environmental and energy impacts that specific mitigation pathways evoke. These impacts have been proved to be key factors in achieving policy targets since they can affect the formulation of the measures to be adopted (TRANSrisk 2017). A variety of impact assessments have already been conducted but without concluding in one specific direction (Ernst and Young 2014; De Vos et al. 2014; Cambridge Econometrics 2013; POLIMP 2012).

The multi-criteria decision analysis (MCDA) has been developed and used to support the evaluation of different scenarios relying on their performance on several criteria under different assumptions (Ribeiro and Ferreira 2013). Hundreds of MCDA methodologies have been proposed (Loken 2007) and applied to a wide range of problems in the energy sector (Goumas and Lygerou 2000; Karakosta and Psarras 2012; Oskari et al. 2017; Papapostolou et al. 2017; Rahman et al. 2015; Volkart et al. 2017; Wang et al. 2009) as well as to explore alternative domestic energy and electricity policy scenarios (Browne et al. 2010). Policymakers need digestible information to design good policies and understand their options and the possible impacts of these options (Karakosta and Flamos 2016). It is towards this direction the herein study attempts to move.

The MCDA approach was selected here due to the fact that this method provides the flexibility and capacity to support the views of many decision-makers or stakeholders (Cherni et al. 2007) and has been applied to a wide range of energy sector's issues, such as for analysis of RES policy scenarios (Papapostolou et al. 2017), for the local energy planning (Marinakis et al. 2017), for energy efficiency measures in buildings (Doukas et al. 2016), as well as in many other decision-making problems (Karakosta and Askounis 2010; Papapostolou et al. 2016). On the other hand, there is a variety of papers that use socioeconomic, environmental and energy impacts in order to perform assessments of the new climate regime and of energy and climate targets (EC 2014; Ernst and Young 2014; De Bruyn and Warringa 2014; POLIMP 2012). To the best of our knowledge, there are only very few papers that deal with energy strategies combined with their socioeconomic, environmental and energy impacts (see Rahman et al. 2015). Thus, the added value of this study is the merging of a multi-criteria decision analysis—in order to assess climate and energy policy scenarios—with their socioeconomic, environmental and energy impacts.

In particular, for the purposes of the herein work, one of the most efficient outranking methods was selected, the Preference Ranking Organization Method of Enrichment Evaluation (PROMETHEE) (Herva and Roca 2013; Kalogeras et al. 2005). PROMETHEE is closely coinciding with the human perspectives, and it determines the preferences among multiple decisions. It is also a suitable approach for an integrated analysis as its flexible algorithm can enable tailor-made enhancements to meet specific requirements for an integrated assessment approach, in particular, the explicit consideration of uncertainty information in the input values (Brans and Mareschal 2005; Kabir et al. 2014; Mohamadabadi et al. 2009).

Given the complexity of the different design options, policymakers and stakeholders need a manageable tool to reduce this complexity. It should be mentioned that the tools which explore the climate change implications, as well as the models' outcomes regarding the cost and benefits, face a high degree of uncertainty. After all there are several mitigation pathways. On the other hand, the policies adapted in order to achieve the desirable socioeconomic, environmental and energy impacts hide many risks, which means that there is the possibility that the outcomes may differ from the expected ones. Risks include the potential damage that derives from uncertainty and the vulnerability to that damage. Towards this direction the study presents a methodology and a variety of criteria through which the risks and uncertainties could be taken into consideration and be addressed.

In this context, the aim of this paper is to evaluate the climate and energy policy scenarios in the EU region exploiting the available data of their environmental, social, economic and energy systems impacts. Examination and suggestion of the most beneficiary scenario in terms of living conditions' improvement have not yet been made using a MCDA and especially the PROMETHEE method. Consequently it is envisaged that this study may contribute to the effort of leading Europe to a low-carbon emission path considering the several implications the proposed actions may have.

The paper unfolds as follows: The second section elaborates the definition of the problem and is comprised of two subsections. The first one presents the proposed alternative actions of the problem which constitute the climate and energy policy scenarios. The second one regards the evaluation system including the definition of the criteria which are the impacts of the scenarios as well as the MCDA method to be used. The third section analyses how the PROMETHEE method for the evaluation of the actions and the Simos procedure for the definition of the criteria weights function, in other words the pilot application. In the last section, the conclusion of the analysis is provided.

2 Defining the Problem

The current situation in EU regarding climate and environment is quite unpropitious and constantly deteriorates creating an increasing number of serious problems. For that reason it is vital for Europe to seek and develop the appropriate climate and energy policies. In order to achieve climate and energy targets for 2030, the adoption of specific actions and measures, regarding emissions and energy management, is considered indispensable. The question is which of these actions and in what way may lead to the most sustainable solutions. The answer of this question will support the policy associated stakeholders to extract key policy conclusions.

As already mentioned, the aim of this paper is to evaluate and classify the abovementioned scenarios in order to be able to assess which one has longer positive effects on the living conditions in the European region. The evaluation will be based on the impacts each scenario has in different areas emphasising the social and economic ones.

2.1 The Scenarios

During the energy and climate policymaking process, multiple possible combinations and ambition levels may be considered. The European Commission (EC) has defined a number of scenarios regarding the achievement of 2030 targets, of which four representative scenarios have been chosen to be evaluated (plus the reference scenario). The basic characteristics of each scenario are presented below in brief (EC 2014).

Reference Scenario The EU Reference Scenario 2013 explores the consequences of current trends, including full implementation of policies adopted by late spring 2012.

GHG40 This scenario achieves GHG emission reductions by tightening the linear reduction factor in the ETS (EU 2015). It presents a medium ambition in terms of GHG emission reduction that meets by 2030 a 40% GHG reduction compared to 1990 levels. It is based on the assumption of equalisation of marginal abatement cost of GHG emissions across the economy driven by increasing carbon prices and simulated carbon values as described for scenario GHG37.

GHG40/EE This scenario presents a medium ambition in terms of GHG emission reduction and is mainly driven by explicit ambitious energy efficiency policies that ensure progress by addressing market imperfections and failures. Beyond concrete EE policies, carbon pricing incentivises fuel shifts, energy savings and nonenergy-related emission reductions.

GHG40/EE/RES30 This scenario presents also a medium ambition in terms of GHG emission reduction and is mainly driven by explicit ambitious energy efficiency policies and pre-set RES target of 30% that ensure progress by addressing market imperfections and failures. Beyond concrete EE policies, carbon pricing continues to incentivise fuel shifts, energy savings and nonenergy-related emission reductions. Moreover EE policies contribute to higher shares of RES as they reduce total energy consumption as well.

GHG45/EE/RES35 This scenario presents a high ambition in terms of GHG emission reduction and is mainly driven by explicit and very ambitious energy efficiency policies and pre-set RES target of 35% that ensure progress by addressing market imperfections and failures. Beyond concrete EE policies, carbon pricing continues to incentivise fuel shifts, energy savings and nonenergy-related emission reductions.

In Table 1 an overview of the scenarios is provided.

Table 1 Scenarios' overview

	Carbon price (€/tonne)	Ambition-level EE policies	Ambition-level RES policies	Indicative policies employed
Reference	35	2020 stand.	2020 stand.	
GHG40	40	Moderate	Moderate	Equalisation of increasing carbon prices and values
GHG40/EE	22	Ambitious	Moderate	• Carbon pricing • Speeding up the buildings' renovation • Energy management systems • Extended and more ambitious energy efficiency obligations • Eco-design regulations • Smart grids
GHG40/EE/RES30	11	Ambitious	30% in 2030	• Similar to those in GHG40/EE plus • The average RES values rise from 49 €/MWh in 2020 to 56 €/MWh in 2030 and decline to 36 €/MWh in 2050
GHG45/EE/RES35	14	Ambitious	35% in 2030	Similar to those in GHG40/EE/RES30

2.2 *The Multi-criteria Evaluation System*

2.2.1 The Criteria

A multi-criteria evaluation model is proposed for the assessment of the different scenarios that could reach Europe's 2030 targets. It comprises 12 evaluation criteria based on four evaluation points of view to support decisions in selecting the most suitable scenario for the previous description (Fig. 1):

1. Environmental impacts
2. Social impacts
3. Economic impacts
4. Energy systems impacts

The evaluation criteria and model were selected after an extensive literature review (Cambridge Econometrics 2013; De Vos et al. 2014; Ernst and Young 2014; POLIMP 2012) and with the support of a panel of experts with experience in the climate and energy sectors (Karakosta et al. 2008; Karakosta and Askounis 2010; Papapostolou et al. 2017), in an attempt to cover all the different aspects.

The definition of criteria is presented below (EC 2014).

GHG reductions vs. 1990 (*g*1) This criterion is a percentage that indicates the total amount of greenhouse gas reduction in the ETS and non-ETS sectors compared to the respective emissions in 1990.

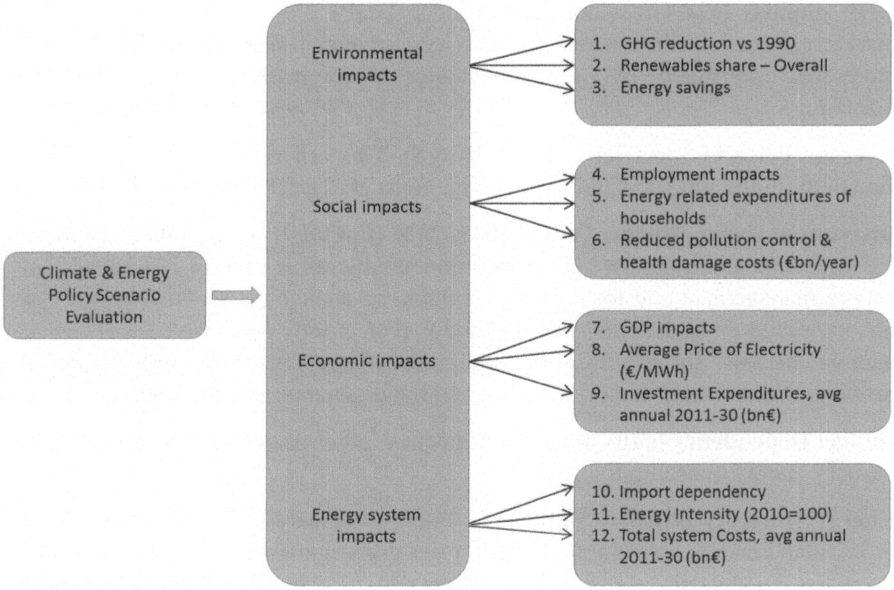

Fig. 1 Evaluation points of view and criteria

Renewable Share: Overall (*g*2) This criterion represents the share of energy from renewable sources (e.g. electricity, energy for heating and cooling and energy in transportation) in gross final energy consumption.

Energy Savings (*g*3) This percentage shows how much less primary energy will be consumed compared to projections made in 2007 for the 2030 as a result of the implementation of proposed measures of the scenarios and not of a decrease of GDP.

Employment Impacts (*g*4) This number indicates the total number of employees in all the sectors such as agriculture, extraction industries, basic manufacturing, engineering and transport equipment, utilities, construction, distribution and retail, transport, communication and publishing, business services and public services.

Energy-Related Expenditures of Households (*g*5) This number stands for the share of energy-related expenditures of households (referring to stationary uses) in average household expenditure. For the Reference Scenario, the corresponding value was 7.5% in 2010.

Reduced Pollution Control and Health Damage Costs (€bn/year) (*g*6) The reduction in air pollution has positive impacts on human health. The number of life year lost is being reduced due to lower harmful emissions' concentrations. The reduction in mortality can also be valued economically. Because of lower emissions and air pollution, costs to control them are lower as well. Consequently this index shows the monetary reduction of health damage costs due to reduced air pollution compared to the Reference Scenario. Valuation uses value of life year lost used for the Thematic Strategy on Air Pollution, ranging 57,000–133,000 € per life year lost.

GDP Impacts ($g7$) Gross domestic product (GDP) is a monetary measure of the market value of all final goods and services produced in a period. This criterion represents how much the GDP will change in comparison with the Reference Scenario.

Average Price of Electricity (€/MWh) ($g8$) This criterion presents the price of electricity in the final demand sectors with constant the 2010 Euro's value.

Investment Expenditures, Annual 2011–2030 (bn€) ($g9$) Direct efficiency investment costs include costs for house insulation, double/triple glazing, control systems, energy management and for efficiency enhancing changes in production processes not accounted for under energy capital and fuel/electricity purchase costs. They also include the cost for the transport equipment. They regard the industry sector; residential and tertiary sector; transport, grid and generation; and boilers.

Import Dependency ($g10$) Import dependency indicates the share of net imports to primary energy consumption.

Energy Intensity (2010 = 100) ($g11$) Energy intensity is the ratio between the gross inland energy consumption (GIEC) and gross domestic product (GDP) calculated for a calendar year. GIEC is calculated as the sum of the gross inland consumption of the five sources of energy: solid fuels, oil, natural gas, nuclear and renewable sources. To monitor trends, GDP is in constant prices to avoid the impact of inflation, with a base year of 2010 (EEA 2016). Gross inland energy consumption is measured in 1000 tonnes of oil equivalent (ktoe), while GDP is expressed in million Euros at 2010 market prices. To make comparisons of trends across countries more meaningful, the indicator is presented as an index.

Total System Costs, Avg Annual 2011–2030 (bn€) ($g12$) Energy system costs from an end user perspective as calculated in the modelling comprise mainly three elements: annuities for capital expenditure on energy using equipment (for energy installations such as power plants and energy infrastructure, energy using equipment, appliances and vehicles); fuel and electricity costs, including the capital expenditure for the production and distribution of electricity; and so-called direct energy efficiency costs incurred (not related to energy equipment itself), such as expenditure for insulation. Capital costs are expressed in annuity payments.

The preference value of each criterion is given in Table 2.

2.2.2 The PROMETHEE Method

PROMETHEE (Preference Ranking Organization METHod for Enrichment Evaluation) is an outranking method, developed by Brans et al. (1986), which is used when someone has to decide which action is better compared to others. Its main characteristic is the pairwise comparison using the inner relationships of each pair of actions based on the respective weights. PROMETHEE I (obtained from the positive and negative outranking flows) is used in cases where only a partial ranking is

Table 2 Criteria preferences

Criterion	Preference
g_1	Max
g_2	Max
g_3	Max
g_4	Max
g_5	Min
g_6	Max
g_7	Max
g_8	Min
g_9	Min
g_{10}	Min
g_{11}	Min
g_{12}	Min

needed, whereas PROMETHEE II (based on the net preference flow) is used for complete ranking of the alternatives (Brans et al. 1986).

The application of the method includes five steps:

1. Determination of the criteria weights, w_j, using the preferable method. The decision-maker has to define the weights so as to counterbalance the importance of each alternative. The sum of these weights has to equal the unit: $\sum w_j = 1$. The methodology used for the calculation of the weights is presented in Sect. 2.2.3.
2. Determination of a particular preference function in order to translate deviation between the evaluations of two alternatives (a and b) on a specific criterion (g_j) into a preference degree. Each criterion has a preference function:

$$P_j(a,b) = F_j\left[d_j(a,b)\right] \tag{1}$$

where F_j is a non-decreasing function of the observed deviations d_j and $d_j(a, b) = g_j(a) - g_j(b)$

The larger the deviation, the larger the preference (0 for no preference, 1 for strict preference). If $d_j \leq 0$, then $P_j(a, b) = 0$; in other case the analyst should choose between multiple types: usual, U shape, V shape, level, linear and Gaussian. Some extra parameters may be used such as indifference threshold or preference threshold in case of using any type of the function except the usual.

3. Calculation of global preference index $\pi(a, b)$ which represents the intensity of preference of alternative a over b taking the weights into account:

$$\pi(a,b) = \sum_{j=1}^{n} P_j(a,b) w_j \tag{2}$$

where $P_j(a,b)$ is the preference function and w_j represents the weight of each criterion.

4. Calculation of outranking flows. Positive outranking flow $\varphi^{+}(a)$ and negative outranking flow $\varphi^{-}(b)$ measure how much the alternative a is outranking or outranked by the other alternatives:

$$\varphi^{+}(a) = \frac{1}{n-1} \sum_{b} \pi(a, b) \tag{3}$$

$$\varphi^{-}(a) = \frac{1}{n-1} \sum_{b} \pi(b, a) \tag{4}$$

5. Calculation of the net outranking flow. Finally, in order to rank actions completely, the net outranking flow $\varphi(a)$ is determined using the equation below:

$$\varphi(\alpha) = \varphi^{+}(a) - \varphi^{-}(a) \tag{5}$$

2.2.3 Simos Procedure

As mentioned above, the first thing analyst has to determine is the criteria weights. The method followed in this paper is the Simos procedure, as an adequate aggregation procedure, where the decision-maker (DM) is able to hierarchise the criteria given a specific context. The gist of this method is that it correlates each criterion with one "card". The DM has to rank the criteria managing the respective cards from the least important to the most important and inserting "white cards" in order to make the distance between them larger if he wants to point out the differences between their importance. No white card means that the criteria have not the same weight and that the difference between the weights can be chosen as the unit for measuring the intervals between weights. Let us denote this unit. One white card means a difference of two times u. Two white cards mean a difference of three times u, etc. (Figueira and Roy 2002; Simos 1990).

3 Pilot Application and Scenario Analysis

In this section the collected data is presented and subsequently the results from the application of the PROMETHEE method.

The values of the criteria for each alternative are presented in Tables 3 and 4 (EC 2014).

During the application of a MCDA method, one has to keep in mind that there are no universal best solutions since results derive upon personal judgement of different criteria (Ribeiro and Ferreira 2013). The general approach was to request several

Table 3 Climate and energy policy scenario impacts (a)

Climate and energy policy scenarios	g_1 (%)	g_2 (%)	g_3 (%)	g_4	g_5 (%)	g_6
Reference	−32.40	24.40	−21.00	231,781,000	9.30	0.00
GHG40	−40.60	26.50	−25.10	232,379,000	9.40	10.35
GHG40EE	−40.30	26.40	−29.30	232,132,000	9.50	26.10
GHG40EERES30	−40.70	30.30	−30.10	232,514,000	9.40	24.95
GHG45EERES35	−45.10	35.40	−33.70	232,075,000	9.70	31.70

Table 4 Climate and energy policy scenario impacts (b)

Climate and energy policy scenarios	g_7 (%)	g_8	g_9	g_{10} (%)	g_{11}	g_{12}
Reference	0.00	176	816	55.10	67	2,067
GHG40	−0.28	179	854	53.60	64	2,069
GHG40EE	0.55	174	875	52.80	60	2,089
GHG40EERES30	0.46	178	879	51.80	60	2,089
GHG45EERES35	0.53	196	909	52.30	57	2,102

experts (DMs) to rank these criteria in order to perform a sensitivity analysis. In this context, eight experts, one governmental representative, two energy experts, one from the civil society, one from NGO, one from the scientific community, one from the industry and one from the general public were asked to use the SIMOS procedure and hierarchise the criteria, thus exporting four slightly different weights for each one. The weights that derived from this procedure are presented for each expert in Table 5.

Moreover, the analyst chose to use the preference function of usual type for every single criterion since there aren't any thresholds. This function serves to transform deviations into 0 or 1 in case a is not preferable from b and a is preferable from b, respectively. The preference function is calculated for each criterion (impacts) and all pairs of the alternative actions (policy scenarios) (Eq. 1) and used to calculate the preference index given in (Eq. 2). The final step is to calculate the net outranking flow (Eq. 5) using the positive and negative outranking flows (Eqs. 3 and 4).

For the implementation of the procedure, the Visual PROMETHEE [http://www.promethee-gaia.net/software.html] was used which provides numerical as well as visual results. Thus, inserting all the data and parameters, the programme exports the results based on the perspective of each expert (Table 6).

The alternative scenarios have been ranked using the net outranking flow that varies between −1 and 1, which is presented in Fig. 2.

Observing the comparison of the scenarios, it is obvious that, in all cases, the fifth and the fourth places are taken by the Reference and the GHG40 scenarios, respectively. This result was expected since the Reference Scenario does not include any measures to further contribute to the greenhouse gases' reduction or to increase the RES; and the GHG40 includes limited and very specific actions mainly for the greenhouse gases' reduction, and it does not enhance any other aspect of life.

Table 5 Criteria weights per expert

Criteria	Exp. 1	Exp. 2	Exp. 3	Exp. 4	Exp. 5	Exp. 6	Exp. 7	Exp. 8
GHG reductions vs. 1990 (*g*1)	0.14	0.15	0.10	0.14	0.15	0.15	0.15	0.04
Renewables share—overall (*g*2)	0.07	0.07	0.06	0.09	0.08	0.01	0.02	0.08
Energy savings (*g*3)	0.07	0.07	0.06	0.11	0.08	0.14	0.10	0.08
Employment impacts (*g*4)	0.14	0.15	0.14	0.14	0.13	0.12	0.13	0.05
Energy-related expenditures of households (*g*5)	0.12	0.11	0.14	0.11	0.13	0.13	0.15	0.02
Reduced pollution control and health damage costs (*g*6)	0.14	0.11	0.14	0.06	0.10	0.10	0.05	0.11
GDP impacts (*g*7)	0.10	0.11	0.14	0.11	0.10	0.06	0.07	0.16
Average price of electricity (*g*8)	0.10	0.11	0.14	0.11	0.13	0.09	0.11	0.06
Investment expenditures, annual 2011–2030 (*g*9)	0.02	0.03	0.02	0.06	0.02	0.04	0.11	0.02
Import dependency (*g*10)	0.02	0.03	0.02	0.02	0.04	0.04	0.05	0.14
Energy intensity (*g*11)	0.04	0.03	0.02	0.02	0.04	0.04	0.02	0.14
Total system costs, avg annual 2011–2030 (*g*12)	0.02	0.03	0.02	0.02	0.02	0.08	0.02	0.11

As for the first place, the prevailing scenario is the GHG40EERES30 in all cases except for the last one, since the expert is a governmental representative. This means that criteria like GDP, import dependency, energy intensity, reduced pollution control and health damage costs and system costs had the highest weights, while energy-related expenditures of households, investment expenditures and GHG reductions had the lowest.

The remaining scenarios, the GHG45RES35 and the GHG40EE, are switched between the second and the third places with the GHG45RES35 taking the second place in five out of the seven cases. The case where the GHG40RES35 takes the third place is when the experts (3 and 7) lent weight to criteria for which this scenario presents the worst numbers. More specifically, it seems that these two experts overrate the importance of energy-related expenditures and average price of electricity. This could be explained somehow since the experts originate from the energy sector.

Sensitivity analysis shows that the prevailing ranking is:

- GHG40EERES30
- GHG45EERES35
- GHG40EE
- GHG40
- Reference

Table 6 Outranking flows and ranking

Climate and energy policy scenarios		Exp. 1		Exp. 2		Exp. 3		Exp. 4		Exp. 5		Exp.6		Exp. 7		Exp.8	
			Rank		Rank		Rank		Rank		Rank		Rank		Rank		Rank
Reference	Φ	−0.469	5	−0.440	5	−0.360	5	−0.394	5	−0.427	5	−0.335	5	−0.225	5	−0.535	5
	Φ^{+}	0.265		0.280		0.320		0.303		0.287		0.333		0.388		0.233	
	Φ^{-}	0.735		0.720		0.680		0.697		0.713		0.668		0.612		0.767	
GHG40	Φ	−0.168	4	−0.153	4	−0.205	4	−0.134	4	−0.174	4	−0.113	4	−0.069	4	−0.327	4
	Φ^{+}	0.401		0.410		0.380		0.419		0.397		0.428		0.446		0.334	
	Φ^{-}	0.569		0.563		0.585		0.553		0.571		0.540		0.515		0.661	
GHG40EE	Φ	0.112	3	0.110	3	0.200	2	0.081	3	0.103	3	0.045	3	0.046	2	0.210	3
	Φ^{+}	0.549		0.548		0.595		0.535		0.544		0.508		0.518		0.574	
	Φ^{-}	0.436		0.438		0.395		0.455		0.441		0.463		0.472		0.364	
GHG40EERES30	Φ	0.332	1	0.338	1	0.295	1	0.331	1	0.346	1	0.313	1	0.304	1	0.290	2
	Φ^{+}	0.643		0.648		0.625		0.647		0.650		0.625		0.628		0.611	
	Φ^{-}	0.311		0.310		0.330		0.316		0.304		0.313		0.324		0.322	
GHG45EERES35	Φ	0.194	2	0.145	2	0.070	3	0.116	2	0.152	2	0.090	2	−0.056	3	0.361	1
	Φ^{+}	0.597		0.573		0.535		0.558		0.576		0.545		0.472		0.681	
	Φ^{-}	0.403		0.428		0.465		0.442		0.424		0.455		0.528		0.319	

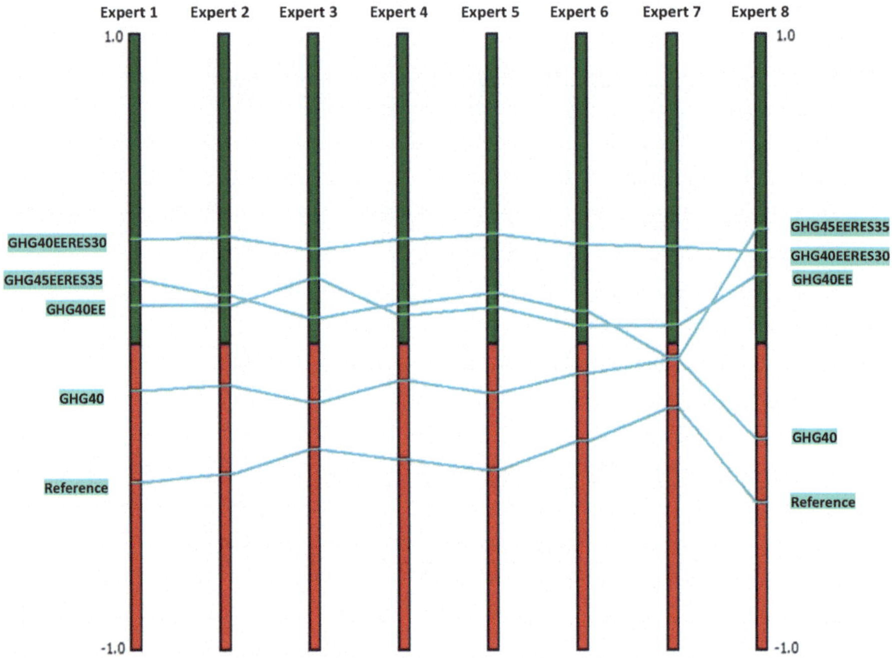

Fig. 2 Comparison of scenario ranking

Table 7 Distances between scenarios for Expert 1

Exp. 1	GHG40EERES30	GHG45EERES35	GHG40EE	GHG40	Reference
GHG40EERES30		−0.14	−0.22	−0.50	−0.80
GHG45EERES35	0.14		−0.08	−0.36	−0.66
GHG40EE	0.22	0.08		−0.28	−0.58
GHG40	0.50	0.36	0.28		−0.30
Reference	0.80	0.66	0.58	0.30	

Negative value means that the scenario on the vertical axis is considered worse (lower in the PROMETHEE ranking) than that on the horizontal axis

In order to verify the above outcome, a further analysis was conducted, which indicated to what degree a specific scenario is better or worse compared to the others. To illustrate this, the distances between the scenarios were calculated as well as the respective averages, and the results showed the degree of preference or not preference between each pair of them (Tables 7, 8, 9, 10, 11, 12, 13 and 14).

The average distances between the scenarios are provided in Table 15. As it can be easily conceived, this analysis's results coincide with the above sensitivity analysis since in the first column the values (distances from the GHG40EERES30) are increasing, as we move to the Reference Scenario and thus the same ranking like the previous derives.

Table 8 Distances between scenarios for Expert 2

Exp. 2	GHG40EERES30	GHG45EERES35	GHG40EE	GHG40	Reference
GHG40EERES30		−0.19	−0.23	−0.49	−0.78
GHG45EERES35	0.19		−0.04	−0.30	−0.59
GHG40EE	0.23	0.04		−0.26	−0.55
GHG40	0.49	0.30	0.26		−0.29
Reference	0.78	0.59	0.55	0.29	

Negative value means that the scenario on the vertical axis is considered worse (lower in the PROMETHEE ranking) than that on the horizontal axis

Table 9 Distances between scenarios for Expert 3

Exp. 3	GHG40EERES30	GHG45EERES35	GHG40EE	GHG40	Reference
GHG40EERES30		−0.23	−0.10	−0.50	−0.66
GHG45EERES35	0.23		0.13	−0.28	−0.43
GHG40EE	0.10	−0.13		−0.41	−0.56
GHG40	0.50	0.28	0.41		−0.16
Reference	0.66	0.43	0.56	0.16	

Negative value means that the scenario on the vertical axis is considered worse (lower in the PROMETHEE ranking) than that on the horizontal axis

Table 10 Distances between scenarios for Expert 4

Exp. 4	GHG40EERES30	GHG45EERES35	GHG40EE	GHG40	Reference
GHG40EERES30		−0.22	−0.25	−0.47	−0.73
GHG45EERES35	0.22		−0.04	−0.25	−0.51
GHG40EE	0.25	0.04		−0.22	−0.48
GHG40	0.47	0.25	0.22		−0.26
Reference	0.73	0.51	0.48	0.26	

Negative value means that the scenario on the vertical axis is considered worse (lower in the PROMETHEE ranking) than that on the horizontal axis

Table 11 Distances between scenarios for Expert 5

Exp. 5	GHG40EERES30	GHG45EERES35	GHG40EE	GHG40	Reference
GHG40EERES30		−0.19	−0.24	−0.52	−0.77
GHG45EERES35	0.19		−0.05	−0.33	−0.58
GHG40EE	0.24	0.05		−0.28	−0.53
GHG40	0.52	0.33	0.28		−0.25
Reference	0.77	0.58	0.53	0.25	

Negative value means that the scenario on the vertical axis is considered worse (lower in the PROMETHEE ranking) than that on the horizontal axis

Table 12 Distances between scenarios for Expert 6

Exp. 6	GHG40EERES30	GHG45EERES35	GHG40EE	GHG40	Reference
GHG40EERES30		−0.22	−0.27	−0.43	−0.65
GHG45EERES35	0.22		−0.05	−0.20	−0.43
GHG40EE	0.27	0.05		−0.16	−0.38
GHG40	0.43	0.20	0.16		−0.22
Reference	0.65	0.43	0.38	0.22	

Negative value means that the scenario on the vertical axis is considered worse (lower in the PROMETHEE ranking) than that on the horizontal axis

Table 13 Distances between scenarios for Expert 7

Exp. 7	GHG40EERES30	GHG45EERES35	GHG40EE	GHG40	Reference
GHG40EERES30		−0.36	−0.26	−0.37	−0.53
GHG45EERES35	0.36		0.10	−0.01	−0.17
GHG40EE	0.26	−0.10		−0.12	−0.27
GHG40	0.37	0.01	0.12		−0.16
Reference	0.53	0.17	0.27	0.16	

Negative value means that the scenario on the vertical axis is considered worse (lower in the PROMETHEE ranking) than that on the horizontal axis

Table 14 Distances between scenarios for Expert 8

Exp. 8	GHG40EERES30	GHG45EERES35	GHG40EE	GHG40	Reference
GHG40EERES30		0.07	−0.08	−0.62	−0.83
GHG45EERES35	−0.07		−0.15	−0.69	−0.90
GHG40EE	0.08	0.15		−0.54	−0.75
GHG40	0.62	0.69	0.54		−0.21
Reference	0.83	0.90	0.75	0.21	

Negative value means that the scenario on the vertical axis is considered worse (lower in the PROMETHEE ranking) than that on the horizontal axis

Table 15 Average distances between scenarios

Average	GHG40EERES30	GHG45EERES35	GHG40EE	GHG40	Reference
GHG40EERES30		−0.18	−0.21	−0.49	−0.72
GHG45EERES35	0.18		−0.02	−0.30	−0.53
GHG40EE	0.21	0.02		−0.28	−0.51
GHG40	0.49	0.30	0.28		−0.23
Reference	0.72	0.53	0.51	0.23	

4 Conclusions

One of the most crucial issues nowadays is dealing with the climate change repercussions and the increasing energy demand. Towards this direction multiple objectives have been set, through the development of different climate and energy policy scenarios, in order to tackle the several implications and achieve a sustainable development. The aim of this study was to examine which of the defined scenarios could respond adequately to the European region's profile and which could affect positively the living conditions based on the socioeconomic, environmental and energy impacts. Consequently a multi-criteria decision analysis was selected to support in the assessment. More specifically, an integrated approach of PROMETHEE II and Simos procedure has been used, where the strengths of both methodologies are combined in a single MCDA tool, which eventually seems to be fully compatible with the overall methodology. Moreover, several experts were asked to participate in the criteria evaluation so as to perform a sensitivity analysis to ensure the robustness of the results.

All things considered, the methodological approach is based on simple and explicit steps, exploiting criteria and scenarios aligned with the EU's strategy imperatives. The developed model supports the decision-maker also in a distinct implementation, so as to extract in a reasonable time the results.

The most beneficiary scenario in terms of living conditions, according to all the experts' preferences, is the GHG40EERES30. This is considered an acceptable and realistic outcome, since it is the most balanced scenario based on literature and stakeholder consultation. This scenario presents a medium ambition in terms of GHG emission reduction through the carbon pricing (11 €/tonne), while it is mainly driven by explicit ambitious energy efficiency policies and pre-set RES target of 30% that ensure progress by addressing market imperfections and failures. Beyond concrete EE policies, carbon pricing continues to incentivise fuel shifts, energy savings and nonenergy-related emission reductions, and EE polices contribute to higher shares of RES as they reduce total energy consumption as well.

The prevailing scenario is followed by GHG45EERES35, and this in turn is followed by GHG40EE, GHG40 and Reference in that order. It also should be noted that even though the scenario with the 45% GHG reduction and 35% RES has the highest targets, it takes the second place because its effects will not be shown up to 2030 but in a 2050 prospect (which is not examined in the present paper). It is also obvious that the "business as usual" will serve no purpose in the next years in view of the increasing energy demand and the intense repercussions of climate change.

As can been seen, the paper has room for improvement by using a different and more integrated approach. The study might be extended by taking into consideration a greater number of alternative scenarios and/or criteria. Moreover a further enhancement could be achieved by involving a greater group of experts to map the diversity of opinions and preferences regarding criteria weights. In that way, it would be possible to extract more concrete and credible suggestions, thus aiding the European effort in the green and sustainable development.

The added value of this analysis to policymakers is its contribution to plan climate and energy strategies towards a low-carbon transition pathway by using the information of this approach and prioritising the enhancement of the living conditions from a socioeconomic, environmental and energy perspective. The final inference is the significant role of RES and the urgent need for energy efficiency measures in order to achieve the EU targets in parallel with sustainable development.

Acknowledgement The current paper was partially based on the research conducted within the framework of the project "SET-Nav: Navigating the Roadmap for Clean, Secure and Efficient Energy Innovation" that received funding from the European Union's Horizon 2020 research and innovation programme under grant agreement no. 691843.

References

Brans, J. P., & Mareschal, B. (2005). PROMETHEE methods. In J. Figueira, S. Greco, & M. Ehrgott (Eds.), *Multiple criteria analysis – state of the art surveys, International series in operations research and management sciences* (Vol. 78, pp. 163–195). New York: Springer.

Brans, J. P., Vincke, P., & Mareschal, B. (1986). How to select and how to rank projects: The Promethee method. *European Journal of Operational Research, 24*(2), 228–238.

Browne, D., O'Regan, B., & Moles, R. (2010). Use of multi-criteria decision analysis to explore alternative domestic energy and electricity policy scenarios in an Irish city-region. *Energy, 35* (2), 518–528.

Cambridge Econometrics. (2013, October). *Employment Effects of selected scenarios from the Energy roadmap 2050*. Final report for the European Commission (DG Energy).

Cherni, J. A., Dyner, I., Henao, F., Jaramillo, P., Smith, R., & Font, R. O. (2007). Energy supply for sustainable rural livelihoods. A multi-criteria decision-support system. *Energy Policy, 35*(3), 1493–1504.

De Bruyn, S., & Warringa, G. (2014, March). *Review of the Impact Assessment for a 2030 climate and energy policy framework*. Delft, CE Delft.

De Vos, R., Van Breevoort, P., Höhne, N., Winkel, T., & Sachweh, C. (2014, March 17). *Assessing the EU 2030 climate and energy targets*. A Briefing Paper. Project number: DESNL14683.

Doukas, H., Siskos, E., Psarras, J., Malamatenios, C., Tournaki, S., & Tsoutsos, T. (2016). Qualification roadmap empowering the Greek building sector workforce in the field of energy. *Renewable and Sustainable Energy Reviews, 65*, 992–1004.

EC-European Commission. (2009, January 15). *Impact assessment guidelines*. SEC 92.

EC-European Commission. (2012). *Energy Roadmap 2050*. Brochure. Accessed from https://ec.europa.eu/energy/sites/ener/files/documents/2012_energy_roadmap_2050_en_0.pdf.

EC-European Commission. (2014, January 22). *Communication from the Commission to the European Parliament, the Council, the European Economic and Social Committee and the Committee of the Regions. A policy framework for climate and energy in the period from 2020 to 2030*. {SWD(2014) 15 final} {SWD(2014) 16 final}. Brussels.

EC-European Commission. (2017a). *Causes of climate change*. Accessed August 28, 2017, from https://ec.europa.eu/clima/change/causes_en.

EC-European Commission. (2017b). *Impact assessments*. Accessed November 6, 2017, from https://ec.europa.eu/info/law-making-process/planning-and-proposing-law/impact-assessments_en.

EEA-European Environment Agency. (2016). *Energy intensity*.

Ernst & Young. (2014). *Macro-economic impacts of the low carbon transition*. © 2014 Ernst & Young et Associés; Studio EY France – 1401sG418.

EU-European Union. (2015). *EU ETS handbook*. Accessed from https://ec.europa.eu/clima/sites/clima/files/docs/ets_handbook_en.pdf.

Figueira, J., & Roy, B. (2002). Determining the weights of criteria in the ELECTRE type methods with a revised Simos' procedure. *European Journal of Operational Research, 139*(1–2), 317–326.

Flamos, A. (2016). A sectoral micro-economic approach to scenario selection and development: The case of the Greek power sector. *Energies, 9*, 77.

Goumas, M., & Lygerou, V. (2000). An extension of the PROMETHEE method for decision making in fuzzy environment: Ranking of alternative energy exploitation projects. *European Journal of Operational Research, 123*(3), 606–613.

Herva, M., & Roca, E. (2013). Ranking municipal solid waste treatment alternatives based on ecological footprint and multi-criteria analysis. *Ecological Indicators, 25*(2013), 77–84.

Kabir, G., Sadiq, R., & Tesfamariam, S. (2014). A review of multi-criteria decision-making methods for infrastructure management. *Structure and Infrastructure Engineering, 10*(9), 1176–1210.

Kalogeras, N., Baourakis, G., Zopounidis, K., & Dijk, G. (2005). Evaluating the financial performance of agri-food firms: A multicriteria decision-aid approach. *Journal of Food Engineering, 70*(3), 365–371.

Karakosta, C., & Askounis, D. (2010). Developing countries' energy needs and priorities under a sustainable development perspective: A linguistic decision support approach. *Energy for Sustainable Development, 14*(4), 330–338.

Karakosta, C., Doukas, H., & Psarras, J. (2008). A decision support approach for the sustainable transfer of energy technologies under the Kyoto protocol. *American Journal of Applied Sciences, 5*(12), 1720–1729.

Karakosta, C., & Flamos, A. (2016). Managing climate policy information facilitating knowledge transfer to policy makers. *Energies, 9*, 454.

Karakosta, C., & Psarras, J. (2012). Fuzzy TOPSIS approach for understanding a country's development priorities within the scope of climate technology transfer. In M. J. Acosta (Eds.), *Advances in energy research* (Vol. 9, pp. 123–149). Nova Science Publishers. ISBN: 978-1-61470-485-0.

Loken, E. (2007). Use of multicriteria decision analysis methods for energy planning problems. *Renewable and Sustainable Energy Reviews, 11*, 1584–1595.

Marinakis, V., Doukas, H., Xidonas, P., & Zopounidis, C. (2017). Multicriteria decision support in local energy planning: An evaluation of alternative scenarios for the Sustainable Energy Action Plan. *Omega, 69*, 1–16.

Mohamadabadi, H. S., Tichkowsky, G., & Kumar, A. (2009). Development of a multi-criteria assessment model for ranking of renewable and non-renewable transportation fuel vehicles. *Energy, 34*(1), 112–125.

Oskari, L., Risto, L., & Pekka, S. (2017). Multicriteria evaluation of sustainable energy solutions for Colosseum. *Sustainable Cities and Society, 35*, 289–297.

Papadelis, S., Flamos, A., & Psarras, J. (2013). A framework to address uncertainties in energy policy formulation. *Energy Sources, Part B: Economics, Planning, and Policy, 8*(3), 270–278.

Papapostolou, A., Karakosta, C., & Doukas, H. (2017). Analysis of policy scenarios for achieving renewable energy sources targets: A fuzzy TOPSIS approach. *Energy and Environment, 28*(1–2), 88–109.

Papapostolou, A., Karakosta, C., Marinakis, V., & Flamos, A. (2016). Assessment of RES cooperation framework between the EU and North Africa: A multicriteria approach based on UTASTAR. *International Journal of Energy Sector Management (IJESM), 10*(3), 402–426.

POLIMP. (2012). *D4.2: Socioeconomic impacts of a new climate regime on the EU*. Deliverable of the POLIMP project.

Rahman, M. M., Paatero, J. V., Lahdelma, R., & Wahid, M. A. (2015). Multicriteria-based decision aiding technique for assessing energy policy elements-demonstration to a case in Bangladesh. *Applied Energy, 164*, 237–244.

Ribeiro, F., & Ferreira, P. (2013). Evaluating future scenarios for the power generation sector using a Multi-Criteria Decision Analysis (MCDA) tool: The Portuguese case. *Energy, 52*, 126–136.

Simos, J. (1990). *Evaluer l'impact sur l'environnement: Une approche originale par l'analyse multicritere et la negociation*. Lausanne: Presses Polytechniques et Universitaires Romandes.

Streatfield, D. R., & Markless, S. (2009). What is impact assessment and why is it important? *Performance Measurement and Metrics, 10*(2), 134–141.

TRANSrisk. (2017). *D4.3: Implications of different mitigation portfolios based on stakeholder preferences*. Deliverable of the TRANSrisk project.

Volkart, K., Weidmann, N., Bauer, C., & Hirschberg, S. (2017). Multi-criteria decision analysis of energy system transformation pathways: A case study for Switzerland. *Energy Policy, 106*, 155–168.

Wang, J.-J., Jing, Y.-Y., Zhang, C.-F., & Zhao, J.-H. (2009). Review on multi-criteria decision analysis aid in sustainable energy decision-making. *Renewable and Sustainable Energy Reviews, 13*, 2263–2278.

Water Stress Implications of Energy Scenarios for the Middle East: An Assessment of Risks and Uncertainties

Bob van der Zwaan, Matthew Halstead, and Tom Kober

Abstract Energy, water, and food systems have so far mostly been studied independently. In this chapter, we argue that it is important to take an "energy-water-food nexus approach" to analyzing these three resource systems. After briefly introducing the emerging literature on the energy-water-food nexus, we inspect the interrelationship between energy and water use in the Middle East. We present results for projected power production, water withdrawal, and water consumption levels until 2050 in the Middle East under both baseline and stringent climate policy scenarios. We also analyze how the use of different cooling techniques for the main power production options in the Middle East can yield water withdrawal and consumption savings in the electricity sector in the region. We end by informing authorities responsible for the implementation of energy-water policies on the risks and uncertainties associated with the water usage of future energy systems.

Keywords Energy-water-food nexus · Climate change · Energy scenarios · Risks · Uncertainties · Middle East

---

B. van der Zwaan (✉)
Energy Research Centre of the Netherlands (ECN), Policy Studies, Amsterdam, The Netherlands

Johns Hopkins University, School of Advanced International Studies (SAIS), Bologna, Italy

University of Amsterdam, Faculty of Science (HIMS), Amsterdam, The Netherlands
e-mail: vanderzwaan@ecn.nl

M. Halstead
Energy Research Centre of the Netherlands (ECN), Policy Studies, Amsterdam, The Netherlands

T. Kober
Laboratory for Energy Systems Analysis, Paul Scherrer Institute (PSI), Villigen, Switzerland

H. Doukas et al. (eds.), *Understanding Risks and Uncertainties in Energy and Climate Policy*, https://doi.org/10.1007/978-3-030-03152-7_6

1 Introduction

In this chapter, we inspect possible water stress implications of energy scenarios developed by an integrated assessment model (IAM). We analyze the potential impacts of developments in the energy sector on requirements from the water sector in a case study for the Middle East, since it is a region in which water stress is likely to become more and sooner apparent than in other parts of the world. We perform our analysis through an assessment of risks and uncertainties. In this chapter, we start by pointing out that energy scenarios are subject to sizeable uncertainties, because they are determined by the parameter values used in, or simulation constraints applied to, the models that generate them. A substantial degree of freedom exists with regard to the choice of parameter values as well as simulation constraints. As a matter of fact, because of these uncertainties, the IAM literature employs the terminology of "scenarios" or "projections," rather than "forecasts" or "predictions."

Below, we show that energy scenarios may involve risks in terms of water usage versus availability and that large disparities exist regarding water use risks between different scenarios. We demonstrate that these risks can be mitigated through technological change and that the water use uncertainties of energy scenarios can be reduced. We quantify these uncertainties and calculate the levels to which the risks can be lowered by developing a set of scenarios for energy and water use in the Middle East with the TIAM-ECN model. TIAM-ECN is an established IAM that has been used for a variety of energy and climate policy analyses over the past decade. Last but not least, we make recommendations for policy-makers, explain that they should be well aware of the possible risks and uncertainties involved in energy and water usage scenarios, and point out the directions in which these risks and uncertainties can be addressed. For the energy and water modeling communities, we make suggestions for how their research can be improved, not only by integrating their respective fields of work but also by explicitly accounting for the risks and uncertainties associated with the scenarios they develop.

The first section of this chapter begins by introducing the published literature on the energy-water-food nexus—often referred to as "the nexus"—in the context of which our work has been performed. The nexus is a broad and still largely undefined and poorly understood concept and constitutes a field of research in statu nascendi. In the following section, we thus narrow down our focus to predominantly inspect the interrelationship between energy and water as applied to the Middle East. In the last two sections, we present and discuss the results of our analysis and formulate a list of conclusions and recommendations, respectively. With our chapter, we intend to inform authorities responsible for the development and implementation of energy-water policies on the risks associated with the water use of future energy systems, notably in the context of efforts to mitigate global climate change. We perform an uncertainty analysis by inspecting various scenarios for future energy demand and deployment of low-carbon energy technologies in the Middle East.

2 The Energy-Water-Food Nexus

Energy, water, and food resource systems are fundamentally interrelated. We need energy to produce food and to treat and move water; we need water to cultivate food crops and to generate essentially any form of energy; and we need food to support the world's growing population that both generates and relies on energy and water services (Halstead et al. 2014). Land availability also constitutes an important element in each of these three resources with important economic consequences, for example, for crop production, for either food or energy purposes (OECD 2017). This mutual relationship is defined as the "energy-water-food nexus." To date, the three individual resource systems of energy, water, and food have mostly been organized and studied independently. In a rapidly developing world with ever more pressing environmental challenges, however, choices and actions in each of these three domains can significantly affect the others, positively or negatively. Therefore, it is important to take a "nexus approach" to analyzing these three resource systems. Conventional policy- and decision-making with regard to each of these domains in isolation is not necessarily anymore the most effective or optimal course of planning or action. A nexus approach—which in our context refers to the multidisciplinary analysis of the relationship between energy, water, and food—can help to reduce trade-offs and to build synergies across these different sectors. In an increasingly complex and interrelated world, this approach can lead to better and more efficient resource use as well as cross-sector policy coherence.

Water scarcity already affects every continent. Around 1.2 billion people, almost one fifth of the world's population, live in areas of scarcity. Another 1.6 billion people, almost one quarter of the global population, face economic water shortage, which means that countries lack the necessary infrastructure to take water from rivers and aquifers. It is estimated that by 2030 almost 50% of people on the planet will be living in areas of high water stress with a likely impact on energy and food security (UN 2014). Even though water is a renewable resource, and sufficient water is available globally to satisfy an expanding and wealthier population, demand for water already exceeds supply in many regions of the world. At present, this supply-demand imbalance is most commonly seen in, for instance, Brazil, China, India, and South Africa, as well as in countries in the Middle East and North Africa (referred to as the MENA region; see SEI 2011).

For pursuing analyses involving water-related issues, one needs to distinguish between three different types of water use: water withdrawal, water consumption, and water discharge. Water withdrawal is the total amount of water taken from a source (groundwater or surface water). Water consumption is the proportion of water that is not returned to its source after it has been withdrawn. Water that is consumed is no longer available because it has evaporated, been transpired by plants, incorporated into products or crops, consumed by people or livestock, or otherwise been permanently removed from its source. Water discharge is the difference between water withdrawal and consumption. In other words, it is water that is not consumed and is returned to the original body of water.

Energy and water are inextricably linked. Non-renewable energy resources currently dominate the global energy generation landscape. These thermal sources of energy generation, mostly derived from fossil fuels, are at present particularly water-intensive, mainly due to the cooling systems that they use. These cooling systems require large amounts of water. A push toward a less carbon-intensive energy sector with a larger share of renewables, stimulated by efforts to mitigate global climate change, requires careful consideration of the potential impacts of such an energy transition on the other nexus sectors. For example, low-carbon biofuels and hydropower are also very water-intensive, sometimes as much as fossil fuels in terms of water use per unit of energy generated. Energy use itself for biomass production may in some cases outweigh the energy that the biomass generates (SEI 2011).

Energy and water are also interconnected to food and agriculture. Agriculture is the largest user of fresh water globally, accounting for approximately 70% of freshwater withdrawals from rivers, lakes, and aquifers. This ratio can rise to up to 90% in some developing countries. An increasing population and shifting dietary trends mean that the demand for food and feed crop cultivation is rising. Food production and its associated supply chain account for approximately one third of the world's total energy consumption (UN 2014). In many regions, rising food production has not only led to agricultural land expansion—largely at the expense of forests—but also to an intensification of agricultural processes on existing land. This simultaneous expansion and intensification place more stress on agricultural input resources such as water, energy, and fertilizers.

The implications on natural resources have been investigated of the baseline food and agriculture projections up to 2050 by the Food and Agriculture Organization (FAO 2006; Bruinsma 2011). Even though the growth rate in agricultural production continues to slow down—as a result of a declining population growth rate and a higher percentage of the world's population reaching medium to high levels of food consumption—agricultural production will need to rise by approximately 70% by 2050 to serve a 40% increase in population and rising average food consumption levels. Approximately 90% of the growth in crop production would be a result of higher yields and increased agricultural intensity, with the remainder being provided via land expansion. Mainly thanks to gradual improvements in water use efficiency, water withdrawals for irrigation would grow more slowly, but still increase by almost 11% by 2050. In terms of the availability of both land and water, both of these resources are more than sufficient globally, but are unevenly distributed throughout the world. Certain regions or countries face scarcity of either land or water for crop production (Bruinsma 2011). Scarcity of these resources could restrict the potential for the expansion of agriculture and intensification of agricultural processes (IEA 2013).

While we recognize that the food and agriculture sectors are an essential part of the nexus, we do not focus on these areas in this chapter. The main aim of this paper is to develop an understanding of how different conventional and innovative energy technologies can be distinguished in relation to their water needs. This is another important part of the nexus. We investigate several energy scenarios and the future water requirements of these scenarios, including some in which climate policy is adopted. We develop a tool to analyze future short-, medium-, and long-term impacts of energy on water and the implications of energy and climate policy on these two resources. Much of what happens in the field of energy is determined by global climate change control; therefore, we inspect low-carbon technologies in particular. For our purposes here, we start by making an inventory of water withdrawal and consumption factors for a set of different fossil fuel-based and renewable energy technologies, much in line with—and relying on data reported in—earlier publications on this topic (Macknick et al. 2012; Meldrum et al. 2013). In the Appendix, we report the results of our findings, summarized in Figs. 7 and 8 (see also Halstead et al. 2014). These figures show how very different the water intensities of distinct energy technologies can be and that all of them are characterized by substantial uncertainty ranges.

3 Case Study on the Middle East

Since a few years, efforts are being undertaken to represent water availability and/or usage in IAMs (see, e.g., Bouckaert et al. 2014; McCarl et al. 2017). Nexus case studies have been performed with IAMs for specific countries, notably several in Africa and the MENA region (Al-Riffai et al. 2017; van der Zwaan et al. 2018). Our purpose here is to inspect with an IAM what the water withdrawal and consumption implications might be of two main types of scenarios for future energy needs and the technologies employed herein to meet these needs: a baseline and a stringent climate policy scenario. We do this for the Middle East, since it is one of the regions in the world where water stress is already today becoming evident (see, e.g., Doukas et al. 2017). Having reported in the Appendix the water intensities of a representative set of different energy technologies, we investigate scenarios that enable us to analyze which of the electricity generation options could potentially be most suitable for this particular region from a water use perspective. We also inspect the uncertainties associated with these scenarios, as well as the risks involved in them. Knowledge of water usage levels of the power sector in the Middle East allows public authorities to match them with data for water availability in this region, which can assist in designing appropriate energy and climate policy.

Our baseline scenario does not include existing and planned climate policies, which means that it does not include greenhouse gas (GHG) emission reduction targets stated by countries in their so-called Copenhagen and Cancun pledges. The baseline scenario does include policy measures on renewable energy that were in place before 2010: these are assumed to remain in effect in the foreseeable future. In the 2 °C climate policy scenario, we assume that enough low-cost options are deployed to reduce GHG emissions so as to reach the global 2 °C target, regardless of where in the world or in what sector the emission reductions take place. This corresponds to a globally harmonized action to mitigate climate change. An example would be a global carbon certificate market that involves GHG emission reduction obligations allocated purely on cost-efficiency criteria. This corresponds to a scenario in which the equilibrium concentration of GHGs in the atmosphere is at most 450 ppm (see, e.g., Kriegler et al. 2013).

The left-hand plot of Fig. 1 shows a possible baseline scenario for electricity generation in the Middle East, which is here defined as the collection of countries in the Levant and the Arabian Peninsula, including also Iran, Iraq, and Turkey. It represents just one of the many ways in which business-as-usual power production could expand over the next several decades, as substantial uncertainty prevails in this respect. This baseline constitutes quite a realistic scenario, given the large role it proffers for the use of natural gas in the power sector. Indeed, this resource is domestically abundant in the region. Our baseline has been developed with the bottom-up energy systems model TIAM-ECN, a tool to make internally consistent long-term energy supply and demand scenarios. For more details and examples of how this model can be used for energy and climate policy analysis, we refer to previous publications (see, e.g., van der Zwaan et al. 2013a, b; Kober et al. 2014; Rösler et al. 2014). As one can see, relatively modest roles are also reserved for coal and oil in power production in the baseline scenario, but by the middle of the century, natural gas remains practically the only fossil fuel left for electricity generation. Hydropower plays a non-negligible role throughout the forthcoming decades, given the potential of this option in countries like Iran and Turkey.

The right-hand plot of Fig. 1 shows how power supply may significantly alter over the next decades if one assumes that in the Middle East, just as elsewhere in the world, stringent climate policy is introduced capable of reaching the global 2 °C target. This scenario is calculated, and this target is met by applying a constraint on emissions of CO_2 and other GHGs in the TIAM-ECN model. TIAM-ECN determines that the cost-optimal transition path involves not only a drastic reduction in the

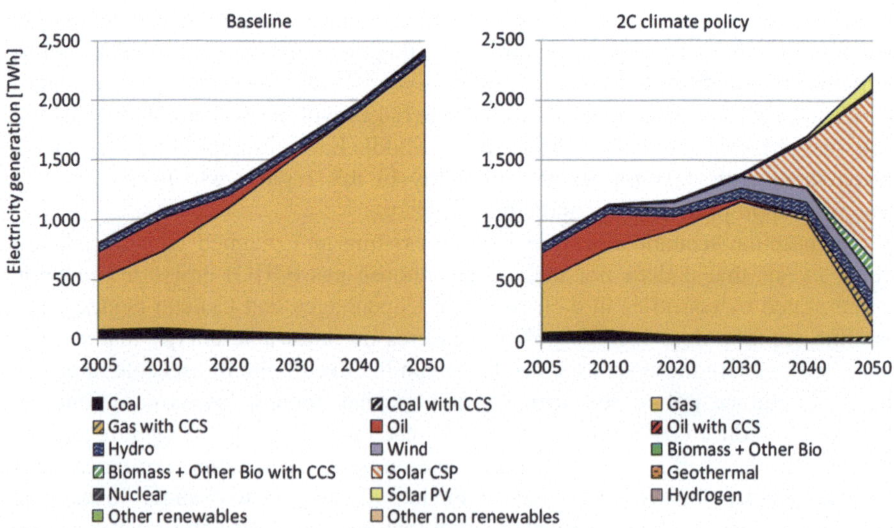

Fig. 1 Baseline and stringent climate policy scenario for power production in the Middle East

role of the single most important fossil fuel, natural gas—while part of its use is subjected to CCS implementation—but also a massive introduction and diffusion of solar energy, in particular CSP. The latter makes sense in view of the large solar irradiation resources of the Middle East. In addition to CSP and some PV, relatively small but non-negligible roles are reserved for power production options such as biomass, wind, and hydropower. It also proves cost-effective to introduce a certain level of energy savings in this climate policy scenario, as evidenced by its lower overall level of power production in 2050 in comparison to that in the baseline. Otherwise, as could be expected, the climate change control stringency necessitates a massive introduction of low-carbon renewables. The precise nature of the renewable energy mix, however, is subject to a large amount of uncertainty, as many different combinations and levels of renewable energy types can achieve the 2 °C target.

4 Results and Discussion

Figure 2 depicts what the water withdrawal levels would be if we superimpose the water intensity factors reported in the Appendix onto the power production patterns of Fig. 1. The color shading of the left plot of Fig. 2 (for the baseline) looks similar to that of the left plot of Fig. 1, except for the fact that the oil and coal shares are slightly fatter in the former. This is an expression of the fact that the water withdrawal intensity of natural gas-based electricity generation is somewhat smaller than that of its fossil fuel counterparts. The color composition of the right plot of Fig. 2, on the other hand, looks very different from that of the right plot of Fig. 1. The reason is that, first of all, CCS—deployed in response to stringent climate policy—is a very

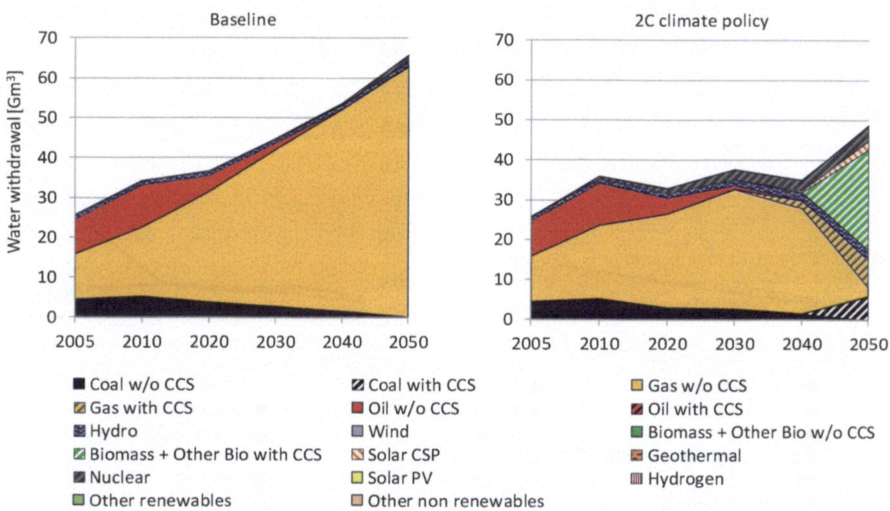

Fig. 2 Baseline and stringent climate policy scenario for water withdrawal in the Middle East

water-intensive technology, as demonstrated by the large water withdrawal shares in Fig. 2 (right plot). This is especially apparent for the use of biomass in combination with CCS, since non-CCS biomass use for power production is also a very water-intensive option. In terms of water withdrawal, CSP possesses a much smaller water footprint, as evidenced by the relatively small contribution of CSP to the right-hand-side graph of Fig. 2. By its very nature, hydropower withdraws large amounts of water. Nuclear power, while hardly discernible in the right-hand-side graph of Fig. 1, occupies a disproportionally large share in the right-hand-side graph of Fig. 2, the explanation for which is the substantial water intensity of nuclear power as thermal electricity generation option.

Over the course of the coming few decades, it can be expected that at least four countries in the region (Iran, Saudi Arabia, United Arab Emirates, and Turkey) consume domestically produced nuclear-based electricity. The right plot of Fig. 3 shows that in terms of water consumption, the stringent climate policy scenario looks quite different from that in terms of water withdrawal. In the former case, CSP is by far the dominant force, since it is substantially more water-consuming than even biomass-based power production complemented with CCS technology. The exponential growth in water consumption in the climate policy scenario depicted in the right graph of Fig. 3 implies a risk in terms of whether sufficient water will be available to meet this demand. This risk needs to be carefully considered.

Underlying the results depicted in Figs. 2 and 3 are the cooling techniques associated with the respective individual electricity generation options, since these cooling technologies are responsible for the vast majority of the indicated water withdrawal and consumption levels. Figure 4 shows what the breakdown is today of different types of cooling techniques in the Middle East for each of the main current power generation alternatives. For calculating the water usage profiles shown in

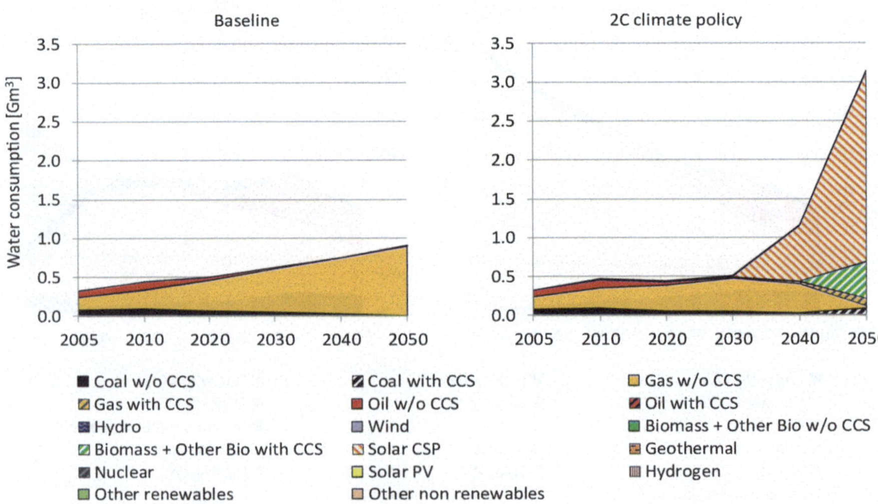

Fig. 3 Baseline and stringent climate policy scenario for water consumption in the Middle East

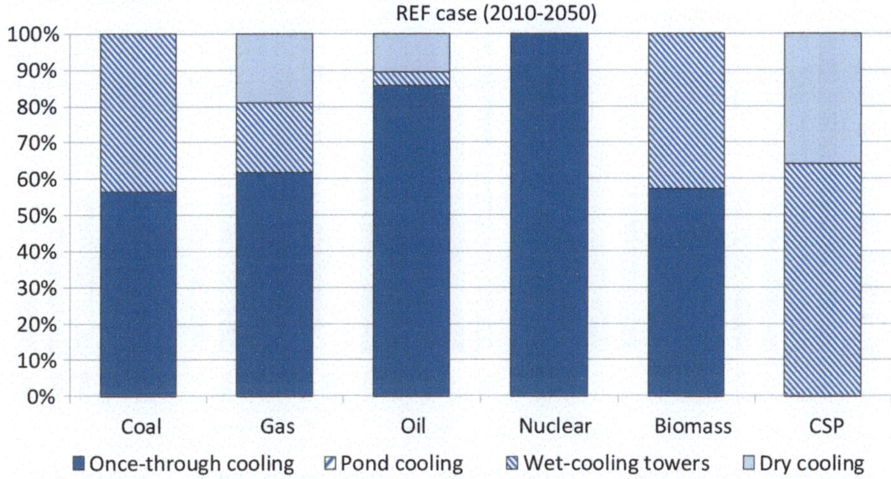

Fig. 4 Current and future shares of cooling techniques for the main power production options in the Middle East in the REF case

Figs. 2 and 3, we have assumed that this breakdown continues to hold until 2050, which we refer to as the reference (REF) case. In other words, for coal usage, for example, we assume that until the middle of the century about 55% of all power plants in this region remain equipped with once-through cooling technology (using either fresh or saline water), while 45% of the coal-based power plants use recirculating methods to cool (either with a cooling tower or pond-based techniques). Likewise, we assume that approximately 60% of natural gas-based power plants remain equipped with once-through cooling, while 20% of these plants use recirculating methods and another 20% dry cooling techniques. For CSP plants, we suppose that the current breakdown of some 65% of recirculating and 35% of dry cooling techniques continues to hold for the forthcoming decades.

Uncertainties abound with regard to the future relative shares of different cooling techniques, which is why we perform a sensitivity test for these shares. Due to serious water constraints in the Middle East, which may intensify over the years to come, it is likely that efforts will be made to reduce the water usage of power plants in the region. This can be achieved by replacing once-through cooling by recirculating cooling and/or substituting the latter with hybrid or dry cooling options. Such replacement will be a gradual process, given the capital intensity of both power plants and cooling technologies and since water constraints will probably gradually emerge in various locations rather than abruptly come to the fore in the region as a whole. A possible scenario for this process is depicted in Fig. 5, in which for all major electricity generation options (based on, respectively, coal, natural gas/oil, nuclear energy, biomass, and CSP) a pattern is supposed for the gradual phase-in of water-saving technologies. This scenario is referred to as the SAVE case. This case engenders additional costs, the overall magnitude of which needs to be considered in joint water-energy policy design. The SAVE case for the evolution of cooling

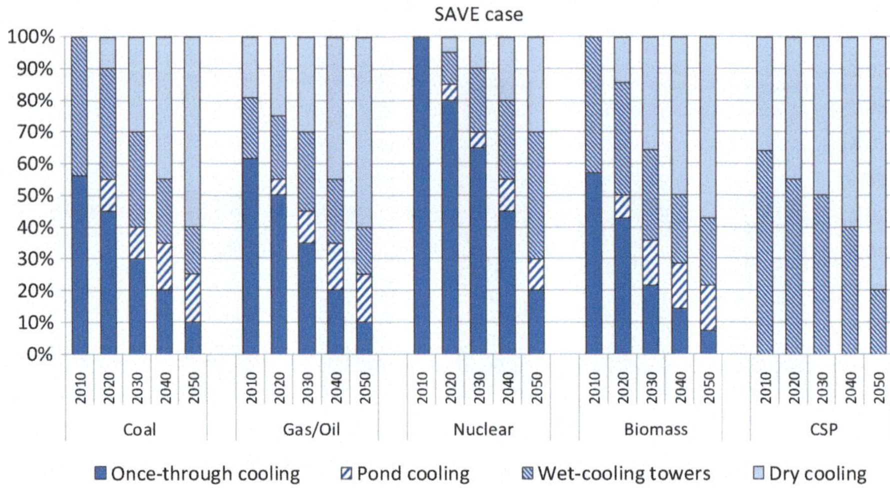

Fig. 5 Current and future shares of cooling techniques for the main power production options in the Middle East in the SAVE case

options has not been developed on the basis of a combined cost minimization procedure for energy and water technologies simultaneously, but on the basis of one for energy technologies only—this is one of the opportunities for improvement that we recommend to be explored in follow-up research. In practice, a SAVE case may thus materialize differently, that is, with other energy and water technologies.

Figure 6 shows how our water withdrawal and consumption projections modify, both in the baseline and stringent climate policy scenario, if one switches from REF-case cooling techniques to those we assume in the SAVE case. The large reduction in water withdrawal in the baseline scenario when switching from the REF to the SAVE case is obvious, which is mostly the result of the gradual phasing out of once-through cooling and the introduction of recirculating, hybrid, and dry cooling systems for natural gas-based power production (see the left-hand plot of Fig. 6). The same plot in Fig. 6 points out that this switch has little effect on water consumption levels during the first couple of decades, which can be explained by the fact that recirculating types of cooling actually possess slightly higher water consumption levels than once-through systems. In the period 2040–2050 though, a reduction in water consumption materializes of about 30%, thanks to the savings introduced as a result of particularly dry cooling systems. These dry cooling systems do not use any water at all, by definition, as air is the principal medium used for cooling.

For the stringent climate policy scenario—see the right-hand plot of Fig. 6—we see a few notable differences. First of all, in terms of water withdrawal, the savings are substantially lower than in the baseline scenario, the explanation for which is the large role played in the climate change control scenario by CCS and biomass-based power production technologies. For water consumption, the difference between the

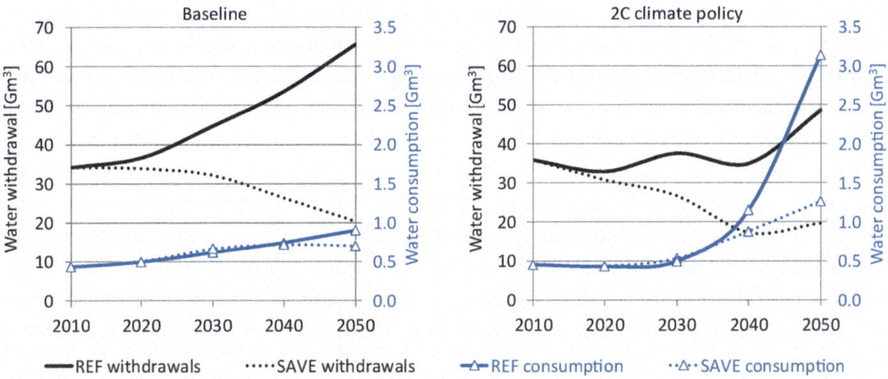

Fig. 6 Water withdrawal and consumption for two electricity sector scenarios under the REF and SAVE cases in the Middle East

REF and SAVE cases is small initially, like in the baseline scenario, but from about 2030, the discrepancy between these two cases becomes very large. The reason is that particularly CSP and, to a lesser extent, biomass are assumed to rely largely on dry cooling systems by the middle of the century in the SAVE case. Figure 6 demonstrates that there is substantial scope for water stress risk reduction, but that uncertainties abound as a result of unforeseeable developments in the application of cooling techniques.

An earlier study performed an investigation of scenarios for energy-water interdependencies in a similar as we have done here (Bouckaert et al. 2014). Its authors added water footprints related to the power system (including cooling systems, gasification, and flue gas desulfurization processes) to their global TIAM-FR energy system model. With their modification, the TIAM-FR model could be used to ascertain whether future energy mixes might be plausible in terms of water availability. The authors evaluated diverse policies concerning water and CO_2 emissions and suggested that the choice for the cooling system and the use of CCS when applying climate policies to the energy system may significantly increase overall freshwater consumption. In our study, we confirm this finding, as can be seen from Fig. 6: if climate policy is implemented and no dedicated water consumption savings strategy is adopted, water consumption may exponentially increase, even in comparison to the baseline scenario. This is a risk that we should hedge ourselves against, certainly in a region like the Middle East.

In regions like the Middle East where water is already scarce today or is likely to become so in the relatively short run, an increase in freshwater consumption levels or withdrawals may not be sustainable for the energy system. Hence, we suggest that in the future, we adapt TIAM-ECN so as to incorporate water use factors, which would allow us to consider limited water availability as a constraint, and evaluate the impact of water scarcity on electricity production in a region such as the Middle East. Even better, in subsequent research, we could improve our model so as to introduce the costs of cooling systems as well as the costs of water employed therein.

Indeed, water usually does not receive a cost price in IAMs—contrary to energy services—so that the impacts on and from water usage normally do not become visible. The inclusion in our type of models of the costs of water and those of cooling systems would allow us to perform a combined cost minimization analysis of energy and water systems simultaneously. It could well prove that the resulting optimal regional energy systems look different from the ones that we obtain with the current version of our model without water cost inclusion, based on cost minimization of the energy system only. Given that the subjects of energy and water are becoming increasingly intertwined in a future climate-constrained world, joint water-energy analysis would not only be an exciting type of new research but may also constitute an essential requirement for any study that attempts to determine the desirable energy system of the future.

5 Conclusions and Recommendations

This chapter has shown that both the energy and water sectors face key challenges over the coming decades. We have demonstrated that in many respects these challenges are interrelated and thus need to be simultaneously addressed. The joint challenges associated with the energy-water nexus, however, are clearly different from those for climate change. The former may pose substantial problems at the local or regional level, but that are often addressable, one way or the other. They may in some cases last only relatively short periods of time, although potential solutions may sometimes come at a high cost. The latter is a truly global *problematique* with challenges that are not easily solvable and are long-lived, that is, span centuries. The costs required to mitigate climate change, although quite uncertain, may amount to percentage points of global world product. In this chapter, we directed our attention primarily to the energy-water nexus, rather than the energy-climate nexus, while addressing—from a scenario analysis perspective—the possible effects of climate change mitigation policy on the former.

The first main conclusion that we draw from our work is that the type of cooling system used for electricity generation is at least as influential for the water needs of power production as the type of energy technology used. This is certainly the case for conventional thermal power generation technologies, such as those based on coal, natural gas, oil, and nuclear energy, but possibly also for other more modern techniques, including a renewable energy technology such as CSP. This does not apply, however, to renewable energy technologies like PV and wind, which do not require water for cooling purposes.

Second, as reported in the Appendix (see also Halstead et al. 2014), we conclude that even when taking the full life cycle into consideration, PV and wind technologies remain the least water-intensive electricity generation options relative to other energy technologies considered in this chapter. This is true even when we only take into account the operational phase for the other energy technologies (thereby making in some sense an unfair comparison), and irrespective of the fact that the water footprint of PV and wind electricity options in the stages of manufacturing and production is often relatively high in comparison to other energy technologies. Indeed, the very high water use of technologies such as based on coal, natural gas, and oil, or nuclear energy and CSP, results predominantly from the operational phase of electricity generation. Overall, technologies like PV and wind thus appear to be clear winners in terms of water savings potential.

Third, it has been suggested in the literature—and we support this claim—that in certain world regions, water availability is becoming as important a concern as security of energy supply. In view of the linkages between water and energy supply, integrated optimization analyses and policies regarding energy and water resource systems are necessary. Rather than first finding least-cost energy systems and subsequently finding the least-cost water supply that these systems require, instead one should attempt to minimize the costs of energy-water systems jointly.

Fourth, due to water constraints, it is likely that further efforts will need to be made to reduce the water usage of power plants in regions such as the Middle East. To achieve these reductions, once-through cooling may need to be replaced by recirculating cooling, which in turn can be substituted by hybrid or dry cooling options. The gradual increase of regional water constraints may make these replacements necessary, but the capital intensity of both power plants and cooling technologies will mean that such replacements will need to take place over decades.

Fifth, in a future that involves more stringent climate policy, a large role may be reserved for CCS and biomass-based power production, which requires large quantities of water. The water withdrawal savings that otherwise could be achieved in a business-as-usual scenario would perhaps be overshadowed by the sizeable water usage of these low-carbon technologies. This is an example of the kind of trade-offs that policy-makers need to consider when designing and implementing climate policies as well as policies related to the energy-water nexus sectors.

Sixth, we conclude that while regions exist today where already substantial water stress risks exist, the problems in principle—from at least a technical perspective—can often be overcome, albeit sometimes at a high cost. In the long run, both water withdrawal and consumption can be reduced significantly if decisions are made—particularly (but not only) in the field of energy production and consumption—that take water issues into account. We thus find that water stress issues, also in those regions where at present they are not yet apparent but may emerge in the future, can often be addressed either by using different energy and water (cooling) technologies or by relocating certain (e.g., industrial) activities to different regions. The real question though is at what cost.

Acknowledgments We would like to acknowledge the feedback from two anonymous referees and the editors of this volume, as well as the comments from colleagues who reviewed earlier versions of our work, as a result of whose expert advice this chapter improved substantially. We thank the TRANSRISK project (EU Horizon 2020 research and innovation program, grant agreement No. 642260) for its financial support.

Appendix

Figures 7 and 8 provide a visual representation of the ranges of water withdrawal and consumption for different energy technologies based on our review of the relevant literature. For more details behind our analysis, literature review, and acquired data, we refer to our report "Understanding the Energy-Water Nexus" (Halstead et al. 2014), available at www.ecn.nl.

We do not include geothermal and hydropower generation sources due to the diversity of technologies used within these two categories that all involve widely diverging water usage factors—deviating from each other sometimes by several orders of magnitude. Technologies within these two categories are also inherently complex, and it is difficult to assess their water withdrawal and consumption factors with a credible degree of accuracy, unless entire studies are dedicated to each of them. We also omit tidal energy, for similar reasons. The impact of electricity generation from tidal power on water resources may be considered minimal as it could be argued that there is no withdrawal or consumption of water during the operational phase. However, we would recommend future studies on the water withdrawal and consumption factors for all these technologies.

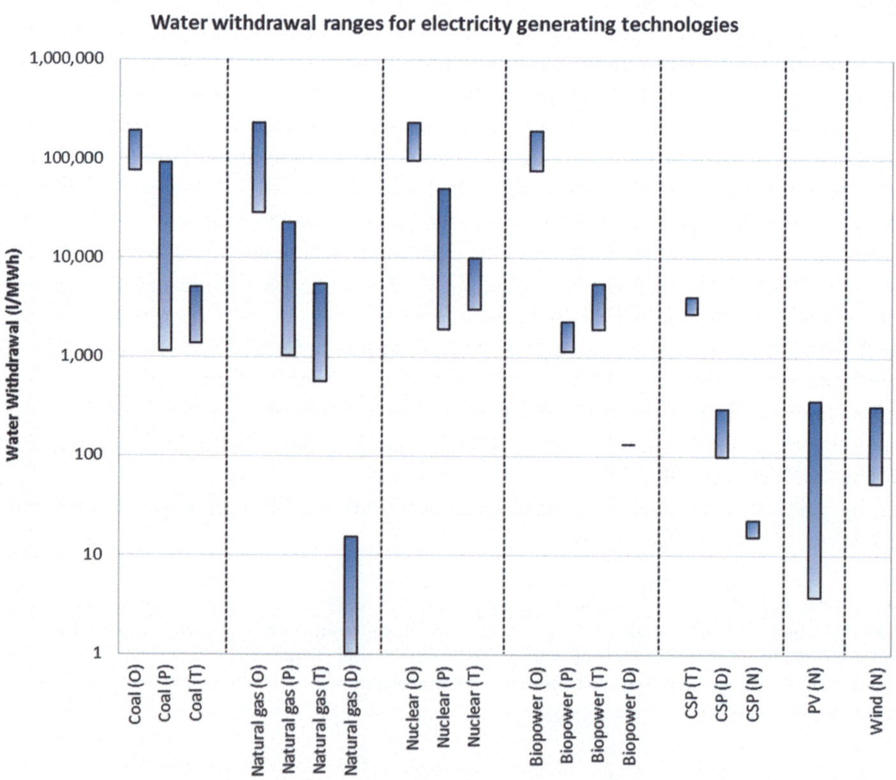

Fig. 7 Water withdrawal factors for electricity generating technologies

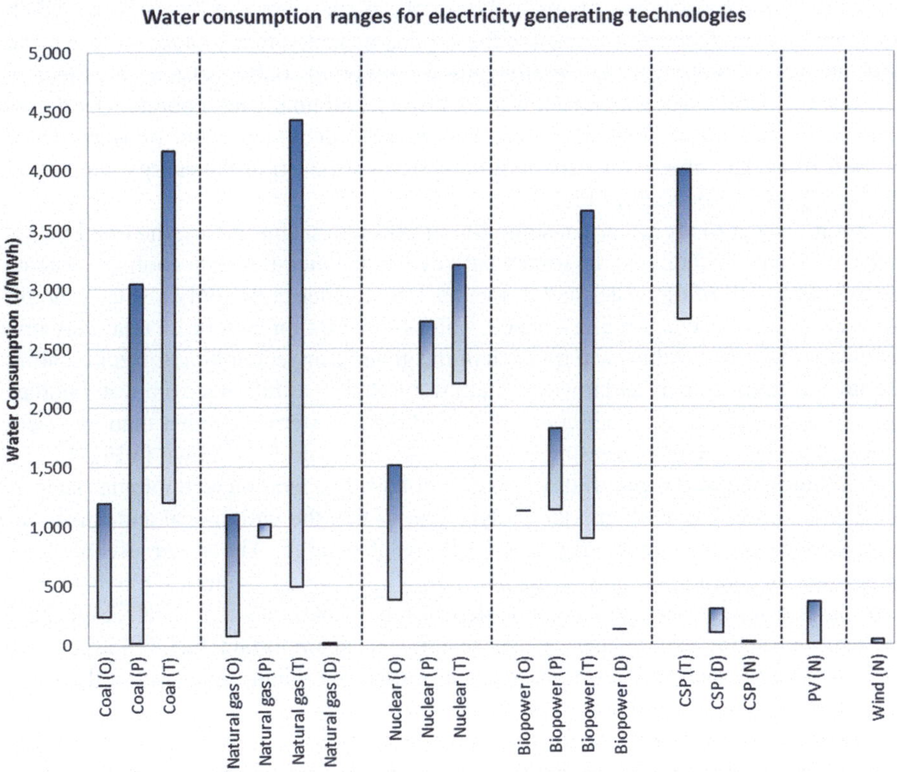

Fig. 8 Water consumption factors for electricity generating technologies. Notes: The different types of cooling system are given in brackets following the energy technology (O once-through, P pond, T tower (both pond and tower are recirculating systems), D dry, N no cooling). The withdrawal and consumption factors for both PV and wind are life cycle estimates which include water withdrawal for power plant procurement and building and fuel extraction, transportation, and recycling. Life cycle data is taken from Meldrum et al. (2013). The data for PV does not include water use of concentrated PV. Withdrawal factors can be approximately 16 times higher for concentrated PV technology. The higher water use of concentrated PV is likely to be because of certain shared operational characteristics with CSP, such as a need for mirror washing (Meldrum et al. 2013)

Our analysis broadly agrees with the previous literature studies by Macknick et al. (2012) and Meldrum et al. (2013) showing that large differences in both water withdrawal and consumption levels exist between not only different types of electricity generating technologies but especially between the cooling systems used. The results show that the cooling system that is adopted often impacts water usage more than the actual electricity generating technology being used. As an example, once-through cooling systems can withdraw between 10 and 100 times more water per unit of electricity generation than cooling tower technologies, but cooling towers can consume typically twice the amount of water of once-through systems (Macknick et al. 2012).

Once-through cooling systems withdraw the highest amount of water per MWh of electricity produced within each of the applicable generation sources (coal, natural gas, nuclear, or biopower). Generally, closed-loop pond cooling systems are the next biggest withdrawers of water, followed by towers, and finally dry cooling which uses minimal water for cooling purposes. This general declining trend of water withdrawal from once-through to dry cooling systems for each of the energy generation technologies can be seen in Fig. 7.

With respect to the water consumption of different technologies, the trend that is seen in water withdrawal is somewhat reversed. Once-through cooling systems return almost all of the water withdrawn back to a water body (only a small amount of water is lost via evaporation); hence, water consumption factors are relatively low compared to water withdrawal for each of the generation technologies. Recirculating cooling systems (ponds and towers) retain water that is withdrawn from water bodies for reuse; therefore, the water consumption of these systems is higher than for once-through systems. This increasing trend for each technology is shown in Fig. 8.

Although the water footprints of fossil fuel-based generation technologies such as coal and natural gas are high, our analysis shows that the withdrawal and consumption factors for both bioenergy and CSP are also large. The water use of these renewable technologies may influence policy-making as countries move toward low-carbon development and begin to deploy renewables on a mass scale, especially in regions of the world where water scarcity is an important factor. Large-scale deployment of renewable energy technologies will be reliant upon, yet at the same time have serious consequences for, water availability.

The data in the two figures above relate predominantly to water withdrawal and consumption during the operational phase of electricity generation. However, from our research, we have identified that the water footprints of both PV and wind in other life cycle phases are relatively significant compared to their footprints during the operational phase. Therefore, we have used water withdrawal and consumption factors for PV and wind that include water usage during the stages of power plant procurement and building, as well as fuel extraction, transportation, and recycling. This is the case in both Figs. 7 and 8. The water footprints of the remaining technologies, during these other life cycle phases, are not included as part of this analysis as they have minimal impact on the data. It can be seen that even when taking the full life cycle into consideration for PV and wind technologies, they remain the least water-intensive electricity generation options. This conclusion is supported by previous work by Meldrum et al. (2013), which incorporates a life cycle analysis of water consumption and withdrawal of different electricity generating technologies. It is important to emphasize here though that despite renewable energy sources generally using less water than fossil fuels, if a full life cycle analysis is performed for bioenergy, then this technology would become by far the most water-intensive option.

Using dry cooling can reduce freshwater usage. However, this may lead to increased costs and decreased plant efficiency. CSP using dry cooling might lead to an annual reduction in electricity output of 2–5% and an increase in levelized cost of electricity of 3–8% compared to wet cooling systems. In the United States, the annual performance loss of switching to dry cooling from wet cooling systems is 6.8% for nuclear facilities, 1.7% for combined cycle plants, and 6.9% for other fossil fuel-based generation plants.

The cooling system chosen is likely to play an important role in our future electricity generation mix. Given future uncertainties surrounding water availabilities and the consequences for power plants, particularly in regions of water scarcity, the use of alternative cooling techniques, such as dry cooling, may be necessary. Utilizing dry cooling or non-freshwater resources avoids some of the risks associated with drought and climate change. By 2035, water withdrawal could potentially increase by 20% and water consumption by 85%, if we shift toward higher-efficiency power plants with more advanced cooling systems that reduce water withdrawal levels, but increase water consumption.

References

Al-Riffai, P., Breisinger, C., Mondal, A. H., Ringler, C., Wiebelt, M., & Zhu, T., (2017). *Linking the economics of water, energy, and food: A nexus modeling approach* (IFPRI, Egypt Strategy Support Program, working paper 4). Washington, DC.

Bouckaert, S., et al. (2014). A prospective analysis of waste heat management at power plants and water conservation issues using a global TIMES model. *Energy, 68*, 80–91.

Bruinsma, J. (2011). The resources outlook: By how much do land, water and crop yields need to increase by 2050? In P. Conforti (Ed.), *Looking ahead in world food and agriculture: Perspectives to 2050* (pp. 1–33). Rome: Food and Agriculture Organisation of the United Nations.

Doukas, H., Karakosta, C., Taoumi, M., Wouters, F., & Makarouni, I. (2017). Analysis of key areas for EU and GCC cooperation in the field of clean energy and the linked sectors of climate and water. *Geopolitics of Energy, 39*(10), 1–16.

FAO. (2006). Food and Agriculture Organization. United Nations, Rome. www.fao.org

Halstead, M., Kober, T., & van der Zwaan, B. (2014). *Understanding the energy-water nexus*. Energy Research Centre of the Netherlands (ECN), ECN-E-14-046, Amsterdam.

IEA. (2013). *Redrawing the energy-climate map: World energy outlook special.*

Kober, T., et al. (2014). Emission certificate trade and costs under regional burden-sharing regimes for a 2°C climate change control target. *Climate Change Economics, 5*, 1–32.

Kriegler, E., et al. (2013). "What does the 2°C target imply for a global climate agreement in 2020?" The LIMITS study on Durban Platform scenarios. *Climate Change Economics, 04*, 04.

Macknick, J., Newmark, R., Heath, G., & Hallett, K. C. (2012). Operational water consumption and withdrawal factors for electricity generating technologies: A review of existing literature. *Environmental Research Letters, 7*, 045802.

McCarl, B. A., Yang, Y., Schwabe, K., Engel, B. A., Mondal, A. H., Ringler, C., & Pistikopoulos, E. N. (2017). Model use in WEF nexus analysis: A review of issues. *Current Sustainable Renewable Energy Reports, 4*, 144–152.

Meldrum, J., Nettles-Anderson, S., Heath, G., & Macknick, J. (2013). *Life cycle water use for electricity generation: A review and harmonization of literature estimates*. Bristol: IOP Publishing.

OECD. (2017). *The land-water-energy nexus: Biophysical and economic consequences*. Paris: OCED.

Rösler, et al. (2014). Electricity versus hydrogen for passenger cars under stringent climate change control. *Sustainable Energy Technologies and Assessments, 5*, 106–118.

SEI. (2011). Understanding the nexus. In *Background paper for the Bonn 2011 Nexus Conference – The Water, Energy and Food Security Nexus: Solutions for the Green Economy.*

UN. (2014). *The United Nations World Water Development Report 2014: Water and Energy.* Paris: UNESCO.

van der Zwaan, B. C. C., Cameron, L., & Kober, T. (2013a). Potential for renewable energy jobs in the Middle East. *Energy Policy, 60*, 296–304.

van der Zwaan, B., et al. (2013b). A cross-model comparison of global long-term technology diffusion under a 2°C climate change control target. *Climate Change Economics, 4*(4), 1–24.

van der Zwaan, B., Boccalon, A., & Dalla Longa, F. (2018). Prospects for hydropower in Ethiopia: An energy-water nexus analysis. *Energy Strategy Reviews, 19*, 19–30.

Evaluation of National Environmental Efficiency Under Uncertainty Using Data Envelopment Analysis

Evangelos Grigoroudis and Konstantinos Petridis

Abstract The evaluation of environmental practices of each country is a very interesting area which has recently gained significant attention. An analysis, which can provide results for comparisons among countries with different economies, is data envelopment analysis. In this chapter, countries' environmental efficiency is estimated using data envelopment analysis. Applying a slack-based model under the consideration of constant returns to scale and variable returns to scale technologies, a composite index is calculated from the efficiency scores of each model. These models consider both desirable and undesirable outputs. However, especially for undesirable outputs, the data collected are not accurate and could potentially be subject to uncertainty. To handle the uncertainty in the undesirable data, a chance-constraint DEA model is applied. Results of the deterministic model show that Australia gathers high values of environmental efficiency. However, in the presence of noise in the undesirable data, the rankings of the countries change.

Keywords Environmental efficiency · Environmental evaluation · Environmental economics · Data envelopment analysis · Slack-based models · Composite index

1 Introduction

The term environment refers to anything that surrounds an object. In natural sciences, as well as in engineering, a system is the part of the world being studied, and the environment is anything outside its boundaries. There can be interactions and exchanges of matter, energy, or information between the system and the environment.

E. Grigoroudis (✉) · K. Petridis
School of Production Engineering and Management, Technical University of Crete, Chania, Greece
e-mail: vangelis@ergasya.tuc.gr

H. Doukas et al. (eds.), *Understanding Risks and Uncertainties in Energy and Climate Policy*, https://doi.org/10.1007/978-3-030-03152-7_7

Human interventions have affected the environment, disturbing the environmental balance, altering its natural processes, and degrading the quality of human life. One of the most dangerous human interventions' side effects is that they jeopardize the sustainability of the planet while creating many problems and serious environmental accidents. The major modern environmental problems known as "global environmental problems" mainly consist of ozone depletion, greenhouse effect, pollution of the environment in general, degradation, and pollution of key environmental resources such as air, water (lakes, seas, and oceans), soil, desertification, and biodiversity loss.

A great time period has passed for the aforementioned risks to be fully understood, and this is evidenced by recent facts of degradation and destruction of the environment. Unreasonable human intervention in the environment and abusive use and exploitation of natural wealth lead to disastrous consequences. Primary goods, which were considered inexhaustible or unaltered, are slowly degraded.

As noted by Grigoroudis et al. (2012), the environment provides the economy with resources (e.g., water, air, fuels, food, metals, minerals, and drugs), services (e.g., the cycles of H_2O, C, CO_2, N, O_2; photosynthesis; and soil formation), and mechanisms to absorb waste. Economic growth is based on these three services, and since the global ecosystem does not grow, economic growth cannot continue indefinitely. The concepts of sustainability and sustainable development have received much attention among policy-makers and scientists as a result of the existence of limits to growth and the dramatic environmental changes of the last decades.

The term of environmental efficiency was devised by the World Business Council for Sustainable Development (WBCSD) in 1992 in its "Changing Course" publication. It is based on the idea of creating more goods and services using less resources and creating less waste and pollution. It is a philosophy that aims to minimize ecological impacts while maximizing the efficiency of the processes of a production unit. The term has become synonymous with a concept oriented toward sustainable development.

Uncertainty, as a term, is inherited in data measurement in the presence of noise, while it is also present when dealing with the environment. The difficulty of uncertainty is its stochastic nature which is approximated with stochastic procedures and models. Nevertheless, the measurement of stochasticity or uncertainty is never accurate. It can, however, be estimated under various assumptions.

The economic activities that are conducted around the world have a direct effect on the environment, which is affected irrespective of the geographical location. A source of pollution, due to excess of emissions from industry or any related production process, affects the total environment. Nevertheless, modern way of living and globalization have led to form the economies without considering the environment, since environmental protection does not add value to economic activities. To reduce this phenomenon, several environmental protocols and directives have been agreed in an effort to regulate terms and undesirable outputs from the production process of national economies. Therefore, a methodological framework is needed to measure the effect at which the environment is affected from the economic activity of each country and try to find an "equilibrium" point at which each country can increase its productivity without putting an extra burden on the environment.

Estimating environmental efficiency gives emphasis on producing the maximum possible economic output, using minimum resources, and at the same time minimizing environmental impacts. Therefore, it is a process different from environmental performance estimation. The estimation of environmental efficiency is a rather difficult problem mainly because the concept does not have a universal definition and usually the applied measurement framework serves also as a definition context. In practice, this makes more difficult the selection of appropriate indicators, while data availability is always a significant shortcoming when estimating environmental efficiency in a national level.

The aim of this chapter is to present a methodological framework for the measurement of environmental efficiency and to highlight the indicators of evaluation of the studied units. The data used in this study refer to a 12-year period (1992–2003) and concern 108 countries from all over the world, belonging to various social, political, and economic categories. A nonparametric method, data envelopment analysis (DEA), is used to estimate national environmental efficiencies. Moreover, a chance-constraint DEA model with desirable and undesirable outputs is applied in the proposed approach, assuming that the undesirable outputs (harmful gas emissions) are subject to noise. This incorporation of uncertainty in the applied DEA models may be considered as the main contribution of this research, while the examination of alternative measurement variables gives the ability to compare how different indicators may affect the environmental efficiency scores.

The chapter is organized in four more sections. Section 2 presents briefly a literature review of environmental efficiency evaluation focusing mainly on DEA models. The methodological background of the proposed approach is given in Sect. 3, where, in addition to the main principles of DEA models, the proposed slack-based model is presented. Moreover, Sect. 3 presents a chance-constraint DEA model in order to handle the uncertainty of environmental data. The results of the proposed approach in a set of 108 countries covering a period from 1992 to 2003 are given in Sect. 4. Section 5, finally, summarizes some concluding remarks and discusses potential extensions of the research.

2 Literature Review

Based on Farrell's original ideas (Farrell 1957), DEA was firstly used in Germany by Brockhoff in 1970 to measure R&D and production efficiency (Brockhoff 1970). Within an environmental context, DEA was first used in 1986 by Färe, in a sample of steam power stations in the United States, to measure the impact of environmental constraints and measures (Färe et al. 1986).

A mathematical programming approach to environmental management and industrial efficiency is outlined in 1994 by Haynes as an alternative to decision support processes related to the monitoring of pollutant reductions. Färe et al. (1996) applied DEA using US data on fossil fuels that are burning power companies, resulting in pollution and efficiency indicators. DEA models have been also used to compare the efficiency of several units of a company or various production units in a sector with specific characteristics and environmental constraints (Tyteca 1996). As a comparison technique, DEA has been used for OECD countries to assess environmental performance based on CO_2 emissions (Zofío and Prieto 2001). Assessment of business participation in sustainable development has been also examined with the use of DEA (Callens and Tyteca 1999). The study of performance indicators with ecological and environmental extensions has been proposed by Dyckhoff and Allen (2001).

The estimations of environmental efficiency measures have also been examined in several studies. For example, Reinhard et al. (2000) compare different efficiency estimation approaches (DEA and stochastic frontier analysis) for the case of Dutch dairy farms. They define environmental efficiency as the ratio of minimum feasible to observed use of multiple environmentally detrimental inputs, conditional on observed levels of output and conventional inputs.

Environmental efficiency through energy consumption and carbon dioxide emissions has been measured using DEA with data from 17 Middle East and North African countries (Ramanathan 2005). The environmental performance of the states of America has been also evaluated under the assumption that air pollution is mainly a by-product of the production process for the years 1972–1983 and 1985–1986 (Zaim 2004). Korhonen and Luptacik (2004) applied a two-stage DEA approach to 24 European power plants. In the first stage, the problem is decomposed in two parts:

(a) The problem of measuring the technical efficiency (such as the ratio of the desired costs to the entrances)
(b) The problem of measuring the so-called ecological efficiency (such as the ratio of the desired costs to the unwanted ones)

The performance indicators of each stage are then combined into one. In the second stage, pollutants and system inputs are treated the same as the aim is to increase the desired costs and reduce pollutants and inputs.

Triantis and Otis (2004) developed a performance measurement model, which examines environmental measures and the harmful ecological consequences of the production process over time. More specifically, they present a pair-based approach that examines production and environmental efficiency and evaluate desirable environmental interactions of the production system through various combinations of inputs and outputs. Zhou et al. (2006, 2008) proposed a non-radial DEA approach for assessing environmental performance. The time period of this analysis spans from 1995 to 1997, while the 26 countries of the OECD are treated as decision-making units.

Prieto and Zofío (2007) presented a network efficiency analysis model that allows possible increases in technical efficiency by comparing technologies that correspond to different economies. Input-output tables represent a network where various nodal factors use primary inputs to produce intermediate inputs and outflows to meet the final demand. The proposed model optimizes the underlying multistage technologies, defining the best financial practice. The model is applied to a total of five OECD countries in the period from 1970 to 1990. Zhou et al. (2007) in the context of environmental efficiency show an extension of the DEA and more specifically the nonincreasing returns to scale (NIRS) model and the variable returns to scale (VRS).

A comprehensive literature review of DEA models applied to energy and environment may be found in Sueyoshi et al. (2017). Their review covers almost 700 articles from the 1980s to 2010s. The authors report an increased number of DEA articles, particularly after the 2000s, and discuss three major future research directions:

(a) *Technology heterogeneities and time lag*: The authors note that different organizations and different regions may have different engineering capabilities, so that they have many different types of technology heterogeneities among them. This, as well as the time lag that is always associated with technology development, should be considered in DEA modeling.
(b) *Statistical inference*: The authors argue that one of the major shortcomings of DEA environmental assessment is that it does not have a statistical inference at the level of statistics and econometrics; therefore, the exploration on the statistical inference on DEA may provide an important future research direction.
(c) *Applications to China*: The authors emphasize that China is the world's largest energy consumer and carbon emission contributor. Thus, it is important, through the application of DEA models, to identify better ways to reduce Chinese energy uses and carbon emissions.

In the same context, Mardani et al. (2017) provide a review, focusing however in energy efficiency. Their review covers 144 published scholarly papers appearing in 45 high-ranking journals between 2006 and 2015 and shows that DEA may be a good evaluation tool on energy efficiency issues, when the production function between the inputs and outputs is virtually absent or extremely difficult to acquire.

Stochasticity is part of the everyday operations; therefore, efficiency measurement calls for probability calculations. In the context of DEA, several stochastic models have been proposed in the literature over the years. The first DEA stochastic model was applied by Charnes and Cooper (1963). Following their paper, the stochastic DEA modeled with chance constraints has grown significantly.

Stochasticity has been incorporated in DEA for measuring technical efficiency and inefficiency indices (Cooper et al. 2002). In the majority of the relevant papers, modeling involved the stochastic nature of CO_2 emissions and the impact on efficiency or energy production. Gutiérrez et al. (2008) presented a methodology to analyze the gradual secular trends present in the time evolution of certain endogenous variables, which are of particular interest in environmental research. Also, several stochastic environmental performance indices have been constructed (Baležentis et al. 2016; Zha et al. 2016).

DEA's applications in environmental efficiency and the assessment of ecological constraints are significantly increasing. DEA gains ground in the preferences of researchers who have been using it extensively especially in recent years.

3 Methodology

3.1 General

In recent years, great emphasis has been placed on efficiency measurement methods. Efficiency is defined as the ability of a unit to efficiently transform, with a generally unknown production mechanism, inputs into outputs. Traditional econometric methods have been used to evaluate efficiency. These techniques were designed to calculate theoretically analytical production functions. One of the shortcomings of these methods is that they cannot handle multiple inputs but only a single output. Data envelopment analysis (DEA) has been introduced to handle multiple inputs and multiple outputs in order to evaluate homogenous units, the decision-making units (DMUs).

DEA technique is widely applied in a series of studies to estimate relative unit efficiency, with respect to a set of similar units that have multiple inputs and outputs. In DEA, DMUs consume inputs to transform them into outputs. Therefore, a DMU includes the activities of many different organizations as mentioned above. Outputs are defined as products or services produced by each unit, while inputs are generally defined as resources used to produce outputs (land, labor, fuel, etc.).

Figure 1 shows a graphical representation of a DEA model in the context of this study, where a DMU corresponds to the production process of a national economy, while the inputs refer to the main resources used in this production process and may include national labor force, available capital, or energy produced. On the other hand, the outputs of the production process include both desirable (e.g., national income) and undesirable results (e.g., emissions, waste).

3.2 Envelopment Models

The envelopment models considered in this study include both input- and output-oriented models. In particular, the input-oriented DEA model has the following form:

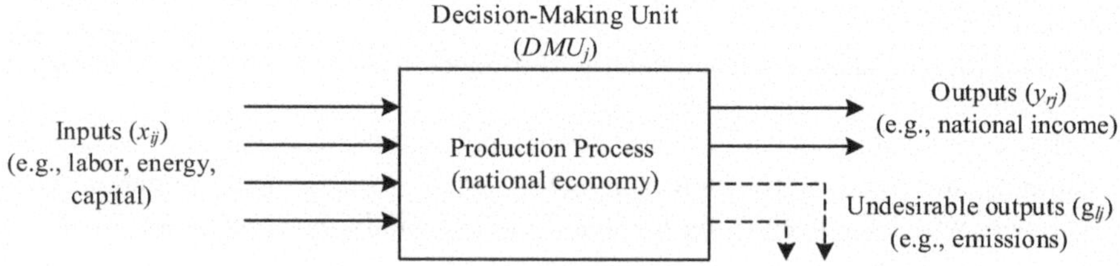

Fig. 1 Graphical illustration of a production process in a national economy

$$\begin{aligned}
&\min \theta - \varepsilon\left(\sum_{i=1}^{m} s_i^- + \sum_{r=1}^{s} s_r^+\right)\\
&\text{s.t.}\\
&\sum_{j=1}^{n} x_{ij}\cdot\lambda_j + s_i^- = x_{io}\cdot\theta, i = 1, \ldots, m\\
&\sum_{j=1}^{n} y_{rj}\cdot\lambda_j - s_r^+ = y_{ro}, r = 1, \ldots, s\\
&\lambda_j, s_i^-, s_r^+ \geq 0, \forall j, i, r\\
&\theta \text{ free}
\end{aligned} \tag{1}$$

The objective function in this model is to minimize free variable θ, which measures the efficiency of each DMU. The evaluation of each DMU's efficiency is conducted upon a predetermined set of i inputs (x_{ij}) and r inputs (y_{rj}) for each DMU_j. As a consequence, the aim of the previous model is to determine the least possible level of available inputs (x_{io}), for a DMU under examination (DMU_o), which are capable to produce the desired level of outputs (y_{io}). Variables λ_j are the peers of each DMU_j; peers are used in order to provide information regarding the proximity of the DMU under investigation with other DMUs. The mathematical formulation (1) represents a *linear programming* (LP) model which is solved for each DMU under examination (i.e., DMU_o). If, for example, DMU_5 has in its reference set DMU_2 and DMU_6, then λ_2,$\lambda_6 \neq 0$. Nonnegative variables s_i^- and s_r^+ are slack variables corresponding to the inputs and outputs, respectively, while ε is a small number. A fully efficient DMU is the one with $\theta^* = 1$ and $s_i^- = s_i^+ = 0$, whereas θ^* is the optimal value of LP model (1) for each DMU under examination. The range of values for input efficiency is $0 \leq \theta \leq 1$.

Similarly, the output-oriented DEA model is defined as follows:

$$\begin{aligned}
&\max \varphi + \varepsilon\left(\sum_{i=1}^{m} s_i^- + \sum_{r=1}^{s} s_r^+\right)\\
&\text{s.t.}\\
&\sum_{j=1}^{n} x_{ij}\cdot\lambda_j + s_i^- = x_{io}, i = 1, \ldots, m\\
&\sum_{j=1}^{n} y_{rj}\cdot\lambda_j - s_r^+ = y_{ro}\cdot\varphi, r = 1, \ldots, s\\
&\lambda_j, s_i^-, s_r^+ \geq 0, \forall j, i, r\\
&\varphi \text{ free}
\end{aligned} \tag{2}$$

The objective function in this case is to minimize free variable φ which measures the efficiency of each DMU, where the variables x_{ij}, y_{rj}, s_i^-, s_i^+, and λ_j are defined similar to LP (1). Model (2) represents also a LP model which is solved for each DMU under examination (i.e., DMU_o). A fully efficient DMU in this case is the one with $\varphi^* = 1$ and $s_i^- = s_i^+ = 0$, whereas φ^* is the optimal values of LP model (2) for each DMU under examination. For output-oriented efficiency, we have $\varphi \geq 1$, and in order to capture the degree of inefficiency of a DMU, the reciprocal is calculated, such that $0 \leq 1/\varphi^* \leq 1$.

Following the discussion about the orientation of input and output DEA models, the optimal values of efficiency variables are of great interest as the projections of inputs and outputs to the efficient frontier are calculated. To do that, the following equations are considered:

$$\begin{cases} \widehat{x}_{ij} = x_{ij} \cdot \theta^* - s_i^{-*} \\ \widehat{y}_{rj} = y_{rj} - s_r^{+*} \end{cases} \quad \text{(input orientation)} \tag{3}$$

$$\begin{cases} \widehat{x}_{ij} = x_{ij} - s_i^{-*} \\ \widehat{y}_{rj} = y_{rj} \cdot \varphi^* - s_r^{+*} \end{cases} \quad \text{(output orientation)} \tag{4}$$

3.3 *Slack-Based Models*

Several environmental performance indicators have been constructed, combining DEA with various types of performance measurement. As mentioned above, in order for a DMU to be fully efficient, $\theta^* = 1$ and $s_i^- = s_i^+ = 0$. However, two DMUs may have the same efficiency (maximum efficiency), despite the fact that their input and output data may indicate a difference in this measure. In terms of DEA, only one of these DMUs is fully efficient. This is an inherent inefficiency of the indicator which is fixed by considering slack-based measure (SBM) models.

The SBM model considering desirable outputs has the following form:

$$\begin{aligned} &\theta_1 = \min \frac{1 - \frac{1}{m}\sum_{i=1}^{m}\frac{s_i^-}{x_{io}}}{1 - \frac{1}{s}\sum_{r=1}^{s}\frac{s_r^+}{y_{ro}}} \\ &\text{s.t.} \\ &\sum_{j=1}^{n} x_{ij} \cdot \lambda_j + s_i^- = x_{io}, i = 1, \ldots, m \\ &\sum_{j=1}^{n} y_{rj} \cdot \lambda_j - s_r^+ = y_{ro}, r = 1, \ldots, s \\ &\lambda_j, s_i^-, s_r^+ \geq 0, \forall j, i, r \end{aligned} \tag{5}$$

Similarly, the SBM model with undesirable outputs can be defined as follows:

$$\theta_2 = \min \frac{1 - \frac{1}{m}\sum_{i=1}^{m}\frac{s_i^-}{x_{io}}}{1 - \frac{1}{s}\sum_{r=1}^{s}\frac{s_r^+}{y_{ro}}}$$

$$\begin{aligned}
&\text{s.t.}\\
&\sum_{j=1}^{n} x_{ij}\cdot\lambda_j + s_i^- = x_{io}, i = 1, \ldots, m\\
&\sum_{j=1}^{n} y_{rj}\cdot\lambda_j - s_r^+ = y_{ro}, r = 1, \ldots, s\\
&\sum_{j=1}^{n} g_{lj}\cdot\lambda_j = g_{lo}, l = 1, \ldots, d\\
&\lambda_j, s_i^-, s_r^+ \geq 0, \forall j, i, r
\end{aligned} \tag{6}$$

The previous model examines both desirable (y_{rj}) and undesirable (g_{lj}) outputs.

Models (5) and (6) are SBM models under constant returns to scale (CRS) technology. The environmental indicator is based on the efficiency scores derived from these models. With the addition of the constraint $\sum_j \lambda_j = 1$, the aforementioned models are solved under the assumption of variable returns to scale (VRS).

Having obtained θ_1^* and θ_2^*, the following slack-based efficiency measure for modeling environmental performance may be defined (Zhou et al. 2006):

$$\text{SBEI} = \frac{\theta_1^*}{\theta_2^*} \tag{7}$$

This index is able to measure the impact of environmental constraints and standards on efficiency. If the index value is equal to the unit, then the two partial indices θ_1^* and θ_2^* are equal. This means that there are no undesirable exits or that the environmental constraints do not affect the entire production process at all. Finally, the degree of impact of these undesirable costs can be estimated by type $1 - \text{SBEI}$.

One of the advantages of the proposed approach is that with the use of DEA, efficiency is calculated based on composite environmental indices (slack-based environmental index, SBEI) which integrate both desirable and undesirable outputs. In addition to SBEI, the opportunity cost can be calculated due to environmental regulations and constraints. On the other hand, the main weaknesses of the proposed approach are based on the limitations of all DEA models. More specifically, DEA models are able to provide relative efficiency scores for the examined DMUs, but they cannot estimate absolute efficiency results. In addition, measurement errors in the assumed inputs/outputs may affect the stability of the provided results, given that DEA is an extreme point method.

3.4 Incorporating Uncertainty

One of the major problems in real-life applications is the uncertainty that lies in the collected data and the examined operations. Since the values that are provided do not represent the "real" image of the data, due to the presence of noise, the data are by nature stochastic. In this study, it is assumed that the undesirable outputs (CO_2, NOx, and SOx emissions) are stochastic. The stochastic variable that models the undesirable emissions is denoted with $\widehat{g}_{lj}$, while $\bar{g}_{lj}$ is the corresponding expected value. In order to measure the stochastic efficiency, the following model is applied (Zha et al. 2016):

$$\begin{aligned}
&\min \theta \\
&\text{s.t.} \\
&\sum_{j=1}^{n} x_{ij} \cdot \lambda_j + s_i^- = \theta \cdot x_{io}, i = 1, \ldots, m \\
&\sum_{j=1}^{n} y_{rj} \cdot \lambda_j - s_r^+ = y_{ro}, r = 1, \ldots, s \\
&\sum_{j=1}^{n} g_{lj} \cdot \lambda_j + s_l^+ = \theta \cdot g_{lo}, l = 1, \ldots, d \\
&\sum_{j=1}^{n} \lambda_j \left[\bar{g}_{lj} + b_{lj} \cdot \Phi^{-1}(\beta_{lj}) \right] = \theta \cdot \bar{g}_{lo}, l = 1, \ldots, d \\
&\lambda_j, s_i^-, s_r^+, s_l^+ \geq 0, \forall j, i, r, l
\end{aligned} \tag{8}$$

The aim of the previous model is to minimize the efficiency taking into account both desirable and undesirable outputs. An additional constraint is integrated in the DEA model to capture the stochasticity of the data. More specifically, b_{lj} denotes the standard deviation of the undesirable output l for each DMU$_j$, whereas $\Phi^{-1}(\beta_{lj})$ is the inverse cumulative distribution of level β_{lj}. The results of the efficiency θ can potentially receive values larger than 1. If $\theta > 1$, then the corresponding DMU is stochastic super-efficient; if $\theta = 1$ and all the slacks equal to 0, then the corresponding DMU is stochastic efficient, while if $\theta < 1$, the corresponding DMU is stochastic inefficient.

3.5 Data and Modeling

In this section, the selected data and the alternative modeling formulations in the examined problem are presented. The selection of appropriate data is based on previous studies, as well as on the rationale of environmental efficiency. Also, the set of countries (DMUs) examined in this study is quite large in order to capture different social, political, and economic conditions.

As already mentioned, the examined DMUs in the presented DEA models correspond to countries, in order to study their national production process in terms of environmental efficiency. The analysis considers a total of 108 countries covering different geographic areas and economies. However, the number of examined countries may differ in the alternative DEA models, due to data availability.

Regarding the inputs, four types of data were used, including the following indicators:

1. Labor force (10^6 people): It comprises people ages 15 and older who supply labor for the production of goods and services during a specified period. It includes people who are currently employed and people who are unemployed but seeking work as well as first-time job-seekers[1].
2. Population (10^6 people): It is based on the de facto definition of population (midyear estimates), which counts all residents regardless of legal status or citizenship[2]. It may be used as a proxy of labor force, since the previous indicator does not include everyone working in a national economy.
3. Gross capital formation (current US dollars): It consists of outlays on additions to the fixed assets of the economy plus net changes in the level of inventories[3]. It is considered as a major resource of a national economy in the context of DEA modeling.
4. Primary energy supply production (10^6 toe): It is defined as energy production plus energy imports, minus energy exports, minus international bunkers, and then plus or minus stock changes[4]. It is also considered as a major resource of any national economy's production process.

All of these resources are used in a general production process by countries and produce desirable and undesirable results.

The outputs of the study are divided into desirable and undesirable outputs. More specifically, the desirable output is the gross domestic product (GDP), while the undesirable outputs are carbon dioxide (CO_2) emissions, sulfur dioxide (SO_2) emissions, and nitrogen dioxide (NO_2) emissions. The definition of outputs is as follows:

1. Gross domestic product (current US dollars): GDP at purchaser's prices is the sum of gross value added by all resident producers in the economy plus any product taxes and minus any subsidies not included in the value of the products[5].
2. CO_2 emissions (kt): Carbon dioxide emissions are those stemming from the burning of fossil fuels and the manufacture of cement. They include carbon

[1]World Bank Indicators (https://data.worldbank.org/indicator/SL.TLF.TOTL.IN?view=chart)

[2]World Bank Indicators (https://data.worldbank.org/indicator/SP.POP.TOTL?view=chart)

[3]World Bank Indicators (https://data.worldbank.org/indicator/NE.GDI.TOTL.CD?view=chart)

[4]OECD Data (https://data.oecd.org/energy/primary-energy-supply.htm)

[5]World Bank Indicators (https://data.worldbank.org/indicator/NY.GDP.MKTP.CD?view=chart)

dioxide produced during consumption of solid, liquid, and gas fuels and gas flaring[6].

3. SO_2 emissions (kt): It arises from the oxidation, during combustion, of the sulfur contained within fossil fuels. Fossil fuels, including coal, oil, and to a lesser extent gas, contain sulfur in both organic and inorganic forms[7].
4. NO_2 emissions (kt): It primarily gets in the air from the burning of fuel, and it forms from emissions from cars, trucks, and buses, power plants, and off-road equipment[8].

The examined alternative models include both slack-based and stochastic DEA approaches, which consider different inputs that are consumed to produce different outputs (desirable and undesirable).

Specifically, the first slack-based model (model A) includes inputs that relate to population of the country and energy supply. These inputs are consumed to produce a desirable output (GDP) and three undesirable outputs (CO_2, SO_2, NO_2). The number of the countries used in this model is 108 from all over the world based on an amalgamation of different socioeconomic factors. The time period of the data spans from 1992 to 2003.

The second slack-based model (model B) is based on the underlying assumptions of the previous model. Three inputs and four outputs are considered here as well. The total number of DMUs (countries) is 104 and consists of countries of the world from various socioeconomic stratifications. In this model, a desirable output (GDP) and three undesirable outputs (CO_2, SO_2, NO_2) are considered which are produced from the consumption of three inputs, namely, labor force, gross capital formation, and primary energy supply. The data for the considered countries cover the period from 1992 to 2003.

Finally, the stochastic DEA model examined in this study is similar to model A, where the inputs include population and energy supply, while the outputs refer to GDP (desirable output) and CO_2, SO_2, and NO_2 emissions (undesirable outputs). A total of 101 countries examined in this model and the time period of the data spans from 1992 to 2003.

4 Results

The results of the three alternative DEA models are presented in this section. Each model is applied separately with the data from time period 1992 to 2003. Since the analysis considers spatiotemporal data regarding inputs and outputs, a geometrical average is presented for each model.

---

[6]World Bank Indicators (https://data.worldbank.org/indicator/EN.ATM.CO2E.KT?view=chart)

[7]European Environmental Agency (https://www.eea.europa.eu/data-and-maps/indicators/eea-32-sulphur-dioxide-so2-emissions-1/assessment-3)

[8]US Environmental Protection Agency (https://www.epa.gov/no2-pollution)

4.1 SBEI Results for Model A

The geometric mean of SBEI for model A is presented in Fig. 2, where the different shades of color represent different values of SBEI (darker shades represent higher SBEI values than lighter shades). The countries that were not considered in the analysis are painted with gray. Based on these results, we may observe that there is a wide selection of countries with average SBEI values in the range [0, 0.2]. Some of these countries are developed with strong economies and big influence (Canada, Germany, France, or Italy), and some are developing countries (Ethiopia, Pakistan, or Tanzania).

The most and the least environmental efficient countries are shown in Table 1. The countries that are fully efficient, in terms of SBEI (i.e., SBEI = 1), are Australia, Cyprus, Hong Kong, and Luxembourg. Especially for Luxembourg, it should be noted that it is a model country for environmental and economic measurements, which is partly ought to its legislation framework. The very small annual energy output (which is an input to our DEA model), coupled with its relatively large GDP (desirable output) and low emissions allow it to dominate the top efficiency. It should be also emphasized that the country's economy is based on the provision of services and not on industry or on the production of products. This is decisive for the country's environmental efficiency. The most important feature of Hong Kong's case is the annual energy supply.

On the contrary, countries such as China, India, or even Russia are in the last places, mainly because of their large population, their annual energy supply (input data), and their pollutants (output). More specifically, China, which is a country with a large population, would be expected to have the room for more emissions. But its

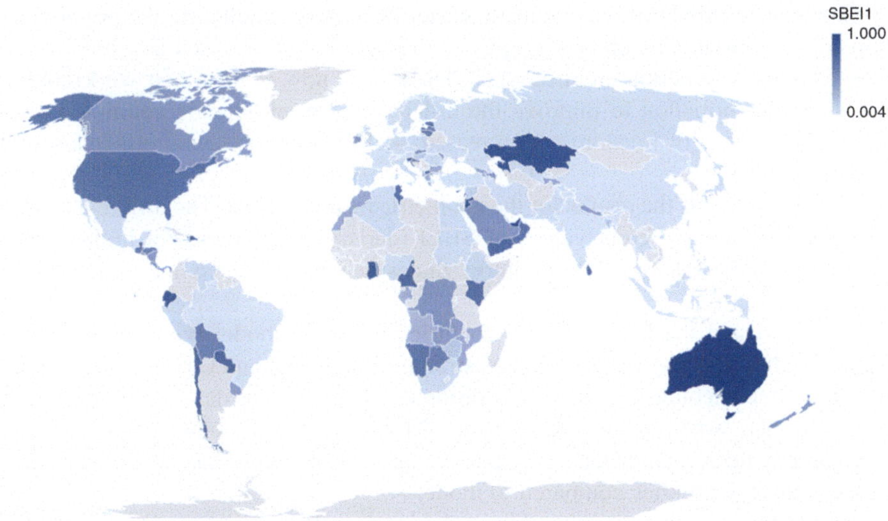

Fig. 2 SBEI results for model A

Table 1 Most and least environmental efficient countries (model A)

Efficiency	Countries	SBEI
Most efficient countries	Australia	1.000
	Cyprus	1.000
	Hong Kong	1.000
	Luxembourg	1.000
	United Arab Emirates	0.915
Least efficient countries	Indonesia	0.012
	Brazil	0.007
	Iceland	0.005
	China	0.004
	India	0.004

annual energy supply does not follow the same figures. In addition, countries such as Russia or the United States show higher amounts of energy supply. Also, as mentioned before, pollutant emission values in these countries are among the highest, which is justified by a large percentage of the large population and energy supply. Nevertheless, a driving factor for this increase in the pollutant emissions is associated with high GDP values (which is considered as a desirable output).

4.2 SBEI Results for Model B

The results of model B are significantly different from those of the model A presented in the previous section. This is attributed to the fact that different inputs are considered in this model. Since the consideration of the examined models was based on the minimization of inputs, the additional input effectively enabled the comparisons of the DMUs to be made easier to achieve. Excluding the population, which is a non-variable entry (except in extremely rare cases such as China where there is a legal limitation on birth rates), DMUs can now change their workforce or gross capital formation to improve the efficiency. By introducing additional variables, the degrees of freedom are increased, and the environmental efficiency of a country increases.

Figure 3 presents the results for the geometric mean of SBEI. The different shades of color represent different values of SBEI (darker shades represent higher SBEI values than lighter shades). The countries that were not considered in the analysis are painted with gray.

The most and the least efficient countries based on model B are presented in Table 2, where we may observe that Australia, Hong Kong, and Luxembourg remain as fully efficient countries, with the addition of smaller or less developed economies (e.g., Estonia, Israel, Jordan, Kazakhstan, Moldova). On the other hand, the least efficient countries include Gabon, Japan, Congo, while Indonesia and China remain as low efficient counties compared to model A.

However, it can be noted that this model has less discrimination power since the majority of the countries have higher SBEI values than model A.

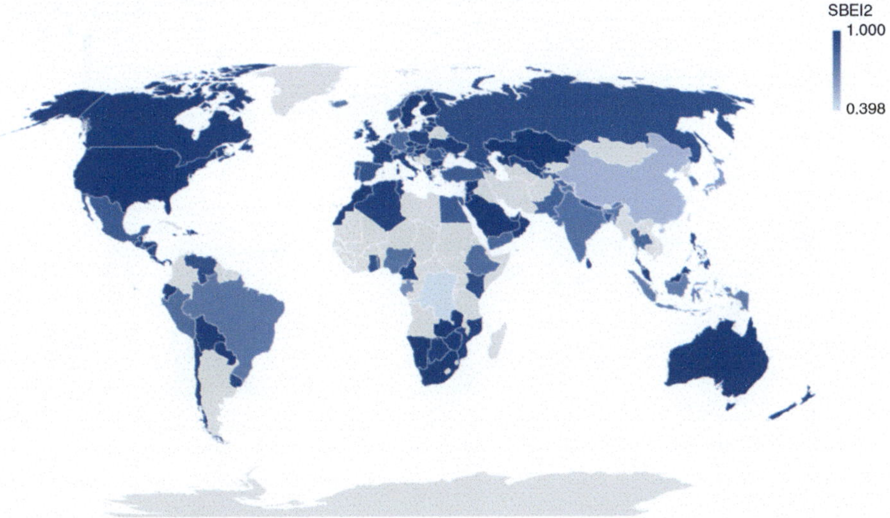

Fig. 3 SBEI results for model B

Table 2 Most and least environmental efficient countries (model B)

Efficiency	Countries	SBEI
Most efficient countries	Australia	1.000
	Estonia	1.000
	Hong Kong	1.000
	Israel	1.000
	Jordan	1.000
	Kazakhstan	1.000
	Luxembourg	1.000
	Moldova	1.000
	Morocco	1.000
	Saudi Arabia	1.000
Least efficient countries	Indonesia	0.737
	Gabon	0.680
	Japan	0.613
	China	0.514
	Congo	0.398

In order to overcome this problem, several ways have been proposed to increase the discriminatory power of DEA, as, for example, applying principal component analysis to reduce the dimensionality of inputs and/or outputs; however, this type of formulation exceeds the scope of the proposed modeling.

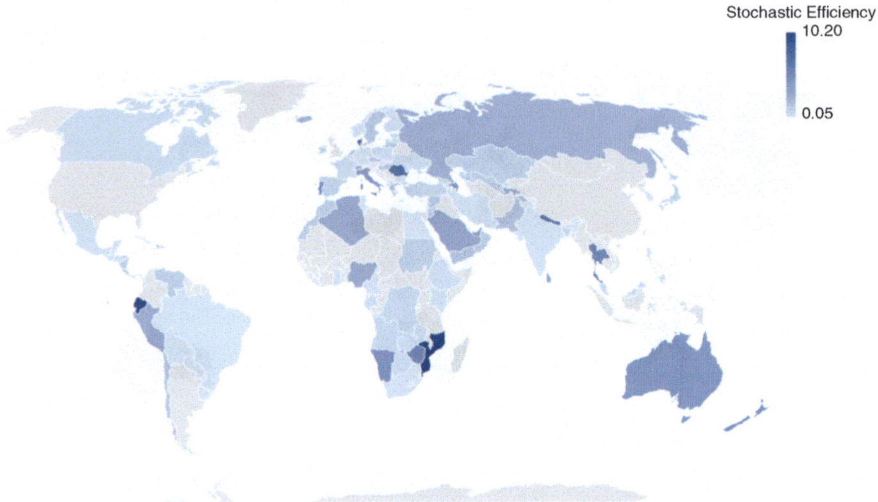

Fig. 4 Efficiency results for stochastic DEA model

4.3 *Stochastic Efficiency*

The stochastic efficiency results of the DEA model presented in Sect. 3.4 are shown in Fig. 4. Similarly to the previous sections, the results refer to the geometric mean of stochastic efficiency, while the different shades of color represent different values of stochastic efficiency (darker shades correspond to higher stochastic efficiency). The countries that were not considered in the analysis are painted with gray.

These results are quite different compared to the previous model due to the large variation of undesirable outputs (i.e., emissions) in some cases. It should be noted that due to its nature, the range of the stochastic efficiency score is larger, as shown in Fig. 4.

The results of the stochastic DEA model give the ability to sort countries based on the estimated efficiency scores. As shown in Table 3, countries may be categorized in three main groups:

(a) Counties with $\theta > 1$: This group refers to countries that are stochastic efficient and includes counties that, based on the previous results, are expected to be environmental efficient, like Australia, Luxembourg, Israel, Switzerland, Saudi Arabia, or the United Arab Emirates. However, additional developed (e.g., Denmark, New Zealand, Russian Federation, Portugal, Italy) and developing countries (e.g., Mozambique, Zimbabwe, Lebanon, Yemen, Pakistan) are included in this group. This phenomenon is ought to the fact that these countries utilize low resources to produce medium values of GDP but higher values of harmful emissions.

Table 3 Categorization of countries based on stochastic efficiency scores

Stochastic efficiency	Countries		
$\theta > 1$	Mozambique	Portugal	Tajikistan
	Ecuador	Italy	Peru
	Denmark	Australia	Iceland
	Romania	United Arab Emirates	Russian Federation
	Nepal	Sri Lanka	Israel
	Thailand	Switzerland	Luxembourg
	Zimbabwe	Saudi Arabia	Lebanon
	New Zealand	Haiti	Greece
	Namibia	Nigeria	Yemen
	Azerbaijan	Algeria	Pakistan
$\theta \simeq 1$	Costa Rica	Czech Republic	Sudan
	Hungary	Venezuela	Jordan
	Netherlands	Syria	Ireland
	Chile	Japan	Kazakhstan
	Sweden	Moldova	Cyprus
	Oman	Morocco	Turkey
	Nicaragua	Spain	Georgia
	Lithuania	Uzbekistan	
$\theta < 1$	Bangladesh	FYROM	Republic of Korea
	Belgium	Latvia	South Africa
	Angola	El Salvador	India
	Congo	Ethiopia	Norway
	Botswana	Guatemala	Bulgaria
	Malaysia	Dominican Republic	Germany
	Honduras	Philippines	Jamaica
	France	Mexico	Iran
	Canada	Ghana	Slovakia
	Ukraine	Cameroon	Tunisia
	Zambia	Albania	Brazil
	Estonia	Finland	Croatia
	Poland	Armenia	Paraguay
	Gabon	Kenya	Bolivia
	Bahrain	Panama	Egypt
	Togo	Uruguay	Tanzania

(b) Countries with $\theta \simeq 1$: Several countries are ranked lower based on stochastic DEA method. Some of them refer to Cyprus or Kazakhstan that appear to have higher efficiency in the previous DEA models. In general, some of the strong national economies are included in this group, like Netherlands, Sweden, and Spain.

Table 4 Comparison of selective counties' ranking based on efficiency scores

Stochastic efficiency	Model A	Model B	Stochastic DEA
Australia	1	1	13
France	100	56	61
Germany	84	92	91
Greece	56	42	28
India	108	97	88
Italy	75	40	12
Japan	76	102	46
Luxembourg	1	1	26
Russian Federation	94	75	24
Spain	91	80	45
Sweden	80	22	35

(c) Countries with $\theta < 1$: This group refers to inefficient countries and includes both developed and developing countries. Countries with strong economies like France, Canada, Germany, or Finland are ranked quite low, as it can be seen in Table 3, since the standard deviation presents large fluctuations, while the probability levels for the undesirable outputs are also high.

The results of the stochastic DEA model, although appear to have some similarities with the slack-based DEA model, in several cases, they provide very different efficiency scores. Table 4, for example, shows the comparison of rankings obtained by the three alternative DEA models for selective countries. As it can be observed, in some cases the stochastic DEA model estimates larger efficiencies compared to slack-based models (e.g., Greece, Italy, Japan), while in other cases, the estimated efficiencies are lower (e.g., Luxembourg, Russia, Spain).

Finally, it should be noted that environmental efficiency should not be confused with environmental performance, since different combinations and levels of inputs may result to an environmental efficient production system.

5 Concluding Remarks

The aim of the present study is to present a methodological framework for the measurement of environmental efficiency and to highlight the evaluation indicators of the production units studied. To this end, a nonparametric method, data envelopment analysis, is applied. The study of environmental efficiency, in the context of DEA approaches, is mainly focused on analyzing how, with a given set of resources (inputs), the outcomes may be maximized (desirable outputs), while at the same time, emissions are minimized (undesirable outputs).

Alternative SBM models are applied in this chapter under CRS and VRS technologies. The considered inputs and outputs have resulted in significant differentiations of the composite index. More specifically, model A assumed to have two inputs and four outputs, providing a meaningful ranking of countries with justified possible sudden fluctuations in the behavior of decision units. Model B, with three inputs and four outputs, provided less reliable results. Several decision units showed maximum efficiency since the model had less discrimination power. The results of this work can be compared to those of similar studies, possibly using a different methodological framework, to give a more general and complete picture of the subject. Finally, a stochastic DEA model is presented to measure stochastic efficiency of countries assuming that the undesirable outputs are stochastic.

According to the results of SBEI of model A, Australia, Hong Kong, and Luxembourg are fully efficient. Due to legislation regarding the emissions (e.g., CO_2) and low input to the process producing large values of GDP, Luxembourg appears as a fully efficient country. This finding can drive other countries to adopt Luxembourg's paradigm and adjust their legislation. This finding indicates that countries should produce high values of GDP with less labor force, capital, and energy production while minimizing undesirable outputs. Based on the results of model B, Australia is fully efficient. In this model, the results of SBEI are closer to 1 compared to the corresponding results of SBEI of model A. This is attributed to the fact that given the inputs and outputs of model B, less countries are inefficient or gather lower values of efficiency leading to less discriminatory power.

Uncertainty is measured with a stochastic DEA model which categorizes countries based on the stochastic efficiency according to three categories: stochastic efficiency greater than 1, equal to 1, and less than 1.

The main limitations of this study refer to the availability of data and the selection of appropriate inputs and outputs. For example, similar to previous studies, the selected indicators are actually proxies of the actual variables that should be included (e.g., GDP is a proxy of the true financial outcome of a national economy, although it is affected by several other factors). Thus, future research efforts may study different combinations of resources and outcomes in the context of the presented DEA approaches.

Moreover, the analysis can be further enhanced in the future with the addition of multiple layers or production processes. In such an approach, efficiency and the corresponding environmental indices can be calculated with the use of network DEA modeling. Also, comparing environmental efficiency and performance (effectiveness) may provide useful results for developing appropriate environmental policies. In this context, SBEI may be compared with alternative environmental or sustainability performance indices (see Grigoroudis et al. 2012 for a review). Finally, combining the presented analysis with the Malmquist index may give additional results regarding the evolution of environmental efficiency in the examined period.

References

Baležentis, T., Li, T., Streimikiene, D., & Baležentis, A. (2016). Is the Lithuanian economy approaching the goals of sustainable energy and climate change mitigation? Evidence from DEA-based environmental performance index. *Journal of Cleaner Production, 116*, 23–31.

Brockhoff, K. (1970). Quantification of marginal productivity of industrial research by estimating a production function for a single firm. *German Economic Review, 8*(3), 202–229.

Callens, I., & Tyteca, D. (1999). Towards indicators of sustainable development for firms: A productive efficiency perspective. *Ecological Economics, 28*(1), 41–53.

Charnes, A., & Cooper, W. W. (1963). Deterministic equivalents for optimizing and satisficing under chance constraints. *Operations Research, 11*(1), 18–39.

Cooper, W. W., Deng, H., Huang, Z., & Li, S. X. (2002). Chance constrained programming approaches to technical efficiencies and inefficiencies in stochastic data envelopment analysis. *Journal of the Operational Research Society, 53*(12), 1347–1356.

Dyckhoff, H., & Allen, K. (2001). Measuring ecological efficiency with data envelopment analysis (DEA). *European Journal of Operational Research, 132*(2), 312–325.

Färe, R., Grosskopf, S., & Pasurka, C. (1986). Effects on relative efficiency in electric power generation due to environmental controls. *Resources and Energy, 8*(2), 167–184.

Färe, R., Grosskopf, S., & Tyteca, D. (1996). An activity analysis model of the environmental performance of firms: Application to fossil-fuel-fired electric utilities. *Ecological Economics, 18* (2), 161–175.

Farrell, M. J. (1957). The measurement of productive efficiency. *Journal of the Royal Statistical Society, Series A (General), 120*(3), 253–290.

Grigoroudis, E., Kouikoglou, V. S., & Phillis, Y. A. (2012). Approaches for measuring sustainability. In P. Olla (Ed.), *Global sustainable development and renewable energy systems* (pp. 101–130). Hersey: IGI Global.

Gutiérrez, R., Gutiérrez-Sánchez, R., & Nafidi, A. (2008). Trend analysis using nonhomogeneous stochastic diffusion processes. Emission of CO_2; Kyoto protocol in Spain. *Stochastic Environmental Research and Risk Assessment, 22*(1), 57–66.

Korhonen, P. J., & Luptacik, M. (2004). Eco-efficiency analysis of power plants: An extension of data envelopment analysis. *European Journal of Operational Research, 154*(2), 437–446.

Mardani, A., Zavadskas, E. K., Streimikiene, D., Jusoh, A., & Khoshnoudi, M. (2017). A comprehensive review of data envelopment analysis (DEA) approach in energy efficiency. *Renewable and Sustainable Energy Reviews, 70*, 1298–1322.

Prieto, A. M., & Zofío, J. L. (2007). Network DEA efficiency in input-output models: With an application to OECD countries. *European Journal of Operational Research, 178*(1), 292–304.

Ramanathan, R. (2005). An analysis of energy consumption and carbon dioxide emissions in countries of the Middle East and North Africa. *Energy, 30*(15), 2831–2842.

Reinhard, S., Lovell, C. K., & Thijssen, G. J. (2000). Environmental efficiency with multiple environmentally detrimental variables; estimated with SFA and DEA. *European Journal of Operational Research, 121*(2), 287–303.

Sueyoshi, T., Yuan, Y., & Goto, M. (2017). A literature study for DEA applied to energy and environment. *Energy Economics, 62*, 104–124.

Triantis, K., & Otis, P. (2004). Dominance-based measurement of productive and environmental performance for manufacturing. *European Journal of Operational Research, 154*(2), 447–464.

Tyteca, D. (1996). On the measurement of the environmental performance of firms: A literature review and a productive efficiency perspective. *Journal of Environmental Management, 46*(3), 281–308.

Zaim, O. (2004). Measuring environmental performance of state manufacturing through changes in pollution intensities: A DEA framework. *Ecological Economics, 48*(1), 37–47.

Zha, Y., Zhao, L., & Bian, Y. (2016). Measuring regional efficiency of energy and carbon dioxide emissions in China: A chance constrained DEA approach. *Computers & Operations Research, 66*, 351–361.

Zhou, P., Ang, B. W., & Poh, K. L. (2006). Slacks-based efficiency measures for modeling environmental performance. *Ecological Economics, 60*(1), 111–118.

Zhou, P., Poh, K. L., & Ang, B. W. (2007). A non-radial DEA approach to measuring environmental performance. *European Journal of Operational Research, 178*(1), 1–9.

Zhou, P., Ang, B. W., & Poh, K. L. (2008). Measuring environmental performance under different environmental DEA technologies. *Energy Economics, 30*(1), 1–14.

Zofío, J. L., & Prieto, A. M. (2001). Environmental efficiency and regulatory standards: The case of CO_2 emissions from OECD industries. *Resource and Energy Economics, 23*(1), 63–83.

Hypothesis for a Risk Cost of Carbon: Revising the Externalities and Ethics of Climate Change

Delton B. Chen, Joel van der Beek, and Jonathan Cloud

Abstract Standard market-based policies for addressing climate change mostly aim to internalize the Social Cost of Carbon (SCC) into the economy with either carbon taxes or cap-and-trade schemes. Standard policies are failing to manage the systemic risk of dangerous-to-catastrophic climate change for a variety of reasons. In this chapter we clarify and expand on a market hypothesis that argues for a second externalized cost of carbon, called the Risk Cost of Carbon (RCC), as the appropriate solution to this risk problem.

The combination of the SCC and RCC creates a new paradigm of complementary market pricing for the dual objectives of improving market efficiency and managing systemic risk, respectively. Introducing the RCC addresses the problem of how to decouple gross world product (GWP) from carbon emissions and how to solve the paradox of time discounting under systemic risk. Subsequently the RCC could have major implications for climate change economics, public policy, and sustainability theory. The hypothesis is novel by taking into consideration both the entropy and the mass of the carbon budget.

The RCC is technically defined as the cost of imposing risk tolerances (%) on climate mitigation objectives, and it has units of USD per tonne of carbon dioxide equivalent (CO_2e) mitigated. The RCC is internalized with a "global carbon reward" that manages a trade-off between market efficiency and climate certainty. The carbon reward is issued as a parallel currency and with an exchange rate that is managed by central banks over a rolling 100-year planning horizon. A key recommendation is to test the hypothesis with experiments.

Keywords Climate change · Systemic risk · Risk management · Carbon price · Biophysical · Thermodynamics · Entropy · Central bank · Monetary policy · Parallel currency · Macroprudential

D. B. Chen (✉) · J. Cloud
Center for Regenerative Community Solutions (501c3 Non-Profit), Basking Ridge, NJ, USA
e-mail: deltonchen@crcsolutions.org

J. van der Beek
EconoVision & EconoLytics, Doorn, Utrecht, The Netherlands

H. Doukas et al. (eds.), *Understanding Risks and Uncertainties in Energy and Climate Policy*, https://doi.org/10.1007/978-3-030-03152-7_8

1 Introduction

The topic of this chapter is the theoretical plausibility of a second externalized cost of anthropogenic greenhouse gas (GHG) emissions, called the *Risk Cost of Carbon* (RCC), whereby the first externalized cost is already established as the Social Cost of Carbon (SCC). Chen et al. (2017) originally postulated the existence of the RCC, which they describe as the cost of managing climate risk with positive incentives guided by cost-effectiveness analysis. The possible existence of the RCC is explained with the *Holistic Market Hypothesis* (HMH), which Chen et al. (2017) introduced using an epistemology of complementary relationships. The HMH could have major implications for economic assessments, climate policy, and environmental law because it posits that the total externalized cost of the market failure is significantly underestimated when the RCC is ignored.

To understand why risk can be quantified as a cost—such as with the RCC—it is necessary to accept that risk is the "effect of uncertainty on objectives" (ISO 2009). The RCC is the cost of reducing the uncertainty of achieving specific climate mitigation objectives, whereby the targeted levels of uncertainty are the acceptable probabilities (%) of success or failure. The targeted levels of uncertainty may also be called "risk tolerances", and the risk tolerances are used to estimate the additional mass of carbon dioxide equivalent (CO_2e) to be mitigated.

The HMH involves an interdisciplinary interpretation of carbon taxes and rewards, including an analysis of their economic objectives and their qualitative effect on the *entropy* of carbon in the environment. The topic of entropy is relevant because it is used to resolve a handful of temporal paradoxes that accompany the standard policy toolkit for climate mitigation. The HMH involves terminology from both neoclassical economics and physics, and this may pose a challenge because economists and physicists typically use different conceptual models, methods, and terminology. The HMH is presented as a macroscopic entropic theory, and if it is cogent and correct, then it should be possible to validate it with experiments and verify it with a statistical-mechanical approach.

To guide the reader, the chapter begins with an introduction to the SCC and RCC (Sect. 1.1). Principles and concepts that support the HMH are described, including climate systemic risk, positive carbon pricing, thermodynamic laws, carbon emissions, and network theory (Sect. 1). The HMH is stated in Sect. 2 in terms of temporal relationships and the RCC metric. In Sect. 3, a policy for a global carbon reward is described that can internalize the RCC into the economy. In Sect. 4, an epistemology of complementary-and-opposite relationships is applied to the carbon tax to verify that the SCC and the RCC are complementary. In Sect. 5, the theoretical utility of the RCC is discussed in terms of solving a temporal paradox, and the practical utility of the global carbon reward is discussed in terms of the Paris Climate Agreement (COP21), net zero emissions, global growth, and ethics. In Sect. 6, some concluding remarks are provided, and finally, in Sect. 7, recommendations are given for future research. For reasons of brevity, the political feasibility of climate policies is not discussed.

1.1 Externalized Cost of Carbon

The Social Cost of Carbon (SCC) is a negative externality created by anthropogenic GHGs, and it is typically defined as the time-discounted marginal loss in economic welfare that results from 1 metric tonne of carbon dioxide equivalent (CO_2e) emitted in a given year (e.g., IAWG 2013). Carbon taxes, which are a kind of Pigovian tax, can be used to internalize the SCC into the economy. Other market-based policies can also be used to incentivize emissions reductions, including cap-and-trade schemes. The ideal carbon tax is related to the SCC under cost-benefit analysis, and consequently the SCC is important in the study of climate change economics (Nordhaus 1991, 2017; Stern 2007).

Despite decades of research into the SCC (e.g., U.S. President 1981), the scholarship on carbon pricing has not associated a specific objective with offering global rewards for climate mitigation. If global rewards for carbon represent an unused price signal, then two fundamentally important questions deserve our attention, namely:

Q1. What is the financial mechanism for a global carbon reward?
Q2. What is the economic objective of a global carbon reward?

Chen et al. (2017) claim that—based on their Holistic Market Hypothesis (HMH)—a global reward for carbon mitigation can (1) be provided with monetary policy and (2) can be used to manage *climate systemic risk.* Aglietta and Espagne (2016) originally coined the term "climate systemic risk" in reference to financial and physical fragilities associated with anthropogenic climate change (refer Sect. 1.3). If these answers to the two questions are correct, then a third question arises:

Q3. What is the total externalized cost of carbon emissions?

Chen et al. (2017) postulate that (3) the total externalized cost of carbon emissions is the SCC plus a second externality, called the Risk Cost of Carbon (RCC). The RCC is the assessed cost of providing a positive externality and it does not substitute for the SCC, which is the assessed cost of a negative externality. They define the RCC as follows:

> ...the market price of each metric tonne of additional CO_2-e mitigation service that is needed to reduce climate systemic risk to an agreed limit. (p. 41)

1.2 Positive Carbon Price

The literature lacks clear terminology for defining a positive carbon price (e.g., Sirkis et al. 2015), and to provide clarity, the terms "positive carbon price," "carbon subsidy," and "carbon reward" are given the following definitions for this exposition:

- A positive carbon price is defined here to be a price signal that offers payment for mitigating carbon emissions—for abating emissions and for removing carbon from the atmosphere[1]—and with the payment being divorced from carbon offsetting schemes.

- A carbon subsidy is defined here to be a positive carbon price that is offered as an ex post or ex ante payment and when the payment is made with a national fiat currency or as a tax deduction.
- A carbon reward is defined here to be a positive carbon price that is offered as an ex post payment for verified carbon mitigation, and when the reward payment is (a) made with a parallel currency denominated in carbon by mass, and (b) provided with conditions for the awardees to maintain an agreed standard of service.

The above provisional terms may help to open up a wider discourse on positive pricing. According to Chen et al. (2017), the global carbon reward should be implemented as a parallel currency so that monetary policy and currency trading can be used to internalize the RCC into the economy. The global carbon reward generates a positive externality because it acts as preventative climate insurance, and this insurance may be classified as a 'public good' because it yields physical benefits that are non-rivalrous and non-excludable. The global carbon reward is a macroprudential responsibility of central banks, and this responsibility should be insulated from political interference. In the following section, we will examine the concept of "climate systemic risk," which is used to define the objective of the global carbon reward and the macroprudential agenda.

1.3 Climate Systemic Risk

Aglietta and Espagne (2016) describe "climate systemic risk" as the ensemble of financial and physical fragilities produced by greenhouse emissions, whereby "fragility" is the possibility of an unacceptable systemic failure. The actual impacts of a systemic failure may not be known with confidence, and so an emphasis is placed on assessing the probability of a systemic failure rather than trying to predict the quantity of the damages. One example of a climate systemic risk is the possibility that 2.0 °C of global warming will be exceeded by the year 2100. The international ambition to limit this climate systemic risk is a major inspiration for the Paris Climate Agreement (UNFCCC 2015).

Aglietta and Espagne (2016) equate the climate systemic risk with the cost of preventative insurance, whereby the cost is "...equivalent of a value that society attributes to mitigation activities" (p. 5). Aglietta and Espagne (2016) also link their insurance proposal to central banks and monetary policy as follows:

[1]The positive carbon price does not address geo-engineering of the solar energy balance.

> The incorporation of some kind of climate signal in monetary policy and financial stability oversight is required, not because the central banks should be a direct actor of the low-carbon transition, but as part of their financial stability mandate. (p. 18)

Aglietta and Espagne (2016) suggest that climate systemic risk can form the basis of a new hypothesis that differs from the traditional theory for externalized costs:

> The climate systemic risk hypothesis radically departs from the premises of standard externality theory. It suggests that we might want to drastically diminish the probability of occurrence of some very bad outcomes for society, which might lead to its quasi-destruction. (p. 13)

Chen et al. (2017) claim that a global reward for carbon mitigation is a kind of preventative insurance against unwanted global warming and that it has similarity with payments for ecosystem services (PES) and is analogous to preventative health insurance. Aglietta and Espagne (2016) and Chen et al. (2017) both justify their collective insurance proposals, but they use somewhat different reasoning. Aglietta and Espagne (2016) emphasize the *Radical Uncertainty Hypothesis* (e.g., Knight 1921; Keynes 1921) and climate fragility as key justifications, whereas Chen et al. (2017) present the Holistic Market Hypothesis (HMH) and the RCC as their justification. The main difference between the two proposals is that Aglietta and Espagne (2016) recommend a break from the standard model for externalized costs, whereas Chen et al. (2017) present the RCC as an augmentation to the standard model. In the following sections, we will review the key biophysical concepts underpinning the HMH.

1.4 Biophysical Economics

Biophysical economics is an emerging school of economic thought that attempts to understand the economy using the laws of thermodynamics (e.g., Cleveland 1987). A key strength of the biophysical approach is that the laws of thermodynamics are applicable to all biophysical systems, including to civilization, the climate system, and living organisms. Despite its solid theoretical foundation, biophysical economics is currently only a minor school because it is philosophically at odds with the classical/neoclassical worldview that economic activity is driven by human agency. For example, a commonly cited definition of classical economics is based on Robbins' (1935) review of the subject, as follows:

> Economics is the science which studies human behavior as a relationship between ends and scarce means which have alternative uses. (p. 16)

A distinctive feature of the Holistic Market Hypothesis (HMH) is its acceptance of both the classical/neoclassical approach of studying human behavior, and the biophysical approach of studying energy dissipation as the universal source of all kinds of agency. Under the HMH, the biophysical economic worldview that supports the SCC and RCC is considered to be more general than the classical/neoclassical economic worldview that only supports the SCC. This hierarchy of worldviews is relevant when addressing carbon emissions, and the underlying reason is that civilization has its own emergent agency because it dissipates energy. This agency can be relatively strong compared with the agency of individuals, social collectives and political groups. The issue of competing agency is especially relevant to the challenge of mitigating carbon emissions, and this is because the primary energy supply, the agricultural sector, and many other economic activities are strongly coupled to carbon. Under the HMH, the expectation is that civilization's agency will undermine attempts to quickly reduce carbon emissions. Under the HMH, it is interpreted that previously successful strategies for mitigating other types of pollution are unlikely to be effective when applied to carbon, and this is because most other pollutants are not strongly coupled to the energy supply (e.g. sulphur in acid rain can be removed, and ozone depleting substances can be substituted).

A unique feature of the HMH is the way that the Second Law of Thermodynamics is used to qualitatively interpret the SCC and RCC as a complementary pair. The two most important laws in thermodynamics are known from the works of Clausius (1854), Maxwell (1860), Boltzmann (1872), and others:

(Law 1) The First Law of Thermodynamics is the conservation law for energy. It states that the energy of a system is conserved because energy cannot be created or destroyed. The First Law is time-symmetric, and energy is an extensive property with units of Joules.

(Law 2) The Second Law of Thermodynamics is the entropy law for matter and energy. The entropy of a classical system is a measure of the "disorder" or randomness of the particles that comprise the system. The Second Law states that the entropy of an isolated system increases monotonically with energy dissipation and eventually peaks at thermal equilibrium. Under the Second Law it is possible for energy dissipation to have reduced the entropy of an open system if the entropy of the open system plus its surroundings has increased. The Second Law is time-asymmetric, and entropy is an extensive property with units of Joules/Kelvin.

The first two laws of thermodynamics were originally derived for gases because gases are relatively simple to study (e.g., Clausius 1867). The First Law may be intuitive to most people, but the Second Law often requires clarification. The Second Law is supported by statistical theories for the randomness of particle interactions (e.g., Maxwell 1860; Boltzmann 1872), and the law explains why heat will flow down a temperature gradient—from hot to cold—but not in the other direction.

The Second Law is vitally important to our understanding of civilization as a type of 'heat engine' that dissipates energy to do the work of producing goods and services while releasing heat and high-entropy waste, such as CO_2. For example, the First and Second Laws are used to derive the maximum theoretical efficiency of a cyclic engine that converts heat into mechanical work (i.e., the Carnot heat engine). The Second Law also applies to living cells in terms of explaining the local decrease in entropy within the cell walls: a product of energy dissipation and an increase in the entropy of the environment. England (2013) used the Second Law to develop a probabilistic theory that explains why biological self-replication is favored—thus giving clues to the origins of life.

Civilization and living organisms are similar, because both persist by maintaining a low-entropy condition far from thermodynamic equilibrium. Raworth (2017)

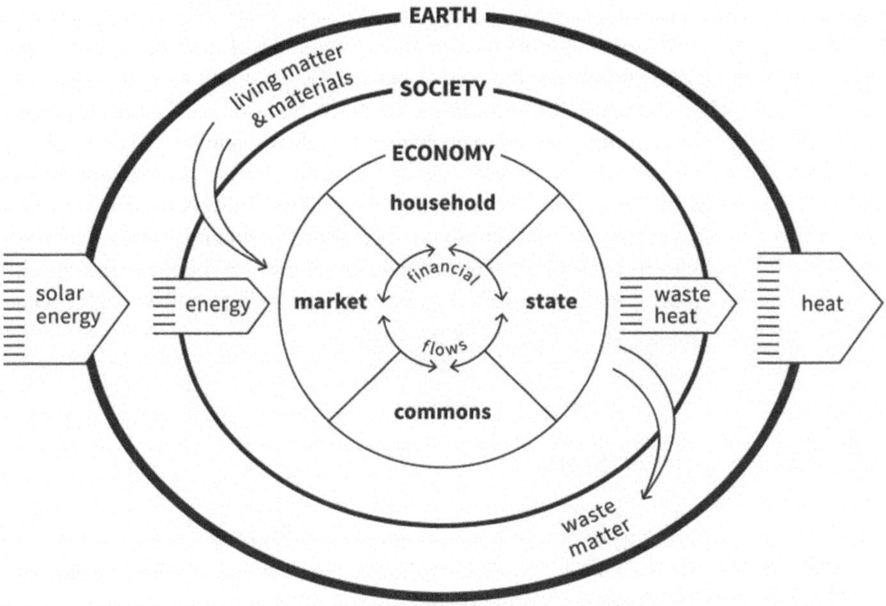

Fig. 1 The embedded economy diagram of Raworth (2017) shows that the economy is part of an open system that dissipates energy and externalizes entropy as waste heat and matter (reproduced with kind permission of the author)

presents the "embedded economy diagram" to illustrate how this condition is maintained in terms of economic processes within civilization (see Fig. 1; note that civilization is synonymous with "society" in Fig. 1). As shown in Fig. 1, civilization is embedded within the Earth's ecosystem. As expected under the Second Law, civilization receives useful energy and low-entropy materials, and it rejects waste heat and high-entropy matter to maintain its low-entropy state. The entropy of civilization is generally too complex to be assessed quantitatively, and in the following discussion entropy is only considered qualitatively.

The economy is an internal process of civilization (refer Fig. 1). Annila and Salthe (2009) present an argument for why the economy is driven by energy dissipation under the Second Law. Garrett (2011, 2012, 2014, 2015) took a major step forward by developing a lumped-parameter model of civilization's energy dissipation that links energy demand and CO_2 emissions to cumulative GDP. Such arguments and models highlight that there are two possible sources of economic agency: (a) human agency and (b) energy dissipation. These two sources of agency are a paradox of the neoclassical and biophysical economic perspectives.

There can be no doubting that civilization and all other biophysical systems adhere to the laws of thermodynamics. Despite this thermodynamic certainty, economic processes are inherently difficult to quantify with the Second Law because civilization is far from thermodynamic equilibrium and also because the financial system is a complex informational system. Despite these complexities, various lines of research have revealed that the economy does express biophysical patterns of behavior. For example, Kümmel and Lindenberger (2014) found that primary energy supply was a dominant factor in economic production in three major developed economies; and Lawrence et al. (2013) found that the global per capita distributions of energy consumption and carbon dioxide (CO_2) emissions are converging on a common exponential distribution without being directed to do so using public policies. The HMH takes a new approach in biophyiscal economics by deriving a policy for a global carbon reward based on a reversal of the carbon tax, whereby the Second Law is used to justify a condition of time-asymmetry in the dominant social agreements/responses that comprise these two policies (refer Sect. 2). Under the Second Law, the change in the entropy of carbon in the environment is expected to be negative as a result of each policy, however the total change in entropy is always positive.

1.5 Future Carbon Emissions

A threat to the well-being of future generations is civilization's strong tendency to emit carbon dioxide (CO_2) and other greenhouse gases while dissipating energy. Raftery et al. (2017) developed a probabilistic model of future CO_2 emissions based on country-specific versions of the Kaya identity and extrapolated future temperature changes to the year 2100. They conclude that the 1.5 °C and 2.0 °C limits of the Paris Climate Agreement have a 99% and 95% chance of failure, respectively (i.e., 1% and 5% chance of success, respectively). Moreover, Raftery et al. (2017) conclude that global warming is trending toward 3.2 °C (2.0–4.9 °C) by 2100.

The study of Raftery et al. (2017) highlights that a major driver of global emissions is a rising total gross domestic product (GDP)—or gross world product (GWP)—which is driven higher by rising GDP per capita and rising population. Capacity to reverse the trend appears limited, because future reductions in the carbon intensity of GDP have limitations and because "Policies to reduce GDP per capita seem unlikely..." (p. 3). The probabilities and macroeconomic trends presented by Raftery et al. (2017) are a stark reminder that climate risk is not well managed.

Garrett (2011, 2012) developed a lumped-parameter model of civilization to assess its primary energy usage and CO_2 emissions. He adopts a biophysical worldview by linking civilization's primary energy consumption to civilization's total wealth, which he represents as the cumulative gross world product (GWP). Garrett (2012) subsequently proposed the following relationship for global CO_2 emissions over time:

$$E(t) \cong c(t)\, \lambda \sum\nolimits_{i=1}^{t} \mathrm{GWP}(i) \tag{1}$$

where E = total mass of CO_2 emissions per year; c = average CO_2 emissions intensity of energy; λ = average power consumption per unit of currency as a time-invariant parameter; GWP = inflation-adjusted gross world product; t = current time in years; and i = time step starting in the pre-industrial period.

Garrett (2012) assessed the historical data for GWP, primary energy, and inflation, to arrive at the following estimate of the mean inflation-adjusted λ:

$$\lambda = 9.7 \pm 0.3 \text{ mW per USD (1990)} \tag{2}$$

Garrett (2012) argues that because of a coupling between economic consumption and primary energy, further improvements in energy efficiency will cause civilization to grow faster and to ultimately consume more energy per unit of time. According to Garrett (2012), this coupling is a driver of CO_2 emissions and puts humanity in a "double-bind," and so he writes: "If civilization does not collapse quickly this century, then CO_2 levels will likely end up exceeding 1000 ppmv..." (p. 1).

In Sect. 5.2 the plausibility of using a global carbon reward to escape from the double-bind problem is examined. The solution involves a currency exchange rate mechanism to create a negative feedback on gross world product (GWP) and dirty patterns of economic growth. Closely related to the problem of unsustainable growth are the problems of energy demand rebound after improvements are made in energy effficiency—called Jevons paradox—and poorly managed human populations. The plausibility of using the global carbon reward to address these two related problems is examined (refer Sects. 5.2.2 and 5.2.3, respectively).

1.6 Network Theory

Network theory involves the study of network organization, topology, and lumped parameters, and it is referenced by the HMH because Chen et al. (2017) consider the social agreements and financial flows of climate policies to be biophysical networks. They represent a policy's authority and market actors as nodes and incentive payments as vectors. The elementary nodes and vectors are shown in Fig. 2 and are used in Sect. 4 to describe policies for negative and positive carbon pricing.

The idea that climate policies can be represented as biophysical networks is supported by the comments of Currarini et al. (2014), who found that network theory has utility in developing environmental policy:

> Recent research in the field of network economics has shown how explicitly modelling the network structure of social and economic relations can provide significant theoretical insights, as well as account for previously unexplained empirical evidence. Despite their critical importance to many environmental problems, network structures and dynamics have been largely disregarded by the environmental economics literature. (p. 1)

Broadbent and Vaughter (2014) similarly claim that social network analysis (SNA) is a technique that is well suited to the interdisciplinary investigation of

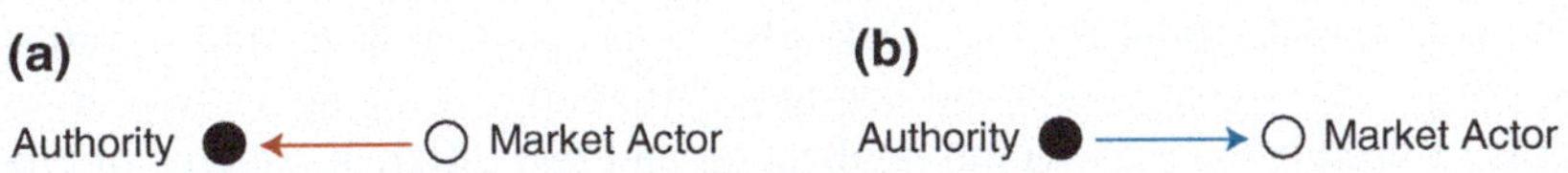

Fig. 2 Elementary nodes and vectors for describing climate policies, such that (**a**) a negative price, such as a carbon tax, results in a financial flow to the authority and (**b**) a positive price, such as a carbon subsidy or reward, results in a financial flow to the market actor

climate policy by integrating ideas from the social and natural sciences. Similar to SNA is actor-network theory (ANT), which considers both human actors and machines as "actants" in the same networks (Law 2009). The thermodynamic analysis of networks was developed independently of SNA and ANT, and according to Oster et al. (1971) "In the [thermodynamic] network approach we 'pull apart' the continuum, revealing the implicit topological relations" (p. 393). They state that network thermodynamics is well defined mathematically but is often neglected, such that "This crucial aspect has been largely neglected in the treatment of thermodynamic systems" (p. 1). The HMH makes use of network theory by referring to climate policies as biophysical networks.

2 Holistic Market Hypothesis

The Holistic Market Hypothesis (HMH) of Chen et al. (2017) is a hypothesis that two externalized costs of carbon emissions are created by carbon emissions—the Social Cost of Carbon (SCC) and the Risk Cost of Carbon (RCC)—and that both costs need to be internalized into the economy to manage the climate problem (see Table 1). The hypothesis is that the SCC associates with the (Type I) carbon tax policy, and that the RCC associates with a (Type II) global carbon reward policy. The hypothesis is that each externalized cost is assessed from a unique frame of reference: the SCC is assessed from a (Type I) neoclassical perspective; and the RCC is assessed from a (Type II) biophysical perspective. Moreover, the Type I and Type II policies, tools, and objectives are complementary because they aggregate benefits and will create a synergy effect of improved social cooperation and economic *resilience* when implemented concurrently. The complementary nature of Type I and Type II policies is described in Sect. 2.3 with a principle called Market Policy Dualism (MPD).

The conceptual model for the HMH requires that the Type I and Type II policies be compared in terms of their common operational objective: to reduce the mass of carbon entering the atmosphere. By restating this operational objective as the reduction of the *entropy* of carbon within the environment, a connection is made between these policies and the Second Law of Thermodynamics. The entropy of carbon increases when it enters the atmosphere as GHGs, because carbon in this gaseous phase has significantly more entropy than the same mass of carbon that is chemically bound in organisms and solids (e.g., in biota, coal, and carbonate rocks), trapped in sedimentary rocks (e.g., in oil and natural gas), and trapped in ice (e.g., in methane clathrates and permafrost).

The conceptual model for the HMH also considers that taxes and rewards for carbon will incentivize goods and services differently, resulting in two different patterns of decarbonization within civilization and two different patterns of energy dissipation within civilization. The conceptual model is that two market prices with opposite sign (Fig. 2) will result in two patterns of energy dissipation that can reduce the entropy of carbon in the environment. This conceptual model includes the radical

Table 1 The Holistic Market Hypothesis (HMH) proposes two externalized costs of carbon emissions and corresponding tools, policies, and objectives

Holistic Market Hypothesis (HMH)	
Social *cooperation* to mitigate carbon emissions is maximized with "carrot" and "stick" carbon pricing, and the economy is more *resilient* with a rebalancing of *good efficiency* with *good inefficiency*[a]. Resilience emerges in a trade-off between the complementary objectives of market *efficiency* and long-term climate *certainty*	
(Type I) Neoclassical perspective	*(Type II) Biophysical perspective*
Negative Externality: The Social Cost of Carbon (SCC) is the time-discounted marginal impact that carbon emissions have on economic welfare. The SCC is a negative externality generated by private producers who do not pay for the spillover damages created by their carbon emissions	Positive Externality: The Risk Cost of Carbon (RCC) is the marginal cost of setting risk tolerances for unwanted global warming and avoiding dangerous tipping points. The RCC is the cost of implementing preventative climate insurance, and this cost is dispersed through the economy via currency trading. The preventative insurance is a positive externality because it is a 'public good' that provides long-term physical benefits that are non-rivalrous and non-excludable
Tool: Carbon tax[b] ("stick")	Tool: Global carbon reward and parallel currency ("carrot")
Policy: The carbon tax is a negative carbon price that is imposed on carbon emissions. Administration of the tax is relatively simple because the tax can be charged on fossil fuels as a carrier of carbon and/or at the points where carbon emissions occur. The tax is a type of fiscal policy and is managed by the government	Policy: The global carbon reward is a positive carbon price that is offered for voluntary reductions in carbon emissions and carbon sequestration. The reward is delivered as a parallel currency, and the administration is relatively demanding because numerous technologies are involved and policing is needed to limit cheating. The reward is a type of monetary policy (an exchange rate mechanism) that is managed by central banks
Objective: The tax price is guided by the SCC and cost-benefit analysis for improving market efficiency (i.e., *good efficiency*). The objective is to maximize economic *welfare*	Objective: The carbon reward price is guided by the RCC and cost-effectiveness analysis for limiting *climate systemic risk* with *good inefficiency*. The long-term objective is to achieve climate *certainty*

Adapted from Chen et al. (2017)
[a]See Sect. 4.3 for an explanation of "good efficiency" and "good inefficiency"
[b]Cap-and-trade schemes also generate a negative price on carbon, but they use different tools and methods to the carbon tax and create a somewhat different set of outcomes in terms of costs and risks

new concept that the dominant social agreements and social responses of the Type I and Type II policies must be time-asymmetric to be consistent with the Second Law of Thermodynamics and the one-directional "arrow of time" (Sect. 1.4).

Chen et al. (2017) further claim that under the HMH "...the SCC and RCC coexist in a cost duality and without paradox" (p. 29). It is argued here that two paradoxes are inadvertently created when the Type I and Type II perspectives are assumed to be incompatible and mutually exclusive. The first paradox is the paradox of agency, which occurs when only one of the two perspectives is believed correct. The first paradox is resolved below. The second paradox is a paradox of time discounting, which is addressed in Sect. 5.1.3.

To resolve the paradox of agency, consider that the (Type I) neoclassical and (Type II) biophysical perspectives differ significantly. The Type I perspective is anthropocentric by assuming that human agency is the primary agency in the economy. The Type II perspective is systemic by assuming that energy dissipation is the primary agency in the economy. The two perspectives appear paradoxical because they both seem plausible but can also result in radically different interpretations. Under the HMH this paradox is resolved by treating the paradox as a misinterpretation: the claim here is that the Type I and Type II perspectives are compatible when human agency is considered a product of energy dissipation. This interpretation of human agency implies that the biomechanics of economic agents is far too complex to be modeled deterministically and that the neoclassical approach of studying human behavior is a pragmatic simplification. Under the HMH, both types of agency—(Type I) human agency and (Type II) energy dissipation—are assumed to operate simultaneously.

The crucial difference between Type I and Type II perspectives is not their source of agency, but rather it is the manner in which these perspectives influence policy choices for managing carbon emissions. The crucial question is this: why do (Type I) neoclassical economists promote policies for a negative price on carbon instead of a positive price? Under the HMH it is posited that Type I economists—with their anthropocentric biases—instinctively seek to maximize economic welfare, and in doing so they have adopted the negative carbon price because the biophysical networks associated with taxes are most suitable for improving market *efficiency*. Under the HMH it is posited that the biophysical networks associated with the global carbon reward—a positive carbon price—are most suitable for improving climate *certainty*. The HMH is self-consistent by providing a unified biophysical model for (Type I) human agency and (Type II) energy dissipation and their associated climate policies (refer Table 1).

2.1 Risk Cost of Carbon

According to Chen et al. (2017), the Risk Cost of Carbon (RCC) is the average market price of voluntary mitigation (USD per t CO_2e) that is sufficient to ensure that a certain level of global warming, ΔT, will not be exceeded within a risk tolerance of R (%). ΔT is defined as a global average surface temperature change (°C) relative to a preindustrial baseline, and ΔT and R are applied over a rolling 100-year planning horizon denoted by the end-year, Y.

A hypothetical example of the RCC is the market price for a global carbon reward that will ensure that the change in average global surface temperature (ΔT) does not exceed 2 °C over the next 100 years, with a risk tolerance of 33% (R). Multiple risk tolerances (ΔT, R) can be defined and addressed simultaneously. For example, four possible (ΔT, R) values are (1.5 °C, 50%), (2 °C, 33%), (3 °C, 15%), and (4 °C, 3%). Actual risk tolerances should be decided in an international forum.

The total cost of carbon (TCC) is a notional measure of the total externalized cost of carbon emissions, and it is defined as the sum of the SCC and RCC over time, as follows:

$$TCC_i(t) = SCC\ (t) + RCC_i(t) \tag{3}$$

where t denotes time in calendar years and subscript i denotes the year of the most recent risk assessment. An alternative to using Eq. (3) is to represent TCC as a two-dimensional vector in a phase space, with the dimensions being the negative externality (SCC) and the positive externality (RCC). The SCC and RCC can have statistical correlation since both values will tend to increase with additional carbon emissions.

The purpose of the rolling 100-year time horizon is to anticipate and avoid unwanted global warming, and it takes into consideration the time lag of warming and the imperfect ability of actors to maintain their service agreements. The 100-year horizon is needed to establish a trade-off between short-term market efficiency and long-term climate certainty (Table 1). The 100-year time horizon is based on findings that (a) 60% of the equilibration surface temperature response occurs 25–50 years after CO_2 emissions (Hansen et al. 2013) and (b) most of the atmospheric CO_2 concentration adjustment occurs 100 years after CO_2 emissions (IPCC 2013). The 100-year time horizon also corresponds to the time standard for the Global Warming Potential (GWP) of greenhouse gases (IPCC 2014c). The application of 100-year agreements has a legal precedent with 99-year leases on property under common law. For example, real estate in the Australian Capital Territory is managed on 99-year leases.[2]

2.2 *Market Policy Dualism*

The HMH rests on a proposed market principle, called *Market Policy Dualism* (MPD), which is inspired by the "carrot and stick" metaphor of combining rewards and penalties to improve social cooperation. Chen et al. (2017) propose Market Policy Dualism (MPD) as follows:

> MPD is a principle that pairs of market policies for environmental mitigation are available based on relationships that are complementary-and-opposite. MPD also includes an implicit assumption that a complementary pair of market-based policies offers benefits, such as rebalancing of social relationships for social feedbacks (e.g., new group dynamics) and policy synergies (e.g., aggregation of price signals). (p. 12)

Evidence that supports MPD is found in a handful of social science experiments (e.g., Andreoni et al. 2003; Hilbe and Sigmund 2010; Chen et al. 2015). Chen et al. (2015) undertook a study of positive and negative incentives based on a public goods game. They found that "...punishment acts as a 'booster stage' that capitalizes on and amplifies the pro-social effects of rewarding..." and that the hybridization of incentives provides "...a surprisingly inexpensive and widely applicable method of promoting cooperation" (p. 1). Andreoni et al. (2003) undertook a study of "carrot and stick" incentives based on a proposer-responder game. They discovered that:

[2]Taylor, G., 2016. Can people own land in the ACT? ABC News. 4 July 2016.

> Thus, while adding rewards only had little effect, adding rewards to punishments has a profound effect. In other words, rewards and punishments seem to act as complements in encouraging proposers to increase their offers. (p. 897)

Andreoni et al. (2003) conclude that (a) rewards alone are relatively ineffective, (b) punishments improve cooperation by eliminating extreme selfishness, and (c) combining rewards and punishments has a "very strong effect" because rewards and punishments can act to complement one another. It also appears that penalties and rewards "…are not merely substitutes in enforcing a fixed objective, but rather that their availability alters the ideals that they enforce" (p. 901). The literature on "carrot and stick" incentives provides solid support for MPD; however a statistical-mechanical model that explains the biophysics of MPD is currently lacking.

2.3 Epistemology of Complementary Relationships

An epistemology is a method or tool that supports the justification of a belief, as opposed to relying on opinions. In this exposition we clarify the epistemology of Chen et al. (2017), which is the epistemology of complementary-and-opposite relationships. The authors define the epistemology as follows:

> The application of MPD begins with the epistemology of defining complementary pairs as two socio-economic relationships that have opposite characteristics and a capacity to aggregate price signals. (p. 12)

The epistemology begins with (Type I) negative pricing and (Type II) positive pricing (refer Table 1). The Type I and Type II policies provide incentives to reduce carbon emissions, but their effects in the marketplace differ because they price carbon in complementary and opposite ways. The epistemology transcends neoclassical economics by considering climate policies as biophysical networks that are designed to reduce the entropy of carbon in the environment (refer Sects. 1.4 and 1.6).

Before attempting to verify the HMH with the epistemology, it is necessary to review the policy for a global carbon reward, including its framework, economic instrument, financial mechanism, and the risk assessment approach for estimating the RCC (see Sect. 3).

3 Global Carbon Reward

3.1 Policy Background

The Holistic Market Hypothesis (HMH) is a market theory that the global carbon reward is inherently suited to the objective of internalizing the Risk Cost of Carbon (RCC) into the economy (Type II in Table 1). Chen et al. (2017) recommend that the global carbon reward be implemented using their *Global 4C Risk Mitigation* policy, which is abbreviated as "Global 4C." Under the Global 4C policy, a parallel currency—which is the reward instrument—is used to internalize the RCC into the economy. The parallel currency is given the generic name *Complementary Currencies for Climate Change* (4C). The Global 4C policy is described below with caveats that the policy has not been validated with models or pilots and the policy may need some adjusting or refining.

3.2 Policy Framework

The Global 4C policy is designed to offer a global carbon reward (Type II in Table 1), whereby the reward payment will be provided with 4C issuance. The reward's financial value will equal the current exchange rate of 4C, and the wealth transferred to market actors is the *seigniorage income* of the 4C as it is issued. The seigniorage income will equal the 4C exchange rate (USD per 100 kg of CO_2e) multiplied by the mass of CO_2e mitigated and less administrative costs that will be deducted as commissions. The Global 4C policy will achieve its macroeconomic objectives by pegging the 4C price to mirror the RCC over time.

The global carbon reward—provided by the 4C parallel currency—will be offered to market actors who voluntarily mitigate carbon, and a wide spectrum of mitigation technologies will be rewarded. The rewards may be weighted to reflect social and ecological co-benefits, and the reward rules will have a scientific basis. The amount of 4C issued will be linked to measurement reporting and verification (MRV) for accountability, and the 4C will be offered conditionally such that actors are required to fulfill service agreements that could last for up to 100 years. Chen (2018a, b) elaborates on how long-lived service agreements can be managed with blockchain ledgers and "smart" digital contracts. A central authority will govern the MRV, but operationally there will be numerous contractors working over the Internet to undertake the MRV. Market actors can be supported with secondary services for information sharing, collaboration, and coinvesting in projects.

Central banks will play a major role in the Global 4C policy, as they will be required to implement unorthodox monetary policies involving quantitative easing (QE) and currency trading—called *carbon quantitative easing* (CQE)—to peg the 4C price to the RCC over the rolling 100-year planning horizon. As is explained in Sect. 3.4, by pegging the 4C price to mirror the RCC, a secular "bull market" in 4C will be invoked in foreign exchange currency markets (the Forex) for the time when the RCC is rising in magnitude. The 4C bull market will attract private purchases of 4C, thereby mobilizing private finance for a high-inertia low-carbon transition.

To maintain public accountability, the total supply of 4C will be maintained in proportion to a global carbon stocktake. The central authority will provide accountability by periodically reconciling defaults on service agreements with 4C chargebacks and 4C demurrage fees. Demurrage fees are equivalent to a negative interest rate charged on 4C holdings. The carbon recorded in the stocktake will *not* be available for trading as carbon offset credits (although 4C will be available for trading); and the retirement of all of the carbon that is recorded in the stocktake will make carbon offsets scarcer and more expensive.

The Global 4C policy will internalize the RCC into the world economy. This internalization process may be described as a deleveraging process for climate risk. The deleveraging of risk will occur as 4C liquidity is provided to low-carbon projects in proportion to their carbon mitigation results. The improved profitability of low-carbon projects will attract private investment to the low-carbon sector of the economy, and it will leverage debt finance for low-carbon projects. Climate risk deleveraging will also occur when market actors respond to the 4C global reward by undertaking long-term R&D programs to develop carbon dioxide removal (CDR) technologies that are scalable and profitable.

3.3 Parallel Currency

As mentioned above, the economic instrument of the Global 4C policy is a parallel currency, called 4C. A justification for adopting a parallel currency is based on the epistemology presented in Sect. 4 and the macroeconomic benefits described in Sect. 5. The chosen unit of account for 4C is smaller than that of the RCC by a factor of 10. The 4C is denominated in 100 kg lots of CO_2e mitigated, whereas the RCC has units of USD per 1000 kg of CO_2e mitigated. A smaller mass was adopted for the 4C unit of account to provide a more convenient exchange rate. The 4C exchange rate and carbon reward price is defined as follows:

$$4\mathrm{C}(t) \cong 0.10 \times \mathrm{RCC}(t) \tag{4}$$

The 4C currency will be a stateless international currency suitable for trading with national currencies in the Forex. Holders of 4C will be subject to domestic laws regarding taxation, peer-to-peer currency trading, banking, and general trade. From an administrative perspective, 4C is a *central bank digital currency* (CBDC) and may be developed using distributed ledger technologies. Special administrative permissions are given to a peak authority for managing the CBDC ledger so that the authority can maintain proportionality between the 4C supply and the carbon stocktake.

Private holders of the 4C parallel currency may trade their 4C holdings with other market actors, and they may exchange their 4C for national currencies, but no amount of carbon will change hands when 4C is traded. This is because the carbon is automatically "retired" and holders of 4C are not granted ownership to the underlying carbon stock. Holders of 4C are only granted ownership of the financial value provided by 4C, and this value is underwritten by central banks. Although 4C is tradable it cannot be used to offset pollution, and with sweet irony 4C will have tangible value because it will help avoid dangerous-to-catastrophic climate change while supporting social and ecological co-benefits.

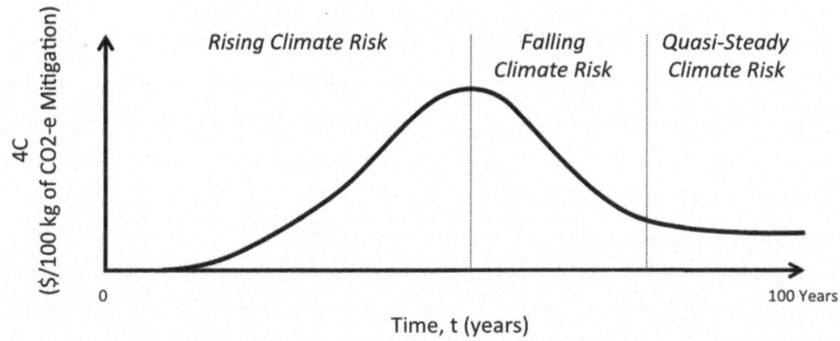

Fig. 3 A hypothetical example of the "100-year advance 4C price alert" that advertises the global carbon reward for climate mitigation. The 4C price mirrors the Risk Cost of Carbon (RCC), and it communicates the level of risk to financial markets (adapted from Chen et al. 2017)

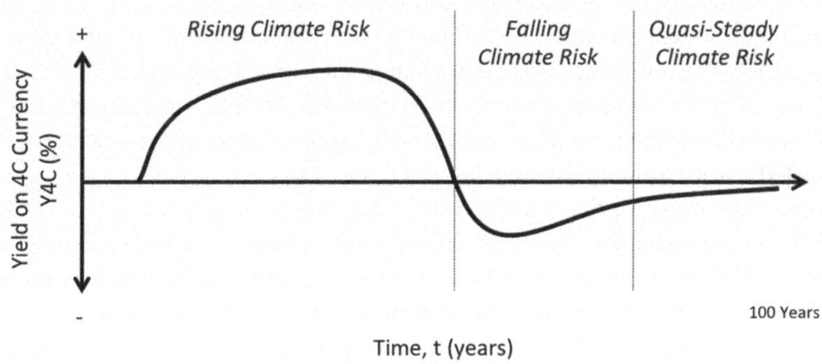

Fig. 4 A hypothetical example of the yield on 4C holdings based on the "100-year advance 4C price alert" shown in Fig. 3. Periods of positive yield represent an opportunity for a risk-free return on 4C holdings, and this is how climate risk is priced into the global financial system

3.4 Financial Mechanism

The financial mechanism of the Global 4C policy is a currency exchange rate mechanism. This mechanism, with its social agreements, may be called the *Carbon Exchange Standard* (CES). The focal point of the CES is the "100-year advance 4C price alert" (see Figs. 3 and 4), which may be described as the 4C floor price. A peak authority will be responsible for administering the CES and guiding central banks when trading 4C so that the 4C floor price is upheld. The 4C price alerts will be published on an annual basis shortly after the annual risk assessments for the RCC are completed, as described in Sect. 3.5.

The "100-year advance 4C price alert" will incentivize private demand for 4C based on the anticipated yield on 4C—Y4C(*t*)—which can be estimated from the annual change in the RCC, as follows:

$$Y4C(t) \cong 100\% \frac{RCC_i(t+1) - RCC_i(t)}{RCC_i(t)} \quad (5)$$

A positive yield on 4C will generate a multi-decade bull market in 4C trading (i.e. promoting 4C saving), and conversely a negative yield will generate a multi-decade bear market in 4C trading (i.e. promoting 4C spending). This exchange rate mechanism has legitimacy based on the estimated RCC. Given that the Forex has high liquidity—with daily currency trading of roughly USD 3 trillion on average[3]—it is assumed that the RCC can be internalized into the global financial system via the Forex. Central banks will act as the "buyers of last resort" whenever private demand for 4C is insufficient to maintain the 4C floor price in the Forex. Central banks will retain their 4C purchases in holding accounts, thereby ensuring that the 4C has the status of both a hard currency and a financial security.

A central authority for the CES will coordinate central bank trading of 4C to target the floor price (refer Eq. 4) and to apportion the financial contributions of each national economy to achieve a uniform marginal change in the exchange rate of each national fiat currency. This will spread the cost of the 4C stimulus as uniform monetary inflation—ceteris paribus. Central banks will employ strategic quantitative easing (QE)—called carbon quantitative easing (CQE)— and currency trading in open markets to negate the need for taxes to fund the policy for a global carbon reward.

The "100-year advance 4C price alert" (see Figs. 3 and 4) simultaneously communicates three important messages: (1) the global reward for mitigating carbon emissions, (2) the 4C investment opportunity with a risk-free yield (e.g., a U.S. treasury bill is considered risk-free), and (3) the long-term climate risk. Over the long run, a period of quasi-steady RCC values and 4C exchange rates may emerge, as suggested in Figs. 3 and 4. A future period of long-term quasi-steady risk is anticipated because there will likely be a long-term need for negative emissions technologies (NETs) to counter residual carbon emissions and to restore the climate with carbon drawdown. The feasibility of NETs is currently uncertain (Fuss et al. 2014; Gasser et al. 2015); however the "100-year advance 4C price alert" will incentivize R&D for new and more effective NETs.

3.5 Risk Assessments

The global carbon reward requires that annual risk assessments be undertaken to quantify the Risk Cost of Carbon (RCC) (refer Sect. 2.1). The RCC should be based on a set of average global temperature rises, ΔT_j, that could occur during the rolling 100-year time horizon as a result of climate sensitivity to anthropogenic GHG emission and possible tipping points (e.g., Lenton 2012). Before the assessments can be undertaken, a political decision is needed to define the risk tolerance, R_j, for

[3]Reuters (March 13, 2017). "Daily FX trade more like $3 trillion than 5: CLS" by Patrick Graham.

each ΔT_j value of concern. For example, (ΔT, R) could be set equal to (2 °C, 33%) over the rolling 100-year time horizon.

The RCC assessments involve three major steps:

(Step 1) Estimating the Systemic Risk of a Climate Mitigation Failure (SRCMF)
(Step 2) Estimating the target mitigation rate, $\Delta Q(t)$, from the SRCMF
(Step 3) Estimating the Risk Cost of Carbon (RCC) from the $\Delta Q(t)$

Step 1 is a risk assessment for estimating the SRCMF$\{\Delta T_j, Y\}$ for each ΔT_j value of concern and over the rolling 100-year planning horizon, denoted by calendar year Y. This should take into consideration the social, political, financial, and biophysical factors that are driving carbon emissions and causing global warming. The SRCMF $\{\Delta T_j, Y\}$ assessments should use adaptable methods that take into consideration quantitative and qualitative knowledge (e.g., Shapiro and Koissi 2015).

Step 2 is a risk assessment for estimating a single target carbon mitigation rate, $\Delta Q(t)$, that can reduce each value of SRCMF$\{\Delta T_j, Y\}$ to below its respective R_j tolerance. A formula that summarizes Steps 1 and 2 is as follows:

$$\Delta Q_i(t) = \text{Function}\ \{\text{SRCMF}\ \{\Delta T_j, Y_i\}, R_j : j = 1, N\} \tag{6}$$

where Function = risk assessment; R = adopted risk tolerance (%); SRCMF = Systemic Risk of a Climate Mitigation Failure (%); N = total number of risk limits (integer); Y = last calendar year of the rolling 100-year planning horizon (year); ΔQ = target mitigation rate (t CO_2e per year); ΔT = global average surface temperature anomaly above a baseline (°C); j = subscript denoting the jth risk limit that is considered concurrently (integer); i = subscript denoting the ith risk assessment (integer); and t = time (year).

Step 3 is a cost-effectiveness analysis that estimates the RCC(t) from $\Delta Q(t)$. This analysis requires the estimation of a *Systemic Risk Abatement Cost Curve* (SRACC) for international markets. The SRACC presents the average cost of abating and sequestering carbon within the context of available technologies and the market's actual capacity and willingness to participate. The SRACC takes into consideration the opportunity costs and the various hidden costs, including the costs of administration, policing, long-term monitoring, free riding, and defaulting. A formula that summarizes Step 3 is presented as follows:

$$\text{RCC}_i(t) = \text{SRACC}_i\{\Delta Q, t\} \tag{7}$$

where RCC = Risk Cost of Carbon (USD per t CO_2e mitigation service); SRACC = Systemic Risk Abatement Cost Curve; ΔQ = target mitigation rate (t CO_2e per year); i = subscript denoting the ith risk assessment (integer); and t = time (year).

4 Analytical Verification

4.1 Premise

The Holistic Market Hypothesis (HMH) of Chen et al. (2017) is a hypothesis that the inherent utility of a global carbon reward is to internalize the Risk Cost of Carbon (RCC) into the economy for the objective of limiting climate risk (refer Table 1). A premise of this hypothesis is that market *inefficiencies* are acceptable as a trade-off for reducing climate systemic risk and avoiding dangerous tipping points. Two key premises are involved in the HMH: (a) market-based climate policies function as biophyiscal networks that dissipate energy and reduce the entropy of carbon in the environment while increasing total entropy, and (b) market-based climate policies that employ positive and negative carbon pricing are time-asymmetric under the Second Law of Thermodynamics. The HMH is verfied by checking for time-asymmetry of the dominant social relationships and dominant social responses of the complementary policies. Market Policy Dualism (MPD) is the inspiration for the epistemology of complementary policy relationships, which is explained in Sect. 4.2. The epistemology is applied in three steps: (Step 1) reversing prices, (Step 2) declaring a currency with biophysical units, and (Step 3) comparing policy objectives based on the "arrow of time." The logic of the three steps is explained below, and the arrow of time refers to monotonic entropy increases under the Second Law (Sect. 1.4). The premise also includes the following five axiomatic statements, which are explained in Sect. 4.3:

Axiom A: Goods and services have embodied energy.
Axiom B: The unit type (unit of account) of money sets a context for the store of value.
Axiom C: Carbon taxes can be used to improve the efficiency of markets.
Axiom D: Actors make risk-reward trade-offs when investing.
Axiom E: Market-based policies that aim to reduce carbon emissions can do so by selectively increasing market *efficiencies* and by selectively increasing market *inefficiencies*.

4.2 Epistemological Translation

The HMH involves a three-step epistemological translation of market policies. The translation is used to derive the policy for a positive carbon price from a policy for a negative carbon price. Perhaps surprising is that market policies have the same dimensionality as money, and this is because they have similar functions (see Table 2). William Jevons (1875) famously defined money as having four functions, and his ideas inspired Milnes (1919) to write the following couplet: "Money's a matter of functions four, a Medium, a Measure, a Standard, a Store" (p. 55). The functions of market policies and money are similar, but only certain types of market policy require a new currency instrument. The global carbon reward is one such

Table 2 The three-step epistemological translation for carbon pricing is compared with the four functions of money to achieve dimensional completeness

Carbon pricing				Money		
Translation[a]	Carbon tax[b]	Carbon subsidy	Carbon reward[c]	Jevons (1875)	Classical	Biophysical
Step 2. Currency	National fiat currency	National fiat currency	Parallel currency	Medium	Medium of exchange	Medium of exchange
Step 2. Units	USD	USD	100 kg of CO_2e mitigation	Measure	Unit of account	Unit of account
Step 3. Arrow of time	Policy of improving market efficiency using carbon taxes guided by the SCC	Policy of reducing carbon emissions	Policy of limiting systemic risk with carbon rewards guided by the RCC	Standard	Social agreement	Entropic relationship
Step 1. Price reversal	Negative price (USD per t CO_2e pollution)	Positive price (USD per t CO_2e mitigation)	Positive price (USD per t CO_2e mitigation)	Store	Store of value	Power (J/s)[d]

[a]See Sects. 4.4, 4.5, and 4.6 for the epistemological translations
[b]Carbon taxes have a negative store of value and are not themselves a currency
[c]The results of the epistemological translation, after Chen et al. (2017)
[d]Garrett (2012); see Eq. (2)

policy, because it declares a parallel currency denominated in carbon. Cap-and-trade schemes declare permits and carbon offset credits for trading in carbon markets, but these are not currencies per se because their social agreements are only designed to address specific markets.

Jevons (1875) discussed the "standard of value" of money in terms of borrowing and lending, but in recent years the term has been replaced with "social agreement," which is a more general description of this function (see Table 2). For example, Eisenstein (2011) writes: "Money is merely a social agreement, a story that assigns meanings and roles" (p. 108).

Table 2 compares Jevons (1875) four functions with a classic description and a biophysical interpretation of these same functions. The "store of value" corresponds to power (J/s) under the biophysical description of money, which is inspired by Garrett's (2012) economic model [refer Eqs. (1) and (2)]. The "social agreement" corresponds to the biophysical function called the "entropic relationship," which is introduced here with Table 2. This biophysical interpretation of the social agreement refers to the influence that social agreements have on future patterns of production, consumption, energy dissipation, and high-entropy waste, such as CO_2 emissions. In Sect. 1.6 it was suggested that market-based climate policies are biophysical networks through which energy dissipation can reduce the entropy of carbon. Step 3 of the translation is the reversal of the social agreement for the carbon tax to achieve a policy objective for the carbon reward: to be consistent with the "arrow of time" under the Second Law (see Table 2).

4.3 Axioms

Axiom A is a statement that goods and services contain "embodied" energy because energy is always needed to produce goods and services. A corollary to Axiom A is that money is indirectly associated with embodied energy, because the value of money grants access to goods and services.

Axiom B is a statement that the unit type (unit of account) of money is fundamentally important for setting the context of value in the economy. The unit type of money must be an *extensive* property, and in the economy there are three principal options: (1) commodities (e.g., 1 oz. of silver), (2) socially or legally declared information (e.g., USD), and (3) services (e.g., 1 hour of education). A fourth option is to adopt a basket of units, but this is not a principal unit type.

A corollary to Axiom B is that service money—money with units defined by a service—is useful for offering rewards to incentivize positive externalities. The corollary is that monetary policy can be used to incentivize the supply of services, assuming that an administrative system is available to couple the currency supply to the observed supply of service. An example is Solarcoin,[4] which is a cryptographic token and reward with units of 1 MWh of solar-derived electricity.

Axiom C refers to an 0established principle in economics that when a carbon tax (a Pigovian tax) is guided by cost-benefit analysis, the tax improves the efficiency of the market by increasing the marginal private cost of production so that the carbon emissions are reduced sufficiently to achieve a social welfare optimum. The optimum in consumption occurs when the marginal social cost equals the marginal social benefit. This kind of efficiency is neoclassical and is described here as "good efficiency."

Axiom D refers to an established principle in economics that when market actors make investments with imperfect knowledge, they invest according to a risk-versus-reward trade-off (or risk-return trade-off). Consequently, market actors take greater risks when they anticipate greater rewards/returns.

Axiom E is a statement that there exist two ethics for reducing carbon emissions. One ethic is to create (a) *good efficiency* (refer Axiom C), and the other ethic is to create (b) *good inefficiency*. The term "good efficiency" is introduced to describe the neoclassical ambition of improving market efficiency to maximize social welfare: the point where the private cost of production plus externalities equals the social benefit. The term "good inefficiency" is introduced to describe the diversion of capital and resources to reduce carbon emissions for the objective of limiting systemic risk and providing climate certainty. This "good inefficiency" is the result of a tradeoff between market efficiency and climate certainty, and it may result in an

[4]https://solarcoin.org

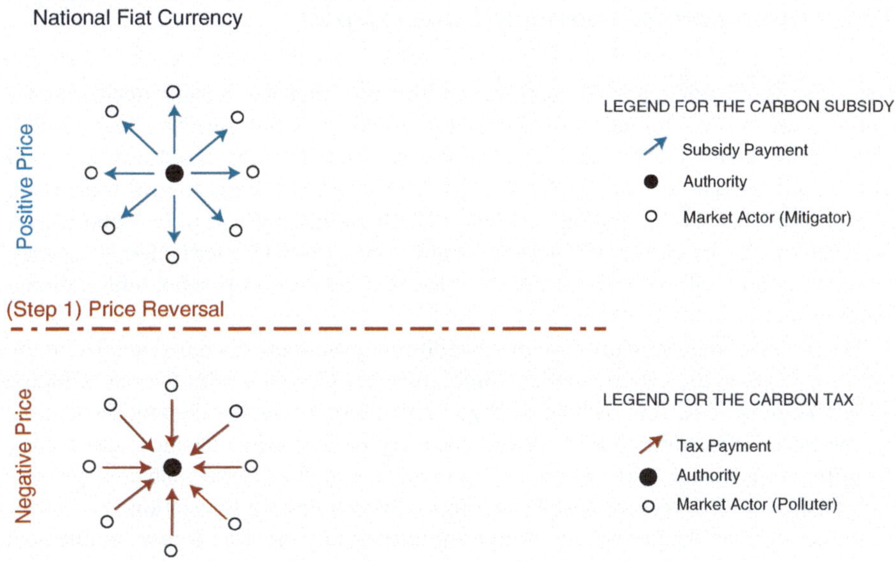

Fig. 5 Climate policies represented as network diagrams: (bottom) the carbon tax and a negative price on carbon emissions and (top) the carbon subsidy and a positive price on carbon mitigation. The horizontal dotted-dashed line denotes the translation for price reversal

increase in other measures of well-being that take into account socio-ecological regeneration and sustainability.

4.4 Translation for Price Reversal (Step 1)

The first epistemological translation compares a negative carbon price with a positive carbon price, as illustrated in Fig. 5. The dotted-dashed horizontal line in Fig. 5 denotes this translation. The translation compares a tax for carbon emissions with a generic subsidy for carbon mitigation. The translation identifies *complementary-and-opposite* pricing and financial flows. The carbon tax and the carbon subsidy are complementary because they both incentivize a reduction in carbon emissions. Tax payments (USD) denoted in Fig. 5 are calculated from the negative price (USD per tonne of CO_2e emissions) multiplied by the mass of carbon emitted, whereas subsidy payments (USD) are calculated from the positive price (USD per tonne of CO_2e mitigation) multiplied by the mass of avoided/sequestered emissions. The avoided emissions are calculated as the difference between a theoretical baseline of emissions (i.e., a rule-based measure of emissions for a single market actor or for an entire market) and the actual carbon emissions over time. In both policies a national fiat currency is used as the medium of exchange for financial payments.

4.5 *Translation for Currency Units (Step 2)*

The term “fiat” refers to any currency with units that are legally declared—ex nihilo—and every national currency in use today is a fiat currency (e.g., USD, EUR, YEN, GBP, CNY, etc.). The carbon tax and the carbon subsidy are paid with a national fiat currency (refer Fig. 5). The second epistemological translation takes the units of the mitigation subsidy—which are biophysical units—and uses it to declare a parallel currency. These units are “100 kg of CO_2e mitigated,” and they become the units of the new currency, which is to be used in parallel with national fiat currencies.

The financial mechanism of the parallel currency was described in Sect. 3, and the currency is given the generic name: Complementary Currencies for Climate Change (4C). The units of 4C (i.e., 100 kg of CO_2e mitigated) are smaller than the RCC units by one order of magnitude as a convenience. By issuing 4C as incentive payments, the policy is termed a “carbon reward,” as distinct from the “carbon subsidy” prior to the translation. The vertical dashed line in Fig. 6 denotes the translation.

Introducing 4C has novel economic implications by creating a new biophysical context for money (Axiom B). Markets and institutions for the parallel currency are represented symbolically in Fig. 6, and these include (a) the mitigation market where actors receive the parallel currency as a reward for mitigating carbon emissions,

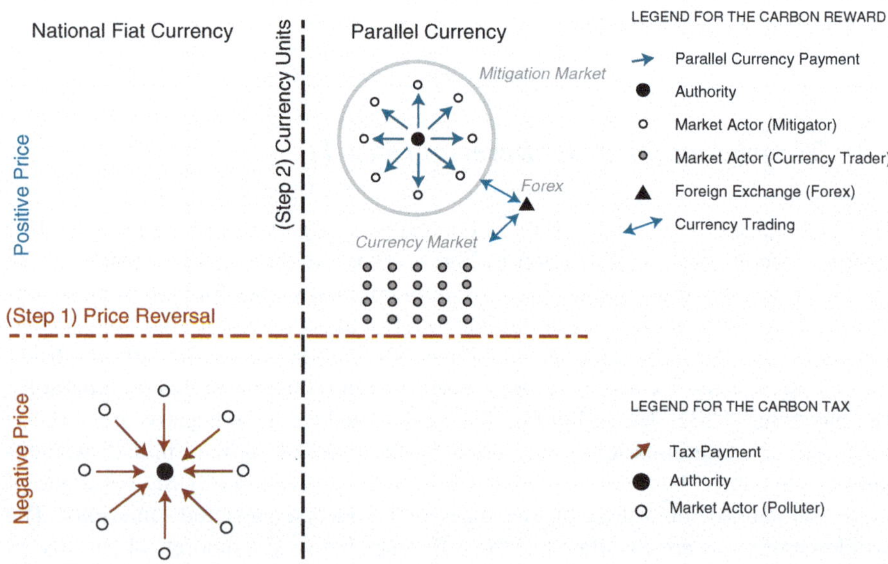

Fig. 6 Climate policies represented as network diagrams: (bottom left) the carbon tax with a negative price on carbon emissions and (right top) the global carbon reward with a positive carbon price on carbon mitigation and using a parallel currency that is traded in the Forex. The vertical dashed line denotes the translation for currency units

(b) the currency market where actors can trade currencies, and (c) the foreign exchange for currencies (Forex) where currency trading is recorded on a ledger.

By trading the 4C parallel currency in the Forex, 4C becomes available in all national economies, thereby establishing 4C as a "global carbon reward." 4C needs to be managed with an official exchange rate mechanism otherwise the price of 4C will likely be too low and volatile. The exchange rate mechanism for 4C is related to the policy's objective or social agreement, and this topic is addressed with the third translation.

4.6 Translation for the Arrow of Time (Step 3)

The third epistemological translation makes a comparison of temporal relationships for negative and positive carbon pricing. The epistemology is used to determine if these policies are complementary-and-opposite based on the "arrow of time." The arrow of time, or time-asymmetry, is coupled to entropy change (refer Law 2 in Sect. 1.4). In this third epistemological translation, there is an expectation that negative and positive carbon pricing will both reduce/limit the mass of carbon entering the atmosphere and consequently both policies should also reduce/limit the *entropy* of carbon in the environment. The translation involves an a priori assumption that the

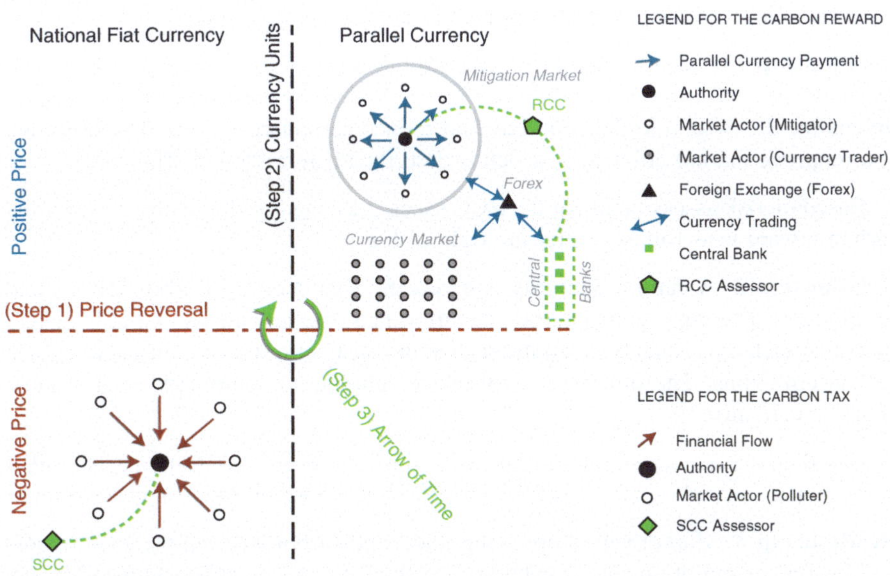

Fig. 7 Climate policies represented as network diagrams: (bottom left) the carbon tax and objectives set by the Social Cost of Carbon (SCC) and (right top) the global carbon reward and objectives set by the Risk Cost of Carbon (RCC). The reward is issued as a parallel currency and managed with an exchange rate mechanism. The circular arrow denotes the translation for the "arrow of time"

global carbon reward can internalize the Risk Cost of Carbon (RCC) (Eqs. 6 and 7). The epistemological translation is a check on this assumption, and the appearance of time-asymmetry in the resulting economic relationships is the anticipated result that will support the hypothesis that the RCC is the second externalized cost of carbon emissions.

The third translation is symbolized in Fig. 7 as a green circular arrow to remind us that time only moves forward with entropic change. The other symbols in Fig. 7 denote the financial mechanisms and social agreements, as explained in the figure's legends. An authority is needed to set the price of the carbon tax to approximate the SCC, thereby setting the objective of the tax policy (bottom left Fig. 7). The a priori assumption is that an authority is needed to set the price of the global carbon reward to approximate the RCC, thereby setting the objective of the reward policy (top right Fig. 7). RCC values are communicated to central banks that are responsible for ensuring that the exchange rate of the parallel currency (4C) matches the RCC over time (refer Figs. 3 and 7, Sect. 3.4). The exchange rate mechanism and the 4C floor price (refer Fig. 3) are major components of the policy's social agreement, which is the "entropic relationship" of the policy (refer Table 2).

To complete the third translation, the following four time-dependent relationships for negative and positive carbon pricing are described: (1) the top-down objective of the carbon tax, (2) the top-down objective of the carbon reward, (3) the bottom-up response to the carbon tax, and (4) the bottom-up response to the carbon reward.

Based on the standard model for the SCC, the top-down objective of the carbon tax is as follows (refer bottom left of Fig. 7 and Axioms C and E):

Relationship 1 A central authority assesses the Social Cost of Carbon (SCC) and responds by imposing a carbon tax that is guided by cost-benefit analysis: to internalize the SCC into the economy and reduce carbon emissions. The objective is to improve market efficiency, which is described here as "good efficiency."

Based on risk assessments for the RCC (Sect. 3.5), the top-down objective for the carbon reward is as follows (refer top right of Fig. 7):

Relationship 2 A central authority assesses the Risk Cost of Carbon (RCC) and responds by offering a global carbon reward (via a parallel currency) that is guided by cost-effectiveness analysis: to internalize the RCC into the economy and reduce carbon emissions. The objective is to reduce "climate systemic risk" and achieve "climate certainty."

Based on the standard model for the SCC, the bottom-up[5] risk-response of market actors under the carbon tax is as follows (refer bottom left of Fig. 7 and Axiom D):

Relationship 3 Market actors operating under the carbon tax can perceive opportunities for consumption and investment, and their response will be to *take risks* in a risk-reward trade-off. Their decisions will be individualistic, and they may respond

[5]The bottom-up responses of market actors may also be called micro-foundational responses.

to the carbon tax by reducing their consumption and production of carbon-intensive goods and services. The aggregate of their responses will improve market efficiency—"good efficency"—but these actions may bear little or no relationship to the "climate systemic risk."

Based on setting the 4C exchange rate to mirror the RCC (Eqs. 4 and 5), the bottom-up risk-response of market actors will be as follows (refer top right of Fig. 7 and Axioms C and E):

Relationship 4 Market actors operating under the carbon reward can perceive opportunities for consumption and investment, and their response will be to take risks in a risk-reward trade-off. Their decisions will be individualistic, and they may respond to the global carbon reward by mitigating carbon emissions to earn 4C, but they may also make *risk-free* investments in 4C according to the advertised yield for 4C. 4C prices will respond to the "climate systemic risk," and the aggregate of the resulting mitigation actions and risk-free investing in 4C will create "good inefficiency" in the market economy.

The next step of the analytical verification is to check the above four relationships for time-asymmetry under the Second Law. This is the preferred verification approach because it is likely the simplest approach.

4.7 Comparative Check for Time Asymmetry

Verification of the RCC as the second externalized cost of carbon is based on the following check of time-asymmetry (the arrow of time) for the above four relationships. By comparing Relationship 1 with Relationship 4, it appears that these two relationships have time-asymmetry because the top-down "good efficiency" objective of the carbon tax is *opposite* to the bottom-up "good inefficiency" response of market actors under the carbon reward. The "good inefficiency" is a result of the risk-free investing in 4C and the way that market actors will respond to the climate risk even if they are unaware of the actual biophysical implications of the 4C price.

By comparing Relationship 2 with Relationship 3, it appears that these relationships are time-asymmetric because the top-down "risk-reducing" objective of the carbon reward is *opposite* to the bottom-up "risk-taking" response of individual market actors under the carbon tax.

By comparing Relationship 1 with Relationship 2, it appears that the two policies have *complementary* objectives, with the carbon tax improving market efficiency and the carbon reward improving climate certainty (i.e., by reducing climate systemic risk).

The policy comparisons are summarized in Table 3. The dominant social agreements and dominant social behaviors that express time-asymmetry are those for efficiency vs. inefficiency (see Relationships 1 & 4) and taking risk vs. reducing risk (see Relationships 3 & 2). These results lend credibility to the interpretation that the

Table 3 Summary of the epistemological translation that compares the (Type I) carbon tax with the (Type II) global carbon reward (after Chen et al. 2017)

	Holistic Market Hypothesis (HMH)	
Translation	(Type I) Carbon tax	(Type II) Global carbon reward
Step 1. Price reversal	The authority for taxation establishes a *negative* carbon price	The authority for rewards establishes a *positive* carbon price
Step 2. Currency units	A national fiat currency provides the store of value and the medium of exchange. A tax instrument provides the *unit of account* and the social agreement for carbon emissions	A parallel currency provides the store of value, the medium of exchange, and the *unit of account*. A currency exchange rate mechanism and service contracts provide the social agreement for carbon mitigation
Step 3. Arrow of time (efficiency vs. inefficiency)	Relationship 1. The policy objective is to promote *good efficiency* in the marketplace by internalizing the SCC into the economy with carbon taxes	Relationship 4. Market actors respond to the 4C price by mitigating carbon and making risk-free investments in 4C. This promotes *good inefficiency* in the marketplace
Step 3. Arrow of time (taking vs. reducing risk)	Relationship 3. Market actors respond to the carbon tax and investment opportunities by *taking risks* under the risk-reward trade-off principle for individuals	Relationship 2. The policy objective is to *reduce climate systemic risk* by internalizing the RCC into the economy with a global carbon reward

(Type I) negative carbon price, and the (Type II) positive carbon price have been correctly assigned time-asymmetric objectives for reducing the entropy of carbon in the environment. The results in Table 3 are further supported with a policy resolution to the temporal paradox of time discounting under systemically risky conditions (refer Sect. 5.1.3) and a possible resolution to the intractible problems of unsustainable GWP growth and unmanaged population growth (refer Sects. 5.2.2 and 5.2.3).

5 Discussion

The Holistic Market Hypothesis (HMH) could have major implications for climate change economics if it is found to be cogent and reliable. The following discussion is offered as a preliminary check on the hypothesis, including an interdisciplinary interpretation of the conceptual model, a proposal for experimental testing, and a solution to the temporal paradox of time discounting under systemic risk (Sect. 5.1). The utility of the global carbon reward is then discussed in terms of its compatibility with the 2015 Paris Agreement and in terms of achieving "net zero" emissions and managing economic growth (Sect. 5.2). The ethics of the new paradigm for externalized costs and "carrot and stick" policies are discussed in Sect. 5.3.

5.1 *Theoretical Cogency*

5.1.1 Interdisciplinary Interpretation

Entropy is an extensive physical property that describes the "disorder" of matter that is the result of energy dissipation. In terms of living organisms and civilization, increasing entropy is often associated with decay, and locally reducing entropy is often associated with life and homeostasis. Stabilization of the climate will require limiting the concentration of GHGs in the atmosphere, and this implies that greater control over the entropy of carbon in the environment is needed. The HMH takes an interdisciplinary approach by proposing a causal link between carbon pricing, energy dissipation, and the changing entropy of carbon. This approach is unusual because few scholars have framed climate change economics in terms of entropy (e.g., Garrett 2012; Guy 2015). The standard narrative on climate change economics is founded on neoclassical assumptions (e.g., Stern 2007), and consequently most attention is given to neoclassical policies such as the carbon tax and cap-and-trade. The global carbon reward policy is unorthodox, however it addresses many common policy issues related to risk and uncertainty associated with climate change (IPCC 2014a).

A reasonable question regarding the HMH is this: given that standard market-based policies and regulatory approaches have successfully limited other types of pollution, such as acid rain (Chan et al. 2012), why should carbon be deserving of special treatment? Under the HMH, carbon is assumed to be a special case because it has exceptional chemical functionality compared with other elements. This functionality includes (a) its *tetravalence* and ability to form a wide diversity of organic molecules, (b) its capacity to store and release chemical energy via long-chain organic molecules, (c) its abundance on the Earth's surface, and (d) its role in storing and replicating large amounts of biological information, such as in DNA. Under the HMH, civilization is interpreted to behave like a heat engine that is currently structured to consume fossil fuels and biofuels as dominant energy sources. This heat engine expresses the two kinds of agency mentioned in Sect. 2: (Type I) human agency and (Type II) energy dissipation. The agency of energy dissipation is overlooked in neoclassical models and policies, which tend to focus on the optimisation of economic welfare. The HMH, on the other hand, points to a need to price climate systemic risk into the financial system to effectively manage climate mitigation in response to dynamic system feedbacks and tipping points of various kinds (e.g. Steffen et al. 2018).

The HMH invites policy makers to reassess climate policies in a thermodynamic context. This may begin with the view that social agreements can influence future energy dissipation and entropy (Table 2). The HMH posits that all climate policies (taxes, fee-and-dividend, cap-and-trade, carbon offset trading, subsidies, etc.) are designed to reduce/limit the entropy of carbon with varying levels of stringency. Under the HMH, the carbon tax and the global carbon reward are complementary policies because they are time-asymmetric and are likely to be optimal for improving market efficiency and climate certainty, respectively (see Table 3).

5.1.2 Experimental Testing

Attempts should be made to experimentally validate the HMH. A variety of experiments and pilot projects should be developed to detect statistical differences in (a) social cooperation, (b) market efficiency, (c) climate certainty, and (d) long-term economic resilience. These tests should apply negative and positive pricing at the decision points of carbon emissions and carbon mitigation, respectively. Experiments should be designed to take into account a variety of factors, including climate sensitivity, climate and social tipping points, economic growth, primary energy supply/demand, energy return on energy invested (EROEI), innovation rates, information sharing, and market sentiment.

5.1.3 Resolution of the Temporal Paradox

Mark Carney (2015, 2016) made obvious the *Tragedy of the Horizon*, which refers to a temporal paradox created by society's inability to address the temporal dilemma of climate change. The dilemma is that society is weakly motivated to avoid climate damages that will occur in the distant future and by the time the damages materialize it will be too late to mitigate them. The 4C exchange rate mechanism can address the dilemma, because the yield on 4C (i.e., the time derivative of the RCC) will generate a secular 4C bull market and preemptive transfers of financial capital into low-carbon sectors of the economy (Figs. 3 and 4). Chen (2018b) surmises that the 4C mechanism will "…convert tomorrow's climate risk into today's profits" to partially resolve the Tragedy of the Horizon. The temporal paradox is further examined in the following discussion of the time discounting of consumption and investments.

Time Discounting of Consumption

According to Nordhaus (2007a), Weitzman (2010), Dietz et al. (2018), and others, a conundrum has emerged concerning the time discounting of the SCC for cost-benefit analysis. The controversy involves the Ramsey formula (after Ramsey 1928) for estimating the social time discount rate, r, for consumption over time:

$$r = \rho + \eta g \tag{8}$$

where r = social time discount rate (%); ρ = rate of pure time preference (%); η = slope of the marginal utility function for consumption (–); and g = average growth rate of per capita consumption (%).

The magnitude of r strongly impacts the time discounting of the SCC which determines the ideal carbon tax. A higher r may satisfy the market's desire for immediate or early consumption, but it also increases the risk of worsening climate change. The paradox is that high r values (more often descriptive) and low r values (more often prescriptive) may be proposed by stakeholders. A related question is this: if a lower r provides a hedge against climate risk, then does a reduction in r constitute a risk premium? Here we argue that reducing r is only a pseudo risk premium, because under the Tinbergen Rule, the number of policy objectives should be matched by the same number of policy tools (Tinbergen 1952). If the objective of the carbon tax is to achieve market efficiency and a welfare maximum, then an additional tool is evidently needed to address the second objective, which is to limit the climate systemic risk (Tables 1 and 2).

A sad irony of manipulating the Ramsey formula to manage climate risk is that the approach will require international consensus on time discounting when economists, politicians, and society have diverse and sometimes contradictory views on the discount rate and taxation. The RCC metric provides an escape from the time discounting quagmire, by allowing policy makers to price risk in global financial markets using the 4C parallel currency. The 4C price and yield—4C(t) and Y4C(t)—are both measures of the climate systemic risk, and these metrics respond reflexively to the climate systemic risk (Sect. 3.5). 4C creates an independent price channel that is suitable for risk communication because 4C is immediately beneficial to market actors. Macroprudential regulation of climate change by central banks may also provide a politically stable method of managing climate risk.

Time Discounting of Investments

A formula for the time discounting of investments is the consumption based capital asset pricing model (CCAPM), after Lucas (1978):

$$R = R_{\mathrm{f}} + \beta\,\pi \tag{9}$$

where R = investment-specific expected return adjusted for risk (%); R_{f} = risk-free rate in financial markets (%); β = elasticity of climate damages with respect to changes in aggregate consumption (–); and π = systematic or market risk premium (%).

The CCAPM formula can be used to value low-carbon projects and provide a theoretical understanding of investors' risk aversion in relation to wealth, consumption, and climate change. For example, a low-carbon project with a β of 0 or 2 may be discounted at 1.6% or 11.2%, respectively, based on a π of 4.8% and an R_{f} of 1.6% (Dietz et al. 2018). The important question is whether a high insurance premium (i.e., low discounting) is justified with Eq. (9) when the investment provides climate mitigation benefits that are correlated to consumption. A difficulty of using CCAPM is that the so-called insurance premium depends on whether society will be richer or poorer in the future and knowing that higher global temperatures positively correlate with being richer. Nordhaus (2007b) acknowledges that "This leads to the paradoxical result that there is actually a negative risk premium on high climate-change outcomes" (p. 113).

Another feature of applying CCAPM, and the traditional capital asset pricing model (CAPM), is that investors have diverse views on climate change and risk, and forcing climate-based standards for β may prove difficult. The RCC metric can bypass these challenges by establishing a risk-free yield on 4C holdings—Y4C(t)—in foreign exchange currency markets. 4C holdings will have low sovereign and exchange rate risks, and so 4C trading can shift the risk-free rate of return, R_f, to a new level based on an internationally agreed tolerances for climate risk (Sect. 3.5). A key benefit of introducing 4C is that the climate systemic risk is automatically factored into investment decisions based on 4C trading and investing.

5.2 *Practical Applications*

5.2.1 The Paris Climate Agreement

The RCC is directly relevant to the Paris Climate Agreement (UNFCCC 2015) because it can be used to set a global risk management objective. Article 2 defines the ambition of staying well below 2.0 °C of global warming and to pursue 1.5 °C and to "...significantly reduce the risks and impacts of climate change." The Paris Agreement lacks a mechanism for setting and enforcing risk tolerances, and so no specific level of climate stabilization is actually guaranteed by the agreement. The global carbon reward can address this shortfall, but this will require a new road map for negotiating risk tolerances for climate mitigation and a concrete plan to implement the 4C exchange rate mechanism. A political agreement will be needed that is either an adjunct to the Paris Agreement or a separate agreement. Regardless of the details, the new policy for enforcing risk tolerances should not result in new direct taxes for citizens and businesses because the effectiveness of the new policy is strongly dependent on offering rewards and not penalties.

A peak authority will be relied upon as the macroprudential regulator of the 4C exchange rate mechanism (Sect. 3.4). This mechanism relates to Article 6 of the Paris Agreement by providing financial rewards anchored in market-based solutions and voluntary actions. The 4C exchange rate mechanism relates to Article 8, by "...reducing the risk of loss and damage..." and by offering "Comprehensive risk assessment and management...." Sections 4 (a), (e), and (f) of Article 8 are most relevant, as follows:

(a) The "100-year advance 4C price alert" can address "early warning systems" (Figs. 3 and 4).
(e) The RCC can address "comprehensive risk assessment and management" (Sect. 3.4).
(f) The 4C exchange rate mechanism can address "risk insurance facilities, climate risk pooling and other insurance solutions" (Sect. 3).

5.2.2 Achieving Net Zero Emissions

Limiting climate risk will require that gross world product (GWP) be decoupled from carbon emissions and that carbon emissions reduce to "net zero" in alignment with an agreed carbon budget (IPCC 2014b) that can avoid critical tipping points (e.g., Lenton 2012). To illustrate a practical application of the RCC and the 4C exchange rate mechanism, the problem of delivering net zero emissions is considered here when the base case is defined by conventional policies. The solution is an extension of Garrett's (2012) empirical relationship for global CO_2 emissions, which assumes that a strong coupling exists between cumulative GWP and primary energy usage (refer Eqs. 1 and 2). An extended relationship is presented below based on the assumption that implementing 4C represents a *structural* change to the economy: effectively establishing a parallel economy that reduces the entropy of carbon in the environment.

When the 4C currency/reward is issued in the marketplace, the supply of 4C is proportional to ΔQ, which is the annual CO_2 mitigation rate that qualifies for rewards. The annual supply of the 4C currency, ΔM, is determined from ΔQ and the unit of account, as follows:

$$\Delta M(t) = \frac{\Delta Q(t)}{100 \text{ kg } CO_2} \tag{10}$$

The revised total mass of CO_2 emissions can be estimated with Eq. (1) by including GWP denominated in 4C (GWP_{4C}) and subtracting ΔQ multiplied by ω, where ω is the fraction of mitigation that is the result of the structural change to the economy for the absolute decoupling of GWP from carbon emissions. If structural decoupling is hightly effective, then ω approaches unity, but if structural decoupling is nearly impossible, then ω approaches zero. The revised mass of CO_2 emissions is as follows:

$$E(t) \cong c(t)\,\lambda \left\{ \sum_{i=1}^{t} \text{GWP}(i) + \sum_{i=k}^{t} \text{GWP}_{4C}(i) \right\} - \omega(t)\,\Delta Q(t) \tag{11}$$

where E = total mass of CO_2 emissions per year; c = average CO_2 emissions intensity of energy; λ = average power consumption per unit of currency as a time-invariant parameter; GWP = inflation-adjusted gross world product for national economies; GWP_{4C} = GWP for the 4C parallel economy; ΔQ = mitigation rate that qualifies for rewards; ω = fraction of mitigation that is decoupled from GWP; t = current time in years; i = time step in years; and k = time step when 4C is introduced.

A point of technical clarification is that currency trading does not register in GWP and the issuance of 4C rewards does not register in GWP_{4C}. If the 4C parallel currency is used to buy goods and services, then that trade with 4C will register in GWP_{4C}, but if 4C is not used to buy goods and services and is only used to trade currencies, then GWP_{4C} reduces to zero. The default policy is that 4C is only used to trade currencies, and so GWP_{4C} is dropped, and Eq. (11) is rearranged to yield the following expression for ΔQ_{zero}, which is the quantity of mitigation rewarded by 4C that achieves net zero CO_2 emissions:

$$\Delta Q_{zero}(t) \cong \frac{c(t)\,\lambda}{\omega(t)} \sum_{i=1}^{t} GWP(i) \tag{12}$$

The required rate of additional mitigation, ΔQ_{zero}, is input into the Systemic Risk Abatement Cost Curve (SRACC) (Eqs. 4 and 7) to determine the 4C price that can achieve net zero emissions:

$$4C_{zero}(t) = 0.10 \times SRACC\{\Delta Q_{zero}, t\} \tag{13}$$

Equations (10, 12, and 13) describe the 4C exchange rate mechanism for delivering net zero CO_2 emissions, and they also illustrate the macroeconomic approach of the global carbon reward. A negative feedback on dirty growth is established because the 4C price is a function of GWP (see Eqs. 12 and 13). Based on this negative feeback between the 4C price and GWP, and an assumed decoupling of mitigation from GWP (i.e. $\omega > 0$), it appears plausible that the 4C exchange rate mechanism could provide a long-term solution to Jevons Paradox (e.g. Brookes 1990) and the Khazzoom-Brookes postulate (e.g. Saunders 1992) in relation to carbon abatement with improved energy efficiency. Additional research is needed to quantify a 4C price schedule that could conceivably achieve net zero emissions.

5.2.3 Managing Global Growth

The 4C reward/currency can be used to manage various aspects of the economy besides the average emissions intensity of energy (refer Eqs. 11 and 12). The policy opportunity is conceptualized here with a modified version of the Kaya identity, as follows:

$$E(t) \cong \text{Population}(t) \times \text{GWP per capita}\ (t) \times CO_2\ \text{intensity of}\ GWP(t) - \omega(t)\,\Delta Q(t) \tag{14}$$

where E = total mass of CO_2 emissions per year (t CO_2 per year); GWP = gross world product (USD per year); ω = fraction of additional carbon mitigation that is decoupled from GWP (–); ΔQ = additional carbon mitigation that is incentivized by the global carbon reward (t CO_2 per year); and t = time (year).

A major advantage of the 4C reward/currency is that it can be used to influence all three Kaya variables to create a negative feedback on carbon-intensive GWP, and it can also provide additional mitigation, $\omega(t)\ \Delta Q(t)$, that is external to the Kaya identity (compare Eqs. 11 and 14). Importantly, the 4C reward/currency could be used to incentivise government agencies to reduce human populations to below statistically determined baselines via socially responsible and cuturally sensitive methods.

Another important feature of the 4C reward/currency is that it can be used to manage average GWP per capita—and GWP growth—in a macroeconomic trade-off between climate certainty and market efficiency. This trade-off, which may be termed the *inefficiency-reward trade-off* after Chen et al. (2017), is based on predictable 4C exchange rates and risk-free investing, and it represents a possible new model for managed degrowth (van den Bergh and Kallis 2012). Returns or yields on 4C holdings/investments are risk free because they are underwritten by central banks [refer Eq. (5), Fig. 4, and Axiom E]. The 4C yield is factored into all investment decisions when the yield influences the official 'risk-free rate of return' in the global financial system (refer Sect. 5.1.3). The resulting increase in 4C holdings will place a negative feedback on spending and GWP per capita. The negative feedback is established because the 4C price is a function of GWP, as defined by Eqs. (12) and (13). The negative feedback on dirty GWP also creates an opportunity to improve economic welfare based on other indices. For example, the 4C reward rules for climate mitigation can be weighted to enhance social and ecological co-benefits that are reflected in the Inclusive Wealth Index (IWI), the Genuine Progress Indicator (GPI), and the U.N.'s Sustainable Development Goals (SGDs). These policy co-benefits give tangible meaning to the term "good inefficiency".

With respect to the first two Kaya variables (Eq. 14), the 4C reward/currency can be used to:

1. Incentivize governments to improve women's education and family planning
2. Incentivize market actors to save 4C, thereby influencing average GWP per capita

With respect to the third Kaya variable and the additional carbon mitigation that is decoupled from GWP (Eq. 14), the 4C reward/currency can be used to:

3. Incentivize reductions in the average CO_2 intensity of GWP
4. Incentivize large-scale CO_2 sequestration using negative emissions technologies (NETs)

5.3 *Philosophy and Ethics*

The Holistic Market Hypothesis (HMH) appears to resolve the Tragedy of the Horizon paradoxes (Carney 2015, 2016). The keystone of the solution is the RCC as the missing complement to the SCC. If the proposed risk management approach (Eqs. 6 and 7) is acceptable to the science and economics community, then a fundamental shift may emerge in the narrative on climate change economics and ethics. This shift could be profound because the RCC may ascribe significantly greater economic value to the protection of the climate, ecosystems, people, culture, and species.

The SCC and cost-benefit analysis emerged from Pigou's (1932) tax as a pragmatic way of improving market efficiency and maximizing economic welfare:

reflecting the ethics of neoclassical economics and anthropocentrism. The RCC and cost-effectiveness analysis described in this exposition (Tables 2 and 3) offer a new road map for managing climate systemic risk: reflecting the ethics of biophsyical economics and naturalism. Synergy provided by the internalization of the SCC and RCC (Eq. 3) might provide civilization with long-term resilience: reflecting an ethic of holism. Geo-engineering the Earth's energy balance was not considered in this exposition, and the economics and ethics of such an approach are not addressed here.

The HMH has important philosophical implications by showing that the externalized cost of carbon is influenced by the observer's worldview and choice of metrics. This is similar to Bohr's (1937) *Principle of Complementarity*, which asserts that an observed quantum state (kinematic or dynamic) is determined by the observer's choice of measurement. In terms of assessing and responding to climate change, an observer could adopted either a (Type I) classical/neoclassical perspective, or a (Type II) biophysical perspective. An observer with a Type II perspective may tend to focus on system dynamics and systemic risks as reported in biophysical studies (e.g., Garrett 2012), probabilistic studies (e.g., Raftery et al. 2017), and Earth system interpretations (e.g., Lenton 2012; Steffen et al. 2018). An observer with a Type I perpsective may tend to focus on standard emissions scenarios, marginal damages, welfare maximization, and mainstream policy options (e.g., Stern 2007; UNFCCC 2015).

6 Concluding Remarks

Climate change is still mostly an unmanaged problem, and this chapter proposes that the Holistic Market Hypothesis (HMH) of Chen et al. (2017) could be a missing link in resolving the climate crisis (refer Tables 1, 2, and 3). The hypothesis is that the Risk Cost of Carbon (RCC) is the positive externality that associates with carbon mitigation services. The RCC is equivalent to the cost of providing preventative climate insurance, and it involves no direct taxes because it is funded through monetary policy and currency trading in open markets. If the SCC and the RCC can be validated as the two externalized costs of carbon, then exciting new endeavors in environmental economics will be possible—including the monetization of climate risk and the improved management of the economy for long-term human prosperity and ecological sustainability.

A technical deduction is that a parallel currency—Complementary Currencies for Climate Change (4C)—can be used to internalize the RCC into the economy. The 4C is used to offer a global carbon reward and to price climate risk in foreign exchange currency markets. If the yield on 4C is risk-free, a higher risk-free rate of return will be created for investors, thereby hedging the climate risk. 4C should be implemented independently of taxes and cap-and-trade markets, because 4C will be used to retire carbon and increase the international price for carbon offset credits. An international agreement on risk tolerances and an international monetary policy are needed to introduce 4C into foreign exchange markets and to create a global price signal for the RCC. The "Global 4C" policy is proposed for this purpose.

7 Research Recommendations

It is recommended that the HMH be intensively researched. This research should involve central banks under their existing mandates or new mandates for the macroprudential regulation of climate risk. The research should be treated with urgency because the window of opportunity for achieving Article 2 of the Paris Agreement is closing (Figueres et al. 2017). Items for research include (1) developing an experimental test for the HMH (refer Sect. 5.1.2), (2) investigating a statistical-mechanical theory for the efficiency/inefficiency and risk-taking/risk-reducing asymmetries of complementary nonequilibrium carbon pricing (refer Table 3), (3) developing a model that shows how economic growth and prosperity are managed with complementary carbon pricing (refer Table 1), and (4) developing 4C reward rules to leverage ecosystem services (e.g., biodiversity) and social co-benefits (e.g., quality of life, employment, peace, and security). Assuming the HMH is correct, the research should include (5) setting a risk management base case for comparative studies, (6) estimating the RCC for the base case, (7) assessing the macroeconomic effects of the base case, (8) undertaking a feasibility study, and (9) pilot testing.

Acknowledgment There are no financial conflicts of interest to disclose. The work carried out by the authors received no financial support from the industry or government. The *Global 4C Risk Mitigation* policy is hosted by the Center for Regenerative Community Solutions 501(c)(3) (non-profit), NJ 07920, USA.

References

Aglietta, M., & Espagne, E. (2016, April). *Climate and finance systemic risks, more than an analogy? The climate fragility hypothesis* (30 pp.). Working Paper CEPII, No 2016-10.

Andreoni, J., Harbaugh, W., & Vesterlund, L. (2003). The carrot or the stick: Rewards, punishments, and cooperation. *American Economic Review, 93*(3), 893–902.

Annila, A., & Salthe, S. (2009). Economies evolve by energy dispersal. *Entropy, 11*, 606–633.

Bohr, N. (1937). Causality and complementarity. *Philosophy of Science, 4*(3), 289–298.

Boltzmann, L. (1872). Weitere Studien über das Wärmegleichgewicht unter Gasmolekülen. *Sitzungsberichte Akademie der Wissenschaften, 66*, 275–370.

Broadbent, J., & Vaughter, P. (2014). Inter-disciplinary analysis of climate change and society: A network approach, Chapter 10. In M. J. Manfredo, J. J. Vaske, A. Rechkemmer, & E. A. Duke (Eds.), *Understanding society and natural resources*. Heidelberg: Springer.

Brookes, L. (1990). Energy efficiency and the greenhouse effect. *Energy and Environment, 1*(4), 318–333.

Carney, M. (2015, September 29). *Breaking the tragedy of the horizon – Climate change and financial stability*. Speech given by the Governor of the Bank of England, Chairman of the Financial Stability Board, Lloyd's of London.

Carney, M. (2016, September 22). *Resolving the climate paradox*. Speech given by the Governor of the Bank of England, Chair of the Financial Stability Board, Arthur Burns Memorial Lecture, Berlin.

Chan, G., Stavins, R., Stowe, R., & Sweeney, R. (2012). *The SO_2 allowance trading system and the clean air act amendments of 1990: Reflections on twenty years of policy innovation*. Cambridge, MA: Harvard Environmental Economics Program.

Chen, D. B. (2018a). Utility of the blockchain for climate mitigation. *The Journal of British Blockchain Association, 1*(1), 1–9.

Chen, D. B. (2018b). Central banks and blockchains: The case for managing climate risk with a positive carbon price, Chapter 15. In A. Marke (Ed.), *Transforming climate finance and green investment with blockchains*. Amsterdam: Elsevier.

Chen, D. B., van der Beek, J., & Cloud, J. (2017). Climate mitigation policy as a system solution: Addressing the risk cost of carbon. *Journal of Sustainable Finance & Investment, 7*(3), 1–42.

Chen, X., Sasaki, T., Brännström, Å., & Dieckmann, U. (2015). First carrot, then stick: How the adaptive hybridization of incentives promotes cooperation. *Journal Royal Society Interface, 12*, 20140935.

Clausius, R. (1854). Ueber eine veränderte Form des zweiten Hauptsatzes der mechanischen Wärmetheoriein. *Annalen der Physik und Chemie, 93*(12), 481–506.

Clausius, R. (1867). *The mechanical theory of heat – With its applications to the steam engine and to physical properties of bodies*. London: John van Voorst.

Cleveland, C. J. (1987). Biophysical economics: Historical perspective and current research trends. *Ecological Modelling, 38*(1–2), 47–73.

Currarini, S., Marchiori, C., & Tavoni, A. (2014). *Network economics and the environment: Insights and perspectives*. Centre for Climate Change Economics and Policy, Working Paper No. 162; Grantham Research Institute on Climate Change and the Environment, Working Paper No. 145.

Dietz, S., Gollier, C., & Kessler, L. (2018). The climate beta. *Journal of Environmental Economics and Management, 87*, 258–274.

Eisenstein, C. (2011). *Sacred economics: Money, gift, and society in the age of transition*. Paperback.

England, J. (2013). Statistical physics of self-replication. *The Journal of Chemical Physics, 139*, 121923.

Figueres, C., Schellnhuber, H. J., Whiteman, G., Rockström, J., Hobley, A., & Rahmstorf, S. (2017). Three years to safeguard our climate. *Nature, 546*(7660), 593–595.

Fuss, S., Canadell, J. G., Peters, G. P., Tavoni, M., Andrew, R. M., Ciais, P., et al. (2014). Betting on negative emissions. *Nature Climate Change, 4*(10), 850–853.

Garrett, T. J. (2011). Are there basic physical constraints on future anthropogenic emissions of carbon dioxide? *Climatic Change, 104*, 437–455.

Garrett, T. J. (2012). No way out? The double-bind in seeking global prosperity alongside mitigated climate change. *Earth System Dynamics, 3*, 1–17.

Garrett, T. J. (2014). Long-run evolution of the global economy: 1. Physical basis. *Earth's Future, 2* (3), 127–151.

Garrett, T. J. (2015). Long-run evolution of the global economy: 2. Hindcasts of innovation and growth. *Earth System Dynamics Discussions, 6*(1), 673–688.

Gasser, T., Guivarch, G., Tachiiri, K., Jones, C. D., & Ciais, P. (2015). Negative emissions physically needed to keep global warming below 2°C. *Nature Communications, 6*, 7958.

Guy, D. (2015). *Energy consumption and entropy release in the biosphere. Encyclopédie de l'énergie*. Accessed June 18, 2018, from http://encyclopedie-energie.org/articles/energy-consumption-and-entropy-release-biosphere

Hansen, J., Kharecha, P., Sato, M., Masson-Delmotte, V., Ackerman, F., Beerling, D. J., et al. (2013). Assessing 'dangerous climate change': Required reduction of carbon emissions to protect young people, future generations and nature. *PLoS One, 8*(12), e81648.

Hilbe, C., & Sigmund, K. (2010). Incentives and opportunism: From the carrot to the stick. *Proceedings of the Royal Society B: Biological Sciences, 277*(1693), 2427–2433.

IAWG. (2013). Technical support document: Technical update of the social cost of carbon for regulatory impact analysis under executive order 12866. 1–22 (US government).

IPCC. (2013). In T. F. Stocker, D. Qin, G.-K. Plattner, M. Tignor, S. K. Allen, J. Boschung, et al. (Eds.), *Climate change 2013: The physical science basis. Contribution of working group I to the fifth assessment report of the intergovernmental panel on climate change* (p. 1535). Cambridge: Cambridge University Press.

IPCC. (2014a). Integrated risk and uncertainty assessment of climate change response policies. In O. Edenhofer, R. Pichs-Madruga, Y. Sokona, E. Farahani, S. Kadner, K. Seyboth, et al. (Eds.), *Climate change 2014: Mitigation of climate change. Contribution of working group III to the fifth assessment report of the intergovernmental panel on climate change*. Cambridge: Cambridge University Press.

IPCC. (2014b). In R. K. Pachauri & L. A. Meyer (Eds.), *Climate change 2014: Synthesis report. Contribution of working groups I, II and III to the fifth assessment report of the intergovernmental panel on climate change* (p. 151). Geneva: IPCC.

IPCC. (2014c). Social, economic and ethical concepts and methods. In O. Edenhofer, R. Pichs-Madruga, Y. Sokona, E. Farahani, S. Kadner, K. Seyboth, et al. (Eds.), *Climate change 2014: Mitigation of climate change. Contribution of working group III to the fifth assessment report of the intergovernmental panel on climate change*. Cambridge: Cambridge University Press.

ISO. (2009). *ISO Guide 73:2009 Risk management – Vocabulary*. Accessed July 28, 2017, from http://www.iso.org/iso/home/standards/iso31000.htm

Jevons, W. S. (1875). *Money and the mechanism of exchange*. New York: D. Appleton. Library of Economics and Liberty. Accessed March 28, 2018, from http://www.econlib.org/library/YPDBooks/Jevons/jvnMME24.html

Keynes, J. M. (1921). *Treatise on probability*. London: Macmillan. Accessed July 28, 2016, from http://archive.org/stream/treatiseonprobab007528mbp#page/n11/mode/2up

Knight, F. H. (1921). *Risk, uncertainty and profit*. New York: Hart, Schaffner and Marx.

Kümmel, R., & Lindenberger, D. (2014). How energy conversion drives economic growth far from the equilibrium of neoclassical economics. *New Journal of Physics, 16*(2014), 125008.

Law, J. (2009). Actor network theory and material semiotics, Chapter 7. In B. S. Turner (Ed.), *Social theory*. Oxford: Wiley-Blackwell.

Lawrence, S., Liu, Q., & Yakovenko, V. M. (2013). Global inequality in energy consumption from 1980 to 2010. *Entropy, 15*, 5565–5579.

Lenton, T. M. (2012). Arctic climate tipping points. *Ambio, 41*(1), 10–22.

Lucas, R. E. (1978). Asset prices in an exchange economy. *Econometrica, 46*, 1429–1445.

Maxwell, J. C. (1860). II. Illustrations of the dynamical theory of gases. *The London, Edinburgh, and Dublin Philosophical Magazine and Journal of Science, 20*(130), 1860.

Milnes, A. (1919). *The economic foundations of reconstruction*. London: Macdonald and Evans.

Nordhaus, W. D. (1991). To slow or not to slow: The economics of the greenhouse effect. *The Economic Journal, 101*(407), 920–937.

Nordhaus, W. D. (2007a). A review of the stern review on the economics of climate change. *Journal of Economic Literature, 45*(3), 686–702.

Nordhaus, W. D. (2007b, July 24). *The challenge of global warming: Economic models and environmental policy*. William Nordhaus Sterling Professor of Economics, Yale University, New Haven, CT.

Nordhaus, W. D. (2017). Revisiting the social cost of carbon. *PNAS, 114*(7), 1518–1523.

Oster, G., Perelson, A., & Katchalsky, A. (1971). Network thermodynamics. *Nature, 234*, 393–399.

Pigou, A. C. (1932). *The economics of welfare, 1920*. London: MacMillan.

Raftery, A. E., Zimmer, A., Frierson, D. M., Startz, R., & Liu, P. (2017). Less than 2°C warming by 2100 unlikely. *Nature Climate Change, 7*, 637–641.

Ramsey, F. (1928). A mathematical theory of saving. *Economic Journal, 38*(152), 543–559.

Raworth, K. (2017). *Doughnut economics: Seven ways to think like a 21st-century economist* (p. 320). White River Junction, VT: Chelsea Green Publishing.

Robbins, L. (1935). *An essay on the nature and significance of economic science*. London: Macmillan.

Saunders, H. D. (1992). The Khazzoom-Brookes postulate and neoclassical growth. *The Energy Journal, 13*(4), 131–148.

Shapiro, A. F., & Koissi, M.-C. (2015). *Risk assessment applications of fuzzy logic* (p. 112). Casualty Actuarial Society, Canadian Institute of Actuaries, Society of Actuaries.

Sirkis, A., Hourcade, J. C., Dasgupta, D., Studart, R., Gallagher, K., Perrissin-Fabert, B., et al. (2015). *Moving the trillions: A debate on positive pricing of mitigation actions*. Nogent-sur-Marne: Centro Brasil no Clima Ed./CIRED.

Stern, N. (2007). *The economics of climate change: The stern review*. Cambridge: Cambridge University Press.

Steffen, W., Rockström, J., Richardson, K., Lenton, T. M., Folke, C., Liverman, D., Summerhayes, C. P., Barnosky, A. D., Cornell, S. E., Crucifix, M., Donges, J. F., Fetzer, I., Lade, S. J., Scheffer, M., Winkelmann, R., & Schellnhuber, H. J. (2018). Trajectories of the earth system in the Anthropocene. *Proceedings of the National Academy of Sciences, 115*(33), 8252–8259.

Tinbergen, J. (1952). *On the theory of economic policy. Volume 1 of contributions to economic analysis*. Amsterdam: North-Holland.

U.S. President. (1981). 'Federal Regulation'. Exec. Order No. 12291 of Feb. 17, 1981, Federal Register 46, Document citation: 46 FR 13193, 3 CFR, 1981 Comp., (p. 127).

UNFCCC. (2015). United Nations. In *Adoption of the Paris Agreement, 21st Conference of the Parties*. Report No. FCCC/CP/2015/L.9/Rev.1. Paris.

van den Bergh, J. C. J. M., & Kallis, G. (2012). Growth, a-growth or degrowth to stay within planetary boundaries? *Journal of Economic Issues, 46*(4), 909–920.

Weitzman, M. L. (2010). Risk-adjusted gamma discounting. *Journal of Environmental Economics and Management, 60*(1), 1–13.

Assessment of Renewable Energy Projects Using a Decision Support System: A Process to Endorse the Social License to Operate

Sotiris N. Kamenopoulos and Theocharis Tsoutsos

Abstract The renewable energy projects involve multiple stakeholders; the social acceptance of these projects may convert into a thoughtful risk for the exploitation of renewable energy. Comprehending the process under which a social license to operate may be granted to a renewable energy project is significant. The "GO" or "NO-GO" decision for a renewable energy project is a vital task. In order to illustrate how renewable energy projects can be assessed from the sustainability point of view, two hypothetical scenarios were constructed. These scenarios describe a conceptual renewable energy project evaluated by five imaginary stakeholders under specific criteria and sustainability pillars. The first scenario is a "NO-GO" scenario: the project is not sustainable due to significant environmental, social, economic, technological and geopolitical negative impacts. In this case, the social license to operate will not be endorsed by the stakeholders; important changes are needed before reassessment. The second scenario is a "GO" scenario: the preferences of the stakeholders are such that the project may be considered sustainable and the social license to operate may be endorsed by the stakeholders. The methodology utilized to assess the two hypothetical scenarios for the same renewable energy project is the multi-criteria decision analysis combined with the multi-attribute utility theory. The quantification of assessment results was conducted with the assistance of a state-of-the-art decision support system ("Acropolis DSS"), which allows decision-makers to evaluate multiple options that offer alternate solutions in "GO-NO-GO" situations.

Keywords Renewable energy · Social license to operate · Multi-criteria decision analysis

---

S. N. Kamenopoulos · T. Tsoutsos (✉)
Renewable and Sustainable Energy Lab, School of Environmental Engineering, Technical University of Crete, Chania, Greece
e-mail: theocharis.tsoutsos@enveng.tuc.gr

H. Doukas et al. (eds.), *Understanding Risks and Uncertainties in Energy and Climate Policy*, https://doi.org/10.1007/978-3-030-03152-7_9

Abbreviations

DSS	Decision support system
MAUT	Multi-attribute utility theory
MCDA	Multi-criteria decision analysis
NGO	Nongovernmental organization
REP	Renewable energy project
RES	Renewable energy sources
SD	Sustainable development
SDF	Sustainable development framework
SLO	Social license to operate
SMART	Simple Multi-Attribute Rating Technique
SP	Sustainable path
UN	United Nations

1 Introduction

Social license to operate (SLO) as a term was perceived by a Canadian mining executive during the 1990s (Fraser Institute 2012; Prno 2013), and since then it has become an important requirement for the sustainability of several projects. The main scope of this chapter is to present a model for the evaluation of renewable energy projects from the standpoint of social license to operate and social acceptance using a decision support system (DSS). The model used is based on multi-criteria decision analysis (MCDA) and the multi-attribute utility theory (MAUT). Subject DSS incorporates different sustainability indicators as these could be applied to renewable energy projects (REPs). For the purposes of this chapter, the term SLO will refer to the level of acceptance and/or approval by local communities and stakeholders of REPs and their operations (Kamenopoulos et al. 2015a). Compliance with legal regulations may not be the only precondition in the case of REPs; societies may be very doubtful when dealing with a REP, and the lack of faith to policymakers is continuously under reassessment. Stakeholders grant an SLO to a REP when they feel that their values/preferences and those of the REP's company are aligned (Govindan et al. 2014). As a result an SLO is not a permanent treaty, but a dynamic social contract which is contingent on the dynamic changes of multiple stakeholders' preferences; it is dynamic because stakeholders' perceptions can change over time for different reasons (Nelsen 2006); also it could be revoked, and it should never be taken for granted (Kemp et al. 2006). Adaptability is required to manage the complexity associated with the founding and preserving an SLO and sustainable operations (Prno 2013); the absence of the SLO downgrades the possibility of sustainable REP operations; this is how the SLO is straightforward interconnected/influenced through the current sustainable development practices. As we may see in Fig. 1, sustainable development (SD) practices are the origins of a valid and constant

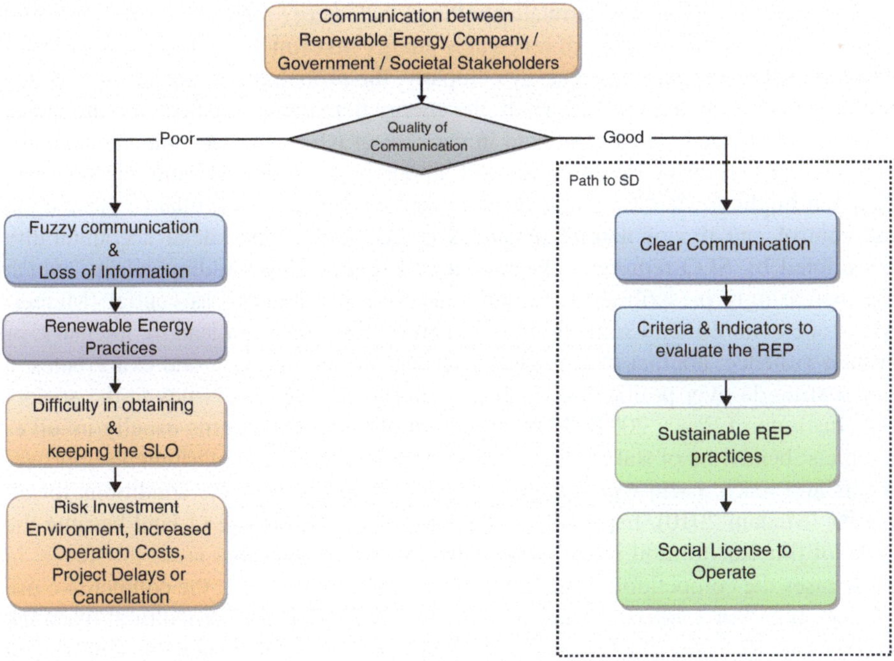

Fig. 1 The modified process that may approve the sustainable exploitation of renewable energy projects and endorse the social license to operate (based on work by Kamenopoulos et al. 2015a, b)

SLO. When a REP is under development, the preferences of involved stakeholders (e.g., local societies, nongovernmental organizations (NGOs), market, government, etc.) are under continuous transformation. Adaptability requires continuous monitoring of stakeholder preferences in every stage of a REP project in order to ensure that the project remains within the sustainable path and the SLO is still valid. Under this continuous "monitoring" process, the decision-makers of a REP may ascertain their decisions on a "GO-NO-GO" frame, adjust policies in accordance with the stakeholders' preferences, and evaluate the project. The notion behind monitoring of stakeholder preferences shall be in such way that preferences are decoded into measurable components within a holistic tool which is flexible and adaptable and continuously supports decision-making.

SLO is straightforward related to the social acceptance of REPs, which is critical for REPs as it could take the form of a risk/uncertainty with severe adverse effects (Lowrance 1976; Kaminaris et al. 2006). These risks may be associated with project delays and/or cancelations due to social turmoil, vandalisms, or increased operational costs (Kamenopoulos and Tsoutsos 2015). These risks may be technological, environmental, social, economic, or geopolitical. As a result, it is very critical to substantiate the reasons why social acceptance comprises a significant risk for which a methodology shall be proposed to address.

The social acceptance is correlated to the global energy market mix of supply and demand through the securing of renewable energy investments: when local societies grant an SLO to a renewable energy company, the probability of social turmoil due to the project is decreased; as a result, the energy demand is satisfied, and the prices may not be affected. The use of land in the case of REPs may become the stake for competition and create potential turmoil among stakeholders (Granoszewski et al. 2011); it might create disagreements and disputes over access similar to those upon the control and use of natural resources (FAO 2000). The social accountability constituted by SLO requires engagement and relationship-building efforts, which are increasingly prescribed to include "meaningful dialogue" as central (Mercer-Mapstone et al. 2017). Competition over natural resources can lead to, intensify, or sustain violence; conflict over natural resources is often part of, and exacerbates, a larger struggle over political, economic, cultural, or religious issues in the society (US Institute of Peace 2007). Managing natural resource systems usually involves conflicts; behaviors of stakeholders, who might be willing to contribute to improvements and reach a win-win situation, sometimes result in worse conditions for all parties (Madani 2010). Especially in the case of REPs, the use of land became the stake of timeless turmoil in the Greek island of Crete and other areas in Greece; in most cases the competition is related to conflict of interests, i.e., the competitive use of land between renewable energy sources (RES) vise the agricultural/livestock breeding or RES vise the touristic sector and the fear of local societies with regard to potential environmental degradation. RES seem to suffer from the "Not in My Back Yard syndrome (NIMBY)" (Smith and Klick 2007) similarly to the mining sector. No man-made project can completely avoid some impact on the environment so neither can REPs (Tsoutsos et al. 2005). A sustainable solution for the exploitation of RES should at least adjust for environmental protection, resource availability, social welfare, and economic viability of the system (Maria and Tsoutsos 2004). Special procedures should be applied prior to the installation of a renewable energy plant especially in countries with a complicated administrative and legislative system (Tsoutsos et al. 2007, 2015; Kokologos et al. 2014). The remainder of the chapter is organized as follows: The methodological framework is presented in Sect. 2. The evaluated hypothetical scenarios and discussion are presented in Sect. 3; the conclusions are presented in Sect. 4.

2 Methodological Framework

Fuzzy communication might create significant negative implications for the acceptance of REPs by stakeholders. If the information is not clearly and transparently shared, the renewable energy firm may be unable to gain an SLO, which refers to the level of acceptance or approval by local communities and stakeholders of renewable energy companies and their operations. This is true regardless of the quality of the renewable energy practices of the firm. The absence of the social license to operate may increase the risks associated with the REP operation. This result is exemplified by the left-hand path of Fig. 1. Effective communication, based on SD's criteria and indicators, when coupled with best practices can ultimately lead to the SLO. This is demonstrated on the right side of Fig. 1 (based on work by Kamenopoulos et al. (2015a, b) and adapted to fit the purposes of this chapter).

A sustainability assessment of a REP typically may involve multiple stakeholders with distinct objectives, preferences, and potential conflicts. The "GO" or "NO-GO" decision for a REP project may become a challenging task due to the interrelated multiple quantitative and qualitative parameters with different positive or negative impacts as expressed by the different stakeholders.

In the past, several frameworks, including multi-criteria decision analysis (MCDA) frameworks, were applied in the field of REP evaluation; tools such as the Spanish method and PROMETHEE were exploited (Tsoutsos et al. 2009a, b). In addition, several related developments are present in the literature. Indicatively, to name but a few, we refer to REP evaluations using linguistic MCDM, TOPSIS, MAUT, and others (Doukas and Psarras 2009; Doukas et al. 2009, 2010, 2012; Doukas 2013a, b).

The decision support system (DSS) presented in this chapter is based on a multi-criteria decision analysis and on the multi-attribute utility theory (MAUT). The DSS was constructed in the Microsoft Excel™ 2013 environment, and its core is based on criteria and indicators; the DSS may assist decision-makers/stakeholders to better assess the impact of a REP from the sustainability point of view. The proposed DSS shall be integrated into a state-of-the-art sustainable development framework (SDF) (Fig. 2) which was previously developed by Kamenopoulos et al. (2015b).

One of the basic characteristics of the MCDA is the contribution of stakeholders at the decision-making through the negotiation process (trade-offs). In addition, the MCDA integrates unique and personal decision-making practices of individual decision-makers (Kamenopoulos et al. 2018).

The multi-attribute utility theory (MAUT) was created by Keeney and Raiffa (1993). In the MAUT the preferences of the decision-makers are expressed by a utility function which expresses the level of preference that a decision-maker has a set of alternatives; the alternatives are compared under specific criteria (attributes), and every criterion has its own weight (Kamenopoulos et al. 2018). The most common MAUT method is the additive model which is represented by the following equation:

$$V_j = \sum w_i p_{ij} \tag{1}$$

where:

V_j is the aggregate score of the jth alternative.
p_{ij} is the score of the jth of the ith criterion.
w_i is the weight of ith criterion.

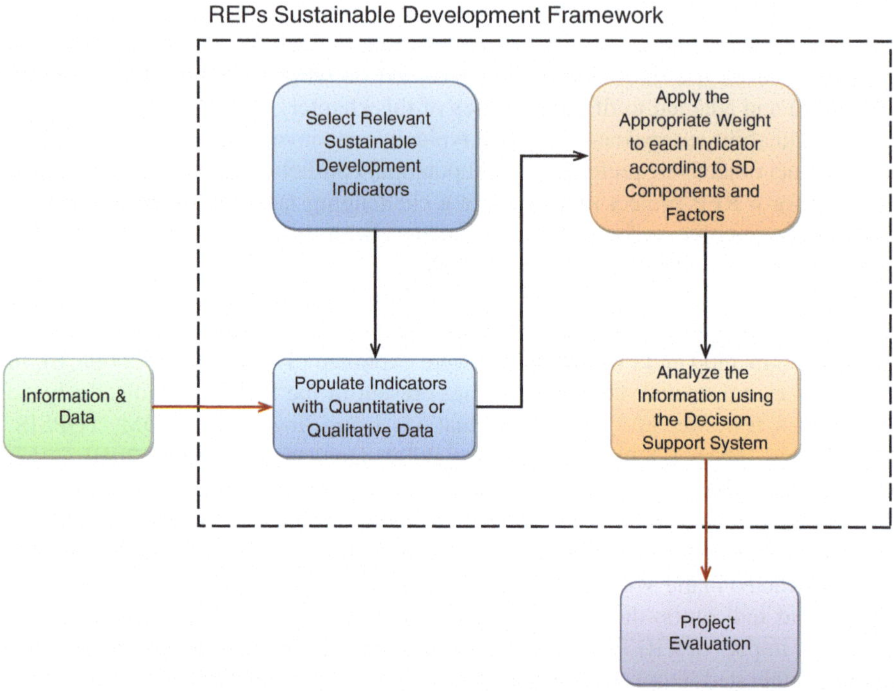

Fig. 2 Modified application of the conceptual framework in the case of renewable energy projects (based on Kamenopoulos et al. 2015b)

A basic constraint of MAUT dictates that the sum of all criteria weights should be equal to one as shown by Eq. (2):

$$\sum_{i} W_i = 1 \tag{2}$$

The strength of multi-attribute utility theory (MAUT) lies on the following (Kumar et al. 2017; Kamenopoulos et al. 2018): (a) it accounts for any difference in any criteria, (b) it simultaneously computes preference order for all alternatives, and (c) it dynamically updates value changes due to any impact. The weakness of MAUT lies on the following: it is difficult to have precise input from the decision-maker, and (b) the outcome of the decision criteria is uncertain. The method which was utilized in order to evaluate the relevant importance of each criterion in terms of weighting factors was the Simple Multi-Attribute Rating Technique (SMART) that was developed by Edwards (1977). The SMART was used in a component part of a hybrid decision support tool for evaluating the sustainability of mining projects, and it was modified for the purposes of this chapter (Kamenopoulos et al. 2018). In that case, the SMART computed the relevant importance of each criterion in two stages (ranking of the criteria and normalization of total score) with the contribution of decision-makers. The modified process allows the formation of a REP's SD index

Table 1 "GO" or "NO-GO" decision based on the evaluation of "ACROPOLIS" index

ACROPOLIS	Decision	Category—color code	Comments
Positive	GO	A—Green	The project is accepted
Negative	NO-GO	C—Red	The project is rejected. Need for critical changes before evaluated
Around zero	On hold	B—Orange	Mitigation measures shall be proposed to offset impacts. The project has the potential to be reevaluated

(Kamenopoulos et al. 2018) that represents the total scoring of decision-maker choices. The following sustainable development index can then be determined (named "ACROPOLIS" index):

$$\text{``ACROPOLIS''} = 100\,(\text{SPA} - \text{SPB})/|\text{SPB}| \tag{3}$$

where:

- The total score of the sustainable path before project start (SPB).
- The total score of the sustainable path after project end (SPA).

A "sustainable path" was defined as "one that allows every future generation the option of being as well off as its predecessors" (US NRC 1994).

The "GO" or "NO-GO" decision for a REP based on the evaluation of the "ACROPOLIS" index is shown in Table 1.

The decision-making process of the proposed DSS is presented in Fig. 3 (based on work by Kamenopoulos et al. (2018) and adapted to fit the purposes of this chapter).

The presented DSS is in a prototype version, and it was designed to include the preferences of five stakeholders, ten quantitative/qualitative indicators, and five sustainable development's pillars (i.e., economy, environment, geopolitics, society, and technology). The model can be modified to incorporate any number of stakeholders, criteria/indicators, or pillars. The proposed model provides the stakeholders with the opportunity for transparent, free decision-making and democratic negotiations, it may quantify/measure the "sustainable path" as this was defined by the US National Research Council, it complies with the UN's SD prerequisites for "...effective citizen participation in decision making and by greater democracy in international decision making..." (UN 1987), and it contributes to gaining and retaining social license to operate. Yet there is no sufficient data yet for testing and it is still in prototype stage (Kamenopoulos et al. 2018).

The process allows several levels of trade-offs between the stakeholders. All criteria/indicators are measured in the same scale for homogeneity reasons; for that reason, a qualitative 5-Likert scale was selected (1, low; 2, below average; 3, average; 4, above average; 5, high). The scale and actual values were selected arbitrarily for the purposes of subject chapter. Since there are no actual data available in order to

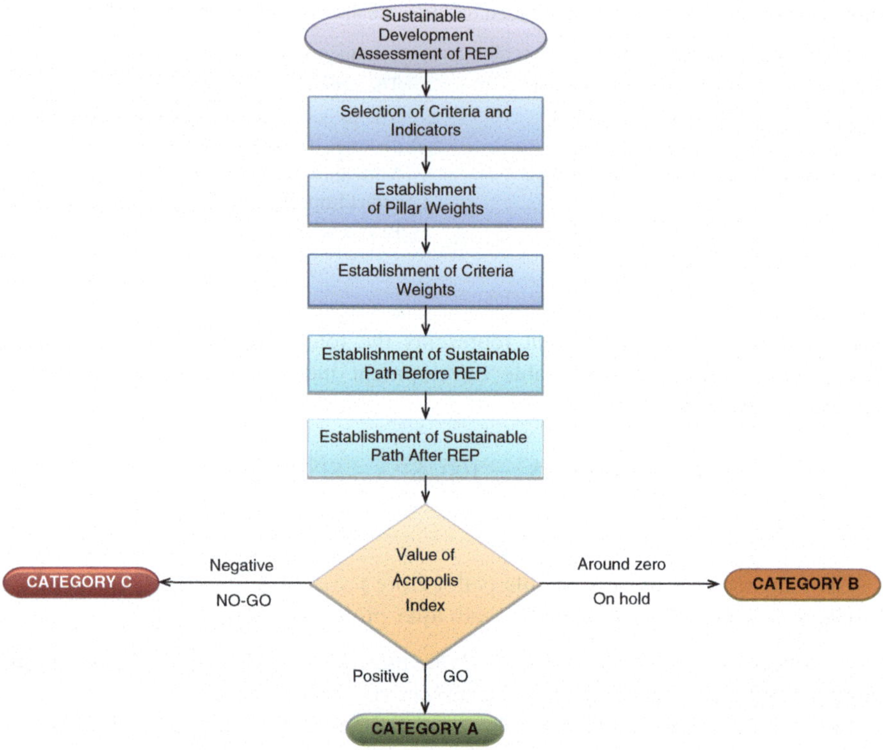

Fig. 3 "GO-NO-GO" decision-making process for REP assessment from the sustainable development point of view (based on work by Kamenopoulos et al. 2018)

test the proposed decision support system, a number of arbitrary assumptions were made (Table 2).

For the purposes of this chapter, ten sustainable development indicators were selected to describe the five pillars of a framework for the evaluation of REPs. The selected indicators and five pillars of sustainability are shown in Table 3.

It should be noted that the proposed model has some advantages as well some disadvantages/limitations that are shown in Table 4 (Kamenopoulos et al. 2018).

3 The Evaluated Hypothetical Scenarios and Discussion

Tables 5, 6, 7, and 8 represent the hypothetical "NO-GO" (red color) and "GO" (green color) scenarios. They show the values of pillars' and criteria weights in accordance to five stakeholder ranking preferences and specific values of indicators. They also show the values of sustainable paths (SP), the values of ACROPOLIS index, and the final result in accordance with five stakeholder ranking preferences and specific values of indicators.

Table 2 Arbitrary assumptions made for the purposes of this paper

Assumption	Explanation
Number of decision-makers	Five
Scenarios	Two scenarios: GO and NO-GO
Selection of indicators	Through bibliographic research and authors' choices
Sustainability pillars	Five pillars: environment, economy, society, technology, geopolitics
Ranking of indicators and sustainable development pillars	Selected by the authors
Values of indicators	Selected by the authors
Criteria/indicators evaluated using 5-Likert scale	5-Likert scale 1 = Low 2 = Below average 3 = Average 4 = Above average 5 = High

Table 3 Ten sustainable development indicators for SD pillars (environment, economy, society, technology, and geopolitics)

Pillar	Indicator
Environment	1. Energy consumption (kW consumed per hectare) 2. Greenhouse emissions (kg of CO_2 per kWh of energy produced) 3. Waste produced (per kW of energy produced)
Economy	4. Average number of people directly employed 5. Total turnover 6. Increase of average annual income in the area
Society	7. Number of lost workdays per 100 full-time equivalent (FTE) employees 8. Regional life satisfaction/well-being (ranking provided as a subindex of the OECD Regional Well-Being Index)[a]
Technology	9. Technology index ranking (subindex of Global Creativity Index by Martin Prosperity Institute, University of Toronto, Canada)[b]
Geopolitics	10. Political stability in the country and absence of violence/terrorism[c]

[a]The OECD regional well-being tool provides information about where regions stand on 11 topics that matter in people's lives: jobs, income, education, health, civic engagement, safety, access to people, housing, community, and life satisfaction. At https://www.oecdregionalwellbeing.org/ (Accessed 22 May 2017)

[b]The GCI is a broad-based measure of advanced economic growth and sustainable prosperity based on the 3Ts of economic development—talent, technology, and tolerance. It rates and ranks 139 nations worldwide on each of these dimensions and on our overall measure of creativity and prosperity. At http://martinprosperity.org/content/the-global-creativity-index-2015/ (Accessed April 4, 2017)

[c]This is an indicator which is included in the World Bank's Worldwide Governance Indicators (WGIs). WGIs are a research dataset summarizing the views on the quality of governance provided by a large number of enterprise, citizen, and expert survey respondents in industrial and developing countries. These data are gathered from a number of survey institutes, think tanks, nongovernmental organizations, international organizations, and private sector firms. The WGIs do not reflect the official views of the World Bank, its executive directors, or the countries they represent. At http://info.worldbank.org/governance/wgi/#home (Accessed April 3, 2017)

Table 4 Advantages and disadvantages of the proposed model

Advantages	Disadvantages
It provides the stakeholders with the opportunity for transparent, free decision-making and democratic negotiations	No sufficient data yet available for testing/prototype stage
It quantifies and measures the US NRC's term of "sustainable path" which was described as "…one that allows every future generation the option of being as well off as its predecessors"	No sufficient data yet available for testing/prototype stage
It complies with the UN's SD prerequisite for "…effective citizen participation in decision making and by greater democracy in international decision making…"	No sufficient data yet available for testing/prototype stage
It contributes to gaining and retaining "social license to operate"	No sufficient data yet available for testing/prototype stage
It is a "value"-oriented tool: stakeholders are encouraged to incorporate and directly or indirectly express their "value" on the stake of the project	No sufficient data yet available for testing/prototype stage
It is designed to support all three stages of a project: before, during, and after project's termination	No sufficient data yet available for testing/prototype stage
If modified, it may include any number of SD pillars	It was built for five SD pillars Need modifications/prototype stage
If modified, it may include an unlimited number of stakeholders	It was built for five stakeholders Need modifications/prototype stage
If modified, it may include an unlimited number of indicators	It was built for ten indicators per each pillar Need modifications/prototype stage
If modified, it could also incorporate financial indicators	Not designed to assess projects from their financial/economic value
It incorporates qualitative and quantitative indicators	No sufficient data yet available for testing/prototype stage
If modified, it may be applicable to any project. Not necessarily in the mining sector	Need modifications/prototype stage
Utilizing parametric analysis the stakeholders may be provided with additional useful information	No sufficient data yet available/prototype stage

In accordance with the stakeholder's ranking preferences with regard to pillars' weights (Table 5), stakeholders #1 and #2 seem to be in favor of the environment and society. The preferences of stakeholders #3 and #4 are balanced equally among pillars at the "NO-GO" scenario; stakeholder #5 seems to be in favor of economy, technology, and geopolitics. Table 6 shows that several trade-offs between stakeholders need to be compromised in order to "move" the project from the "NO-GO" status (red) to "GO" status (green): stakeholders #1 and #2 need to compromise and slightly lessen their preferences with regard to the weight of environment. Congruently, stakeholders #3, #4, and #5 need to slightly lessen their preferences with regard to the weights of economy, technology, and geopolitics; respectively, stakeholders #3, #4, and #5 will need to compromise and upsurge their preferences with regard to the weights of environment and society. Equivalent trade-offs need to be

Table 5 "NO-GO" (red color) and "GO" (green color) hypothetical scenarios: Pillars' weights in accordance with five stakeholder ranking preferences and specific values of indicators

Pillars	Stakeholder 1	Stakeholder 2	Stakeholder 3	Stakeholder 4	Stakeholder 5	Average pillar weight factors
Environment	0.40 – 0.25	0.40 – 0.29	0.20 – 0.29	0.20 – 0.15	0.06 – 0.30	0.252 – 0.254
Economy	0.10 – 0.19	0.10 – 0.14	0.20 – 0.12	0.20 – 0.15	0.29 – 0.21	0.180 – 0.163
Society	0.30 – 0.30	0.30 – 0.36	0.20 - 0.29	0.20 – 0.40	0.07 – 0.21	0.210 – 0.313
Technology	0.10 – 0.13	0.10 – 0.14	0.20 – 0.18	0.20 – 0.15	0.29 – 0.14	0.179 – 0.150
Geopolitics	0.10 – 0.13	0.10 – 0.07	0.20 – 0.12	0.20 – 0.15	0.29 – 0.14	0.179 – 0.120
Sum	1 - 1	1 - 1	1 - 1	1 - 1	1 - 1	1 - 1

compromised on the average criteria weights (Table 6): in accordance with the results of Table 5 when the average weights of environmental and societal criteria increase, then the project may move from "NO-GO" status to "GO" status; this result seems to be reasonable since the society general is willing to provide the social license to operate to REPs when only they fulfill environmental and societal standards. The sustainable path (SP) in the case of "NO-GO" scenario seems to be rationally valued (Table 7): if the SP after a REP (−2.40) has a lower value than the SP before the project (1.01), then the society is not willing to grant the social license to operate (i.e., the social acceptance). This is obvious since no society is willing to abate a good social and environmental condition for a worst one. In that case, the value of ACROPOLIS index (Table 8) has a negative value (−337.62%). Correspondingly, the sustainable path (SP) in the case of "GO" scenario seems to be also rationally valued (Table 6): if the SP after a REP (2.90) has a higher value than the SP before the project (1.14), then the society may grant the social license to operate (i.e., the society is willing to accept the project). This could be obvious since society is probably willing to abate a relatively good social and environmental condition for a better one under specific prerequisites; in that case there may be a bargain/trade-off between stakeholders, and if the economic/technological/geopolitical benefits for the society seem to overcome the expected/future social and environmental returns, then the society may grant the social license to operate and accept the project. In that case, the value of ACROPOLIS index (Table 8) has a positive value (154.38%). Of course, it has to be reminded that each project is unique and the above scenarios are hypothetical.

4 Conclusions

In this chapter, a model was demonstrated for the evaluation of REPs from the perspective of SLO using a decision support system and sustainability criteria. The model can assist decision-makers and stakeholders to evaluate a REP and make

Table 6 "NO-GO" (red color) and "GO" (green color) hypothetical scenarios: Average criteria weights in accordance with five ranking stakeholder preferences and specific values of indicators

Pillar	Criteria	Average criteria weight factors
1. Environment	P1C1. Energy consumption	0.134 – 0.140
	P1C2. Gas emissions	0.134 – 0.140
	P1C3. Waste	0.134 – 0.140
2. Economy	P2C1. Employment	0.082 – 0.060
	P2C2. Turnover	0.082 – 0.060
	P2C3. Area annual income	0.082 – 0.060
3. Society	P3C1. Number of accidents per 100 FTE employees	0.094 – 0.150
	P3C2. People's happiness & well being	0.094 – 0.150
4. Technology	P4C1. Education and social programs	0.082 – 0.060
5. Geopolitics	P5C1. World Bank's political stability ranking	0.082 – 0.040
Sum		1 - 1

"GO-NO-GO" decisions from the sustainable development's (SD) point of view. The model was based on multi-criteria decision analysis and the multi-attribute utility theory. The model provides the capability to visualize the SD perceptions of *citizen participation* and *greater transparency* in a comprehensive mode as recommended in earlier years by the United Nations and quantify the notion of the *sustainable path* as this was defined by the US National Research Council. The model incorporated criteria and indicators under which two hypothetical scenarios for REPs were evaluated: a "GO" scenario and a "NO-GO" scenario. The two hypothetical scenarios

Table 7 "NO-GO" (red color) and "GO" (green color) hypothetical scenarios: Values of sustainable paths (SP) in accordance with five stakeholder ranking preferences and specific values of indicators

Sustainable paths	Value NO-GO	Value GO
SP before project (SPB)	1.01	1.14
SP after project (SPA)	-2.40	2.90

Table 8 "NO-GO" (red color) and "GO" (green color) hypothetical scenarios: Values of evaluation ACROPOLIS index and final result in accordance with five stakeholder ranking preferences and specific values of indicators

Evaluation Index	Value for NO-GO Scenario	Value for GO Scenario
ACROPOLIS	(-2.40-1.01)/1.01 * 100 = - 337.62 %	(2.90-1.14)/1.14 * 100 = 154.38 %
	- 337.62% << 0 NO-GO	154.38% >> 0 GO

involved personal preferences of five decision-makers (stakeholders). Five sustainability pillars were selected: environment, society, economy, technology, and geopolitics. The hypothetical scenarios were subjective and were built under specific assumptions; in addition, it should be noted that each REP has unique characteristics. Indicative results showed that the presented model may provide outcomes that consist of stakeholder's preferences and relatively good judgment.

References

Doukas, H. (2013a). Modelling of linguistic variables in multi-criteria energy policy support. *European Journal of Operational Research, 227*(2), 227–238.

Doukas, K. (2013b). Linguistic multicriteria decision-making for energy systems: Building the "RE2S" framework. *Wiley Interdisciplinary Reviews Energy and Environment (WIREs), 2*(5), 571–585.

Doukas, H., & Psarras, J. (2009). A linguistic decision support model towards the promotion of renewable energy. *Energy Sources, Part B: Economics, Planning, and Policy, 4*, 166–178.

Doukas, H., Karakosta, C., & Psarras, J. (2009). A linguistic TOPSIS model to evaluate the sustainability of renewable energy options. *International Journal of Global Energy Issues, Special Issue "Energy Efficiency, Environmental Performance and Sustainability", 32*(1/2), 102–118.

Doukas, H., Karakosta, C., & Psarras, J. (2010). Computing with words to assess the sustainability of renewable energy options. *Expert Systems with Applications, 37*(7), 5491–5497.

Doukas, H., Marinakis, V., Karakosta, C., & Psarras, J. (2012). Promoting renewables in the energy sector of Tajikistan. *Renewable Energy, 39*, 411–418.

Edwards, W. (1977). How to use multiattribute utility measurement for social decision making. *IEEE Transactions on Systems, Man, and Cybernetics, SMC-7*, 326–340.

Food and Agricultural Organization – FAO (of United Nations). (2000). *Conflict and natural resource management*. Accessed September 12, 2017, from http://www.fao.org/forestry/21572-0d9d4b43a56ac49880557f4ebaa3534e3.pdf

Fraser Insitute. Miningfacts.org. (2012). *What is the social license to operate (SOL)?* http://www.miningfacts.org/Communities/What-is-the-social-licence-to-operate/

Govindan, K., Kannan, D., & Shankar, K. M. (2014). Evaluating the drivers of corporate social responsibility in the mining industry with multi-criteria approach: A multi-stakeholder perspective. *Journal of Cleaner Production, 84*, 214–232.

Granoszewski, K., Reise, C., Spiller, A., & Musshoff, O. (2011). The role of land use competition in farmers' engagement in renewable energy production. In *VII International Agribusiness PAA-PENSA Conference, "The Multiple Agro Profiles: How to Balance Economy, Environment and Society"*, November 30–December 2, 2011, Buenos Aires, Argentina.

Kamenopoulos, S. N., & Tsoutsos, T. (2015). Assessment of the safe operation and maintenance of photovoltaic systems. *Energy, 93*(2), 1633–1638.

Kamenopoulos, S., Shield, D., & Agioutantis, Z. (2015a). Sustainable development criteria and indicators for the assessment of rare earth element mining projects, Chapter 6. In I. B. de Lima & W. L. Filho (Eds.), *Rare earths industry: Technological, economic and environmental implications*. Amsterdam: Elsevier. isbn:978-0-12-802328-0.

Kamenopoulos, S., Agioutantis, Z., & Komnitsas, K. (2015b). A framework for sustainable mining of rare earth elements, Chapter 6. In I. B. de Lima & W. L. Filho (Eds.), *Rare earths industry: Technological, economic and environmental implications*. Amsterdam: Elsevier. isbn:978-0-12-802328-0.

Kamenopoulos, S., Agioutantis, Z., & Komnitsas, K. (2018). A new hybrid decision support tool for evaluating the sustainability of mining projects. *International Journal of Mining Science and Technology, 28*(2), 259–265.

Kaminaris, S. D., Tsoutsos, T. D., Agoris, D., & Machias, A. V. (2006). Assessing renewables-to-electricity systems: A fuzzy expert system model. *Energy Policy, 34*(12), 1357–1366.

Keeney, R. L., & Raiffa, H. (1993). *Decisions with multiple objectives – Preferences and value tradeoffs* (p. 569). Cambridge: Cambridge University Press.

Kemp, D., Boele, R., & Brereton, D. (2006). Community relations management systems in the mineral industry: Combining conventional and stakeholder-driven approaches. *International Journal of Sustainable Development, 9*(4), 390–403.

Kokologos, D., Tsitoura, I., Kouloumpis, V., & Tsoutsos, T. (2014). Visual impact assessment method for wind parks a case study in Crete. *Land Use Policy, 39*, 110–120.

Kumar, A., Sah, B., Singh, A. R., Deng, Y., He, X., Kumar, P., & Bansal, R. C. (2017). A review of multi-criteria decision making (MCDM) towards sustainable renewable energy development. *Renewable and Sustainable Energy Reviews, 69*, 596–609.

Lowrance, W. (1976). *Of acceptable risk* (p. 180). Los Altos: William Kaufmann.

Madani, K. (2010). Game theory and water resources. *Journal of Hydrology, 381*, 225–238.

Maria, E., & Tsoutsos, T. (2004). The sustainable management of reneable energy sources installations: Legal aspects of their environmental impact in small Greek islands. *Energy Conversion and Management, 45*, 631–638.

Mercer-Mapstone, L., Rifkin, W., Moffat, K., & Louis, W. (2017). Conceptualizing the role of dialogue in social license to operate. *Resources Policy, 54*, 137–146.

Nelsen, J. L. (2006). Social license to operate. *International Journal of Mining, Reclamation and Environment, 20*(3), 161–162.
Prno, J. (2013). An analysis of factors leading to the establishment of a social license to operate in the mining industry. *Resources Policy, 38*(4), 577–590.
Smith, E.R., Klick, H. 2007. *Explaining NIMBY opposition to wind power.* Accessed September 12, 2017, from http://www.polsci.ucsb.edu/faculty/smith/wind.pdf
Tsoutsos, T., Frantzeskaki, N., & Gekas, V. (2005). Environmental impacts from the solar energy technologies. *Energy Policy, 33*, 289–296.
Tsoutsos, T., Maria, E., & Mathioudakis, V. (2007). Sustainable siting procedure of small hydroelectric plants: The Greek experience. *Energy Policy, 35*, 2946–2959.
Tsoutsos, T., Drandaki, M., Frantzeskaki, N., Iosifidis, E., & Kiosses, I. (2009a). Sustainable energy planning by using multi-criteria analysis application in the island of Crete. *Energy Policy, 37*, 1587–1600.
Tsoutsos, T., Tsouchlaraki, A., Tsiropoulos, M., & Serpetsidakis, M. (2009b). Visual impact evaluation of a winf park in a Greek island. *Applied Energy, 86*, 546–553.
Tsoutsos, T., Tsitoura, I., Kokologos, D., & Kalaitzakis, K. (2015). Sustainable siting process in large wind farms. Case study in Crete. *Renewable Energy, 75*, 474–480.
United Nations (UN). (1987). *World Commission on Environment and Development: Report of the World Commission on Environment and Development: Our Common Future.* Accessed September 14, 2017, from http://www.un-documents.net/wced-ocf.htm
US Institute of Peace. (2007). *Natural resources, conflict and conflict resolution.* Accessed September 12, 2017, from https://www.usip.org/sites/default/files/file/08sg.pdf
US National Research Council (NRC). (1994). *Assigning economic value to natural resources.* Washington, D.C.: National Academy Press. Accessed September 14, 2017, from http://www.nap.edu/catalog/4844.html

A Unilateral Climate and Supply Market Model

Eike Blume-Werry, Martin Koller, and Martin Everts

Abstract In the midst of its respective energy transitions, the European power sector faces several challenges. Low levels of both European Union Emissions Trading Scheme (EU ETS) allowance prices and wholesale power prices fuelled concerns over drivers for decarbonisation and long-term generation adequacy. Whilst some countries have introduced capacity remuneration mechanisms to ensure generation adequacy, reforming the EU ETS has proven to be difficult. This paper proposes a unilateral approach by a state introducing a CO_2 levy that internalises and prices CO_2 at a national level. The suggested climate and supply market model thereby incentivises and rewards production from CO_2-neutral sources during times when it does not cover the targeted share of production. Prior to describing the model in detail, this paper discusses the theoretical policy steering background and the problems associated with current energy policies. For a broader picture, other CO_2 taxation models are briefly presented. This is followed by a discussion on the legal aspects of the model, its compatibility with international as well as EU law. The model is explored further by using Switzerland as an example, showing that a cross-sector carbon price can be implemented at acceptable costs for consumers. Last but not least, the paper examines varieties of the model and the adaption potential for European countries.

Keywords Border tax · Carbon tax · Climate policy · CO_2 levy · Emission trading · Energy policy · EU ETS · Security of supply

---

E. Blume-Werry (✉)
TU Wien, Vienna, Austria

M. Koller · M. Everts
Axpo Holding AG, Baden, Switzerland

H. Doukas et al. (eds.), *Understanding Risks and Uncertainties in Energy and Climate Policy*, https://doi.org/10.1007/978-3-030-03152-7_10

1 Energy Policy and the Concept of Direct and Multiple Steering

It is the objective of modern energy policies to decarbonise without jeopardising other policy objectives. Within this context, one usually refers to the energy policy objective triangle of (1) security of supply, (2) affordability and (3) sustainability. In order to tackle climate change and fight global warming, governments strive to reduce emissions and mitigate respective risks (Doukas et al. 2008, p. 362; Nikas et al. 2018). Carbon dioxide (CO_2) in particular has become a common standard to measure the sustainability of a given system and quantify emission growth or reduction. CO_2 is emitted by a variety of sources from different sectors, the power and transport sector being amongst the most prominent. Depending on the sector, there are different energy policies and strategies to initiate emission reductions.

Focusing on the power sector, one can differentiate between two theoretical policy concepts: direct and multiple steering. The former centres around the idea of pricing and internalising CO_2 or, more generally, greenhouse gas emission costs. In practice, this can be put into effect by introducing a CO_2 tax and/or an emission trading system such as the European Union Emissions Trading Scheme (EU ETS). Given the declared objective of decarbonisation, this approach can be seen as plausible and straightforward. Putting a price on CO_2 emissions should reduce the overall output and thus give rise to further decarbonisation. This approach directly targets CO_2 emissions, which is why we refer to it as 'direct steering'.

In contrast, an indirect steering method would indirectly work towards achieving the same objective, for example, by setting other incentives to reduce emissions or promote alternatives, which entail less or no emissions. Examples would include taxes on emission-intensive fuels or support for renewables.

If multiple policy measures coincide and interact with the common objective of reducing emissions, one speaks of a multiple steering model (Everts et al. 2016, pp. 119–120).[1] For example, a CO_2 tax and trading system can be a part of a multiple steering model. Indeed, most governments nowadays have a range of policies in place that not only price CO_2 but also incentivise emission reduction in other ways. It is an advantage of this multiple steering model that policymakers have greater control over emission reduction contributions of different sectors, technologies used and general concepts. There are, however, several problems with those designs which currently prevail.

[1]Multiple steering models usually include direct and indirect measures.

2 Current Issues of the EU ETS

There is an extensive body of literature on the weak performance of the EU ETS and the continuous drop in price of allowances since the financial crisis, from about 30 EUR/tCO_2 to below 5 EUR/tCO_2.[2] The low allowance price level has led to circumstances in which there exists great uncertainty as to whether the EU ETS is still and can be a main driver for decarbonisation, despite being designed to fulfil exactly this role (Marcu et al. 2016, pp. 7–11). In general, scholars regard the price level as too low to fulfil the intended functions (Abrell et al. 2016, p. 2; Rogge et al. 2011; European Commission 2014). It has been argued that at the current price level, costs of negative externalities stemming from carbon emissions are no longer properly internalised (Carbon Market Watch 2015, p. 4). Further, the price of allowances is too low to trigger investments in low carbon technologies and facilitate innovation (Carbon Market Watch 2015, p. 4; Hepburn et al. 2016, p. 1; Rogge et al. 2011).

This points towards the prevailing problems of the EU ETS and a necessity for reform. However, reforming the EU ETS has proven a difficult political matter, given the diverse interests of the stakeholders and parties involved. Currently, there is an ongoing reform process of the EU ETS for which the European Parliament voted and on which the European Council recently agreed a position (European Council 2017). However, observers remain sceptical that the proposed changes will be sufficient to end the run of very low allowance prices (Boffey 2017; Rattay 2017).

This run of very low allowance prices (on average 6.2 EUR/tCO_2 during the last 5 years)[3] has also had subsequent effects. It is part of the concept of the multiple steering model to maintain some level of price neutrality with respect to wholesale power prices. But CO_2 pricing systems such as the EU ETS have a price-increasing effect on wholesale power prices, whilst the deployment of low marginal cost renewables has a price-decreasing effect, which is why both policies together have the potential of retaining overall price neutrality. Power prices have fallen drastically since the financial crisis, and the price decline of CO_2 allowances and the deployment of low marginal cost renewables have been found to be the most contributing factors to this wholesale power price drop in Germany (Bublitz et al. 2017, p. 330; Everts et al. 2016, p. 122; Hirth 2016, p. 11).

This development of wholesale power prices has in turn increased the focus on the so-called missing money problem[4]—a problem which might threaten the long-term security of supply. If there is no investment rationale for investments in flexible

[2]Koch et al. provide a concise overview of current state of research on the EU ETS and the causes of the price drop (Koch et al. 2014).

[3]Average ICE EUA futures 2012–2016 (Intercontinental Exchange 2017).

[4]The 'missing money problem' describes a situation in an energy-only market where low power prices and few price spikes do not provide sufficient (long-term) investment incentives in new (flexible) generation capacity. For a closer examination of the missing money problem, see Crampton and Stoft (2006), Joskow (2008) and Newbery (2016).

generation capacity to balance intermittent renewable production or in conventional backup capacity, security of supply is threatened. As a consequence, many governments have recently introduced mechanisms to provide market participants with more incentives for building or maintaining generation capacities in order to guarantee that power demand can be met at all times. In theory, there should be no need for such market interventions, as the energy-only market should provide sufficient incentives for new capacity with scarcity price spikes. Therefore some see capacity remuneration mechanisms[5] as market distortions (European Commission 2016; Hancher et al. 2015).

It should be noted at this point that the investment cycles in the energy industry are generally characterised by their long-term nature and high capital costs. Power plants have long life times with high upfront costs. The high capital intensity amplifies the impact of investment cycle changes and raises the risk of excessive or insufficient capacities (Lu et al. 2015, p. 3242). A low-price outlook as well as regulatory uncertainties may exacerbate this effect. This, combined with technology innovations and doubts over future market designs, has made traditional investments in power plants with an expected lifetime of half a century or longer rather complex.

The dearth in investment incentives has also been identified by the European Commission as a market failure. It is commonly acknowledged that further policy measures targeting emission reductions and addressing concerns on (long-term) investment incentives are needed (European Commission 2016, p. 4). Electricity supply is now more than ever a vital good in modern societies and is also in liberalised markets regarded as a public good[6] (Abbott 2001, pp. 31–33; De Vries and Hakvoort 2003, p. 2; Finon and Pignon 2008, p. 3). Insufficient investments to guarantee long-term security of supply can thus be seen as a market failure, and capacity mechanisms represent a regulatory market intervention to address the issue.

3 Existing Carbon Taxation Models

The following section briefly presents and examines other CO_2 taxation models, namely, a differential taxation model, the use of border tax adjustments and the current CO_2 taxation model of the United Kingdom.

[5]Whilst capacity remuneration mechanisms can take different forms, they generally provide monetary payments towards generators for available generation capacity.

[6]A public good is commonly defined by the characteristics of nonrivalry and non-excludability, i.e. additional consumers add no additional costs (zero marginal costs) or reducing the good's availability for others and people cannot be excluded from consuming the good.

3.1 Differential Taxation

Cottier et al. propose a CO_2 taxation model described as differentiated electricity tax (Cottier et al. 2014b). It aims to replace renewable support schemes and act as a steering system. They suggest different tax rates based on the technology used to generate electricity, aiming not only to reduce consumption but also to promote renewables. Renewable energy sources would profit from exemptions, and tax rates for electricity produced from non-renewable sources would depend on their carbon intensity (Cottier et al. 2014b, p. 3). Guarantees of origin (or alternatively specifically designed renewable energy certificates) occupy a central role by determining the corresponding tax rate. The scholars discuss four main different varieties of the model in detail with respect to the legal considerations they entail.[7]

They find that offering exemptions in their model only for domestic renewable electricity—effectively treating domestic and foreign production differently—would most likely constitute discrimination under the GATT (Holzer et al. 2017, pp. 380–381). One approach which could circumvent this is introducing additional requirements and constraints for imported electricity eligible for tax exemptions. Yet, even in that case, compliance with WTO law remains uncertain, depending on the exact criteria (Holzer et al. 2017, p. 382). A central issue associated with also offering unrestricted exemptions for foreign renewable production is the significant availability of guarantees of origin at very low prices in the EU, especially from Nordic hydropower. Producers could simply purchase those guarantees of origin instead of paying the—it is assumed—costlier carbon tax.

In contrast to the proposed climate and supply market model with one uniform CO_2 levy, the differentiated electricity tax aims to tax electricity at different rates depending on the electricity source. Even though both models use guarantees of origin, their functions differ. In the proposed climate and supply market model, the guarantees act as a source of additional income for renewable producers due to their—at times—significant increase in value. Cottier et al.'s legal analysis of their model regards a limitation of tax exemptions only for domestic renewable electricity as problematic, whereas the structure of the proposed climate and supply market model enables a lawful increase in value of exclusively domestic guarantees of origin from CO_2-neutral production.

3.2 Carbon Tax with Border Tax Adjustments

Border tax adjustments (BTA) have in recent years been increasingly discussed in relation to enforcing an environmental tax on imported goods within WTO law. A particular focus of research has been the interplay of emission trading schemes and

[7] See Holzer et al. (2017) and Cottier et al. (2014a, b).

BTA and how BTA can be used to avoid carbon leakage.[8] Regarding the power sector, one could introduce an environmental CO_2 tax on power generation. BTA could then be used to tax imports in order to prevent competitive disadvantages for domestic production. If the tax shall depend on the CO_2 intensity of the technology used to generate the electricity, one effectively applies a differential taxation model such as the aforementioned model.

In practical terms, one aspect which comes hand in hand with the problem is that the origin of electricity is not always known and guarantees of origin are often only issued for renewables. Guarantees of origin would need to be introduced for all technologies and, possibly, also certify the CO_2 footprint of the electricity. It remains an open question how imports without guarantees of origin would be handled. A subsequent question in this cross-border context is if one can legally differentiate between electricity produced from CO_2-neutral sources (green electricity) and electricity produced from unknown or fossil energy sources (grey electricity). As a matter of fact, there exists significant legal debate amongst scholars as to whether green and grey electricity are considered 'like' or 'unlike' products under WTO law—thus far this question has not been subject to WTO jurisprudence (Holzer et al. 2017, p. 373; Kreiser et al. 2015, p. 167).

If all electricity was considered a 'like' product independent of its method of production and origin, equal treatment is required from a legal perspective. In this case, imported grey electricity could not be treated less favourably than domestically produced green electricity.[9] A flat tax on all electricity generation would fail to set incentives for low or carbon-free production. In case grey and green electricity are deemed 'unlike' products, a taxation model taxing them at different rates and using BTA seems theoretically feasible. However, a unilateral implementation of such a market design would face the aforementioned practical issue of a significant availability of renewable guarantees of origin at comparably low prices in the EU. Treating domestic electricity from CO_2-neutral sources differently than that from foreign CO_2-neutral sources would—as discussed—most likely contravene GATT rules.

3.3 United Kingdom: Carbon Price Floor

In 2013, the British government's Department of Energy and Climate Change (DECC) established a carbon price floor for electricity generation taxing fossil fuels used to generate electricity. It can be described as a 'top-up' of the EU ETS. The carbon price floor was initially introduced at 16 GBP/tCO_2 and was supposed to

[8]See Ismer and Neuhoff (2007), Kuik and Hofkes (2010), Panezi (2015).

[9]It should be noted that Art. XX GATT on exceptions may leave some room for policy measures, potentially enabling a justification of violation of the non-discrimination rules. See Cottier et al. (2014c, pp. 34–37), for a detailed discussion.

reach 30 GBP/tCO_2 in 2020 and 70 GBP/tCO_2 in 2030 (in real 2009 prices). To this end, the government charges power generators a top-up of the EU ETS called Carbon Price Support (CPS) making up the difference between the floor price and the EU ETS allowance price. The amount of this CPS is announced with budget statement by the British Treasury 2 years in advance.

Due to the lower than expected EU ETS allowance prices, the CPS was frozen at 18 GBP/tCO_2 until 2021 in order to limit the competitive disadvantage faced by businesses and prevent electricity bills from rising (HM Revenues and Customs 2014, p. 1; HM Treasury 2016).

The carbon price floor was introduced with the intention of correcting the market distortions created by the low EU ETS prices. It was supposed to underpin the price of carbon at a level that drives low carbon investment (Ares and Delebarre 2016, p. 3). The mechanisms contributed to a significant drop in electricity generation from coal-fired power plants due to higher costs associated with such generation, which in turn helped to reduce emissions (Clark 2017).

Analysing the design of the carbon price floor, it is essential to highlight that the carbon price floor affects only producers in Great Britain. Electricity imports to Great Britain are not subject to the carbon price floor; charging producers abroad would most likely violate the General Agreement on Tariffs and Trade (GATT) (in particular Article III).[10] The subsequent competitive disadvantages for electricity producers in Great Britain compared with those abroad are rather limited due to the relatively low interconnector capacity of Great Britain with other countries (4GW) as a result of its natural geographically isolated island location (Ofgem 2017). In a more interconnected market, the design of the carbon price floor that only charges national producers would not be sensible as it would disadvantage national production disproportionately. This is also the reason why Northern Ireland is exempted from the carbon price floor (Foster in Sync Ni 2013).

Currently there are ongoing discussions about the future of the carbon price floor.[11] There are both calls to phase out the mechanism as well as calls to maintain it. Of particular interest, there are studies that project the carbon price floor to lead to an overall *increase* in emissions. An increased interconnection capacity with planned interconnectors to mainland Europe and Iceland might, together with the carbon price differential, cause European-wide emissions to rise as gas-fired generation in Britain might be undercut by coal-fired generation in mainland Europe (Aurora Energy Research 2016, p. 2). In light of Brexit, there remains policy uncertainty regarding the future of interconnectors and the United Kingdom's participation in the EU ETS (Howard 2016, p. 11). The UK Government has, however, signalled its intention to maintain the carbon price floor mechanism at its current rate until 2021 (HM Treasury 2016).

[10]The role and classification of electric power as a good under WTO law and the subsequent legal implications are subject to substantial legal debate. Cottier et al. (2014c, pp. 34–37) provide an overview of handling electricity under WTO law, and Horn and Mavroidis (2011) discuss the legality of Border Tax Adjustments for climate purposes. It is generally agreed that CO_2 taxation regimes not limited to domestic production may easily violate WTO (GATT) rules.

[11]For an overview see the recent research note by Policy Exchange (Howard 2016).

The example of CO_2 taxation in Great Britain and the model proposed by Cottier et al. illustrate the economic and legal constraints of CO_2 steering models. Reforms on a European level have proven difficult, levying only domestic power producers results in competitive disadvantages, and taxes on imports are likely to violate international law. For countries with a low share of interconnector capacity, a carbon price floor such as the British model—levying only domestic production—effectively counters low EU ETS allowance prices and can (re)establish carbon prices as a driver for decarbonisation. The networks of most European countries are, however, much more deeply integrated and connected, which is why the British model does not represent an appropriate solution.

In the absence of effective European reforms, the suggested climate and supply market model can therefore offer an alternative way of pricing CO_2 on a national basis, internalising the costs of CO_2 emissions and clearing market distortions originating from EU ETS allowance prices which are too low without violating international trade agreements or EU law.

4 Unilateral Climate and Supply Market Model

It has been shown that, regarding energy policy objectives, further policy measures targeting emission reductions as well as addressing concerns over (long-term) investment incentives are needed. The authors propose a climate and supply market model that tackles not only the aforementioned climate issue but also the insufficient investment incentives provided by the energy-only market.

Given the difficulties in reforming emission trading systems such as the EU ETS at a multilateral level, the climate and supply market model suggests a unilateral approach in which a state introduces a CO_2 levy that internalises the external costs of CO_2 emissions at a national level. It aims to correct the aforementioned market distortions caused by the low EU ETS price level and its subsequent effects by reintroducing a significant carbon price for the power sector. The model exempts the consumption of CO_2-neutral electricity from the new CO_2 levy, using national guarantees of origin (which already exist in many countries). The levy is introduced for suppliers who pass the costs on to the final electricity consumers (Fig. 1). The EU ETS allowance price is taken into account, whereby the proposed CO_2 levy decreases when the EU ETS allowance price rises. The government, relevant ministry or institution in question sets the amount of the national CO_2 levy. This amount could be related to the social costs of carbon[12] or, if available, existing CO_2 taxation on fuels.

[12]The social costs of carbon are a scientific approach to measure the marginal costs of emitting an additional unit of CO_2 or CO_2 equivalents at a given time. The comprehensive scientific approach tries to incorporate climate change-related costs to estimate the social costs of carbon. See Committee on Assessing Approaches to Updating the Social Cost of Carbon et al. (2017) for a recent overview of social costs of carbon estimates.

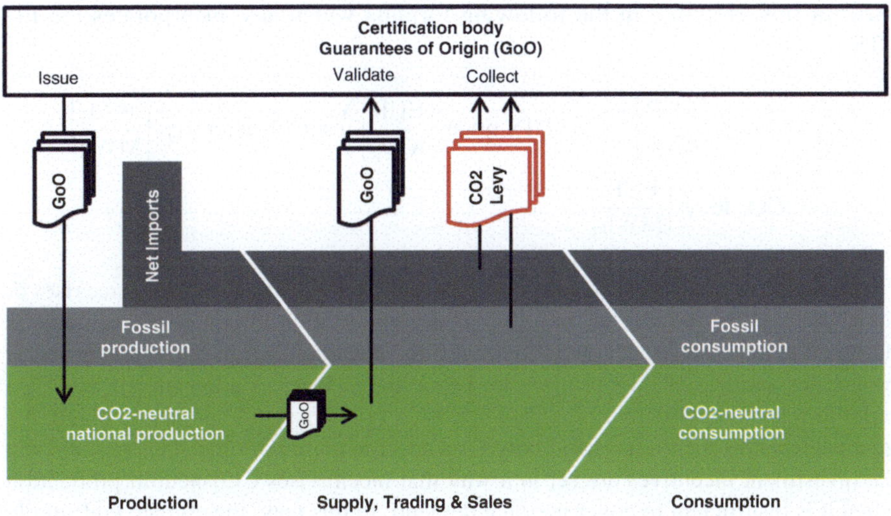

Fig. 1 Climate and supply market model

In the suggested model, producers report their production to a certification body (in many cases already existing and responsible for handling guarantees of origin) and receive corresponding guarantees or certificates. Suppliers report to the certification body their deliveries to the final consumers, and they pay the corresponding CO_2 levy or present national guarantees of CO_2-neutral origin instead.[13] Guarantees of origin are traded, and as long as national CO_2-neutral production does not surpass the national demand, the guarantee price should roughly equal the CO_2 levy, since suppliers have either to pay the CO_2 levy or present the guarantees of CO_2-neutral origin.[14] It should be noted at this point that the certificate market remains independent and separate from the energy-only market on which electricity is traded.

In order not to discriminate against any form of production (thus adopting a principle of non-discrimination), the CO_2 levy applies generally to every unit of electricity consumed within the country and does not differentiate between technologies—apart from the fact that only guarantees of origin from CO_2-neutral sources offer an exemption from the levy. Costs associated with CO_2 (such as the social costs of carbon) are usually expressed per tonne of CO_2. For the purpose of the climate and supply market model, a calculation to express the costs per unit of electricity (i.e. MWh) is necessary. The CO_2 intensity of either the national power sector or a European Economic Area (EEA) average can be

[13]It is assumed that foreign guarantees of origin cannot offer exemptions as long as there is no such agreement.

[14]If national production exceeds final national consumption, the guarantees of origin have no additional value.

used for this, resulting in the following formula which also incorporates the EU ETS:

$$\left(CO_2 \text{ levy} \left[\frac{EUR}{tCO_2}\right] - \text{EU ETS price} \left[\frac{EUR}{tCO_2}\right]\right) * CO_2 \text{ intensity} \left[\frac{tCO_2}{MWh}\right] = CO_2 \text{ levy} \left[\frac{EUR}{MWh}\right].$$

Consequently, the final CO_2 levy decreases if the EU ETS allowance price rises or the CO_2 intensity is reduced. In terms of decarbonisation, this means the model is designed to eventually become redundant if the decarbonisation of the power sector progresses and/or EU ETS reforms lead to a higher level of allowance prices.[15] In order to reach the objective of a CO_2-neutral supply covering national demand, the guarantees of origin remain valid only for a specific period of time. This ensures that the investment incentives are set in a way that incentivises CO_2-neutral production when it is insufficient to cover national demand. In this way, the climate and supply market model strengthens security of supply and makes a country less dependent on imports. As soon as CO_2-neutral sources satisfy the national demand during the chosen time periods, the model becomes obsolete. In the case of renewable power plants that are currently supported through some form of renewable support mechanism such as feed-in tariffs or quota schemes, the support scheme administrative authority (rather than the plant operator) should receive the certificates.[16] Those guarantees can then be sold on the market and used, for example, to finance the costs of the existing renewable support scheme.

Currently, several governments in Europe have some form of CO_2 levy in place usually concerning fuels used in the transport or heating sector.[17] The emissions originating from electricity generation for electric radiators or vehicles is, in contrast, often only charged through EU ETS, and there is no true cross-sector approach. CO_2 levies for fuels could be used and applied to other sectors, in order to implement a cross-sector carbon price. Such a cross-sector approach would help to clear market distortions and effectively reward the most carbon-efficient approach in fields such as heating or mobility. Introducing the climate and supply market model inevitably leads to higher electricity bills for consumers, due to its financing structure. The

[15]The EU ETS allowance price used in the formula can be determined by either using past (ex post) or future (ex ante) allowance prices.

[16]This is done to prevent windfall profits for operators of renewable power plants already supported through some form of support mechanism as the value of the guarantees of origin would increase significantly through the introduction of the climate and supply market model. Those operators will have to be compensated in some way as they will not be able to sell their guarantees of origin. A direct financial compensation and an introduction of new distinct certificates serving the same purpose as guarantees of origin yet not eligible for the climate and supply market model would be possible ways of doing so.

[17]Examples of carbon taxes can be found in Switzerland, Sweden, Finland and Denmark (see World Bank et al. 2016).

exact costs would be country specific and depend amongst other things on the set CO_2 levy, the country's electricity generation mix, the import/export balance, the potential exemptions of some customers and the final consumption of electricity. In this regard, the test case presented later shows an acceptable level of costs for consumers.

Aside from the additional cost burden that comes with the climate and supply market model, one should also address the drawbacks that a uniform CO_2 levy entails. In order not to discriminate against certain forms of production, the CO_2 levy of the climate and supply market model is uniform and applicable to all forms. Guarantees of origin offer exemptions only for electricity produced from CO_2-neutral sources, which in turn means that all electricity produced from fossil sources faces the same uniform CO_2 levy. Electricity from emission-intensive lignite- or coal-fired power plants is thus not treated differently than that produced from less emission-intensive gas-fired and combined-cycle power plants. This consequently makes the climate and supply market model an ill-fitted model for countries that seek to replace coal-fired generation capacities with gas-fired capacities through the introduction of a CO_2 levy.

5 Legal Considerations with Respect to International and EU Law

The concept of pricing CO_2 on a unilateral level is not new, and there have been various studies regarding its legal status in the past.[18] In order to initiate a public debate on a concept such as the suggested climate and supply market model, it is necessary to clarify the legal considerations such a market design entails. Any potential market design has to be compatible with existing legislation and in particular with WTO law, EU law and other trade agreements, since CO_2 taxation regimes have to be carefully designed in order not to violate them. A professional legal assessment of the suggested climate and supply market model by a leading Swiss law firm found the model to be in line with all relevant international and EU legislations.

Fundamental for the legal compatibility of the suggested model is the twofold approach of putting a levy on the final electricity consumption (rather than production) and the clear separation of the national guarantees of origin market on the one hand and the energy-only market on the other. Generally, guarantees of origin are not considered goods or products under internal trade law since they are not tangible and have no customs tariff number in the Harmonized Commodity and Coding Systems (short Harmonized System [HS]) of international law (Petsonk 1999,

[18]See, for instance, Cottier et al. (2011, 2014a), Holzer (2014), Panezi (2015).

pp. 199–200).[19] They are also generally not considered as services under the General Agreement on Trade in Services (GATS) even though the subsequent trading of the certificates may be regarded as such (Delimatsis and Mavromati 2009, p. 251).

Also as concerns EU law, guarantees of origin are not handled as goods which fall under the free movement of goods within the single market. Court rulings by the European Court of Justice underline that green certificates or guarantees of origin are not treated as goods and that guarantees do not have to be recognised by other member states. One can refer at this point to the prominent case of Åland Vindkraft and a following similar case relating inter alia to Belgian guarantees of origin.

The Åland Vindkraft court ruling (which gained prominence by confirming the national character of renewable support schemes) clarified that member states only have to recognise foreign guarantees of origin to a limited extent (European Court of Justice 2014a, p. 11). In a similar case, one of the involved parties explicitly argued that the intangible nature of guarantees of origin prevents their categorisation as goods. The court refrained hereby from ruling definitely, answering that 'even if it were accepted that guarantees of origin … constitute "goods"' it would not change the question at stake (European Court of Justice 2014b, p. 15). Accordingly, it is fair to say that thus far, guarantees of origin have not been treated as goods in international law and EU law, which is crucial for the legal compatibility of the suggested climate and supply market model.

Finally, as regards this legal perspective, it is worth taking a brief look at the EU state aid law vis-a-vis the suggested model. The support of companies or industries with state resources is considered state aid by the European Commission. The legal assessment asserts that the climate and supply market model does not constitute state aid. The associated costs arising through the CO_2 levy of the suggested model are borne by consumers, and the amounts paid for the CO_2 levy are not received by producers. The fact that the value of guarantees of origin experiences a substantial increase through a state intervention does not change this principle. Strictly speaking, the model does therefore not constitute state aid in this context. However, even if the model was considered state aid, it would likely be considered proportionally and approvable (similar to other permitted measures). This should be seen in the broader context of approved measures regarding support for renewables and mechanisms strengthening the security of supply.

[19]It should be noted that there is no clear definition of what exactly constitutes a good in international law and, thus far, guarantees of origin or certificates as part of renewable support schemes (often referred to as quota obligation or green certificates) have not been regarded as goods or services (Buchmüller 2013; Delimatsis and Mavromati 2009; Howse 2009; Petsonk 1999).

6 Climate and Supply Market Model Example: Switzerland

One can use Switzerland as an example to illustrate the proposed climate and market model. Switzerland's domestic power production is virtually CO_2 neutral. Yet during the winter months, Switzerland relies on imports to meet its demand. The validity for guarantees of origin could therefore be set for 1 month. During summer months, when Switzerland is a net exporter of electricity, the guarantees of origin from CO_2-neutral sources would be without additional value, and no CO_2 levy applies. However, in winter months when Switzerland imports electricity, the value of guarantees of origin from CO_2-neutral sources would rise to approximately that of the CO_2 levy to be paid for non-CO_2-neutral production (i.e. imports). Existing Swiss laws include a CO_2 levy on thermal fuels of 84 CHF/tCO_2 ($\approx$78 EUR/tCO_2) (Federal Office for the Environment 2016; Bundesrat 2011). The suggestion is to apply the same carbon price level to the power sector. Consequently, in winter months when imports are necessary, suppliers would need to pay the CO_2 levy for the electricity that cannot be exempted with guarantees of origin (imports).

For the calculation of the applicable CO_2 levy, an estimated average EEA[20] power generation carbon intensity of 0.23 tCO_2/MWh[21] can be used. To take the EU ETS price into account that has already been paid, one can deduct the EU ETS allowance price off the Swiss carbon price. With a carbon price of 78 EUR/tCO_2 and a 2016 average EU ETS allowance price of about 5.3 EUR/tCO_2, the suggested CO_2 levy would thus equal:

$$\left(78\left[\frac{\text{EUR}}{\text{tCO}_2}\right]-5.3\left[\frac{\text{EUR}}{\text{tCO}_2}\right]\right)*0.23\left[\frac{\text{tCO}_2}{\text{Mwh}}\right]=16.7\left[\frac{\text{EUR}}{\text{MWh}}\right].$$

Swiss suppliers would therefore have the choice of either buying national guarantees of origin from CO_2-neutral sources or paying the CO_2 levy. Given the certificate scarcity in import months (i.e. when Swiss national CO_2-neutral production does not cover demand), the price for the guarantees of origin would equal that of the CO_2 levy of 16.7 EUR/MWh—since for every MWh delivered suppliers have to provide either a guarantee of origin or pay the CO_2 levy.

An increase of the EU ETS allowance price level or a decrease of the EEA power generation carbon intensity leads to a lower CO_2 levy without further adjustments. In

---

[20]An EEA plus Switzerland average rather than EU average is used here for legal reasons. The economic area of the EEA plus Switzerland encompasses a wider European market and does not discriminate specific countries or on the basis of political union.

[21]Estimation based on a published EU CO2 emission intensity of 0.276 tCO_2/MWh for 2014 (European Environment Agency 2016). Assuming a similar CO_2 emission intensity reduction as in previous years and taking the virtually CO_2-free Swiss, Norwegian, Icelandic and Liechtenstein production into account to form an EEA + Switzerland average, one can use 0.23 tCO_2/MWh as a rough estimation for 2016 (Amt für Statistik 2016, p. 18; European Environment Agency 2016; Norwegian Water Resources and Energy Directorate (NVE) 2017; Orkustofnun, 2017, p. 1; Swissgrid 2017).

summer months, when Switzerland is traditionally an exporter of electricity and its domestic CO_2-neutral production exceeds consumption, the guarantees of origin have no additional value.

This example shows that the climate and supply market model would reward CO_2-neutral production in times when the national CO_2-neutral production is not sufficient to cover the demand and set incentives for expanding capacities that produce CO_2-neutral electricity during those times.

A subsequent step is to calculate the economic impact and overall costs arising from a CO_2 levy of 16.7 EUR/MWh. Over the last 5 years (2012–2016), there were on average 4 months per year during which the final electricity consumption was greater than the total energy production in Switzerland (Swissgrid 2017).

The average final electricity consumption over those 4 months equalled 21,119,886 MWh (21.12 TWh). A CO_2 levy of 16.7 EUR/MWh would hence result in costs of around 353 million EUR per year. It should be noted at this point that this sum may vary depending mostly on the number of months during which the final electricity consumption exceeds the production of CO_2-neutral electricity. The total annual Swiss final electricity consumption has been just below 60 TWh in the last couple of years (Bundesamt für Energie 2017, p. 4). The costs of 350 million EUR per year would thus equal additional costs of around 0.6 cents/kWh, which can arguably be seen as an acceptable level of additional costs.

7 Variations of the Climate and Supply Market Model

The model can be altered in many ways to incorporate specific requirements or change its effects. The most straightforward steering instrument is the government-set CO_2 levy in EUR/tCO_2, which together with the EU ETS allowance price and the carbon intensity translates into the final levy expressed in EUR/MWh. Policymakers can choose whichever price is perceived as appropriate to work towards the given policy objectives. This might result in a cross-sector carbon price as in the described example or one that relates to the social costs of carbon.

One recurring aspect of CO_2 taxation regimes are exemptions or special conditions for energy-intensive industries. Such measures can be included in the suggested climate and supply market model in case policymakers choose not to place additional burdens on the energy-intensive industry.

It may also be in the interest of a government to choose which particular CO_2-neutral generation technologies should profit from the guarantees of origin value increase. It could, for example, be restricted to CO_2-free rather than CO_2-neutral technologies thereby excluding technologies such as biomass or landfill power plants.

Another central element of the suggested model is the time period for which guarantees remain valid. In the presented example, the time period is set to 1 month, but different time frames are possible. Depending on the country's requirements and circumstances, the administrative body could set a longer time frame (i.e. quarters or seasons) or shorter one (i.e. weeks, days or even hours). A shorter time frame represents a more precise steering instrument but increases potential market power abuses.[22] Additionally, it comes at higher administrative costs and efforts for market participants and the administrative body.

In the suggested version of the model, suppliers have to provide one guarantee of origin from CO_2-neutral production for every MWh delivered or pay the CO_2 levy. But the administrative body could also require suppliers to provide more or less than one guarantee per unit of energy delivered. This way, one could steer the demand for guarantees and it enables setting targets of a desirable share of production from CO_2-neutral sources for the chosen time period.

8 Adaption Potential for the Climate and Supply Market Model

The concept of the climate and supply market model was developed in a European context, and the focus of this research rested on European countries. The general principle of a CO_2 price component with exemptions for CO_2-neutral production using guarantees of origin can theoretically also be applied elsewhere. Naturally this would entail some necessary adaptions such as the removal of the EU ETS pricing in the formula of the CO_2 levy and further legal assessments.

The model is particularly attractive for countries with a high share of traditional CO_2-neutral production (e.g. hydro, biomass) and looking for ways to maintain or increase it. Newer forms of CO_2-neutral/free production such as photovoltaics (PV) and wind are commonly supported through some form of feed-in scheme. As mentioned, operators of plants receiving subsidies through such a scheme would not be eligible for support through the climate and supply market model, since it would result in windfall profits.

With compensation levels of support schemes continuously decreasing due to the cost reductions of renewables technologies, learning curves and further innovation, one may question for how long governments will continue their feed-in support schemes. Recent competitive tenders have again led to a decrease in compensation levels, and the first German offshore wind auction gained prominence as companies

[22]If in a chosen time period domestic CO_2-neutral production is projected to be only marginally greater than the final consumption, producers might abuse their market power to transform it into an import period by withholding generation capacity. This could increase their income as the value of the guarantees of origin from CO_2-neutral sources would rise to that of the CO_2 levy. The shorter the time period, the smaller the generation capacity necessary to do so, as long as market data are available to project if domestic production will exceed or fall short of final consumption.

bid to build offshore wind parks without a guaranteed minimum strike price.[23] In this light, it seems plausible that policymakers discuss the future of support schemes with the climate and supply market model comprising a potential option.

Since the climate and supply market model aims to incentivise CO_2-neutral power production during times when the targeted share of this production type is not met, it seems most apt for countries that already have a relatively high share of CO_2-neutral production in their electricity mix. Aside from the aforementioned example of Switzerland, the following European countries have a large CO_2-neutral share (over 65%) in their gross electricity production (according to data published by the European Commission 2017): Austria, Croatia, Denmark, Finland, France, Slovakia, Slovenia and Sweden.

Given the functioning of the climate and supply market model described herein, the composition of the power sectors of these countries make them ideal candidates for a potential adaption of the model or discussion thereof. This is not to say that the climate and supply market model is not a useful fit for other countries. If policymakers set a lower amount of required guarantees of origin for suppliers per MWh delivered (e.g. 0.5 rather than 1), the model can also be attractive for countries with a lower share of CO_2-neutral production. As mentioned earlier, the fact that the CO_2 levy of the model is uniform means that all electricity is treated the same independent of exact CO_2 intensity of the production and only guarantees of origin from CO_2-neutral sources offer exemptions. Therefore, the model does not represent an adequate instrument for countries that seek to introduce a CO_2 levy to lift the economic viability of gas-fired power generation over that of coal-fired generation.

9 Conclusion

It has been argued that the power sector is facing a series of difficulties in light of energy transitions and decarbonisation. Low EU ETS allowance prices have led to a state of uncertainty regarding its role as a driver for decarbonisation. Previous reform processes have had limited success, and some countries have looked towards unilateral action to tackle the issue.

A run of very low wholesale power prices in Europe, partly caused by low EU ETS allowance prices, has led to concerns over long-term generation adequacy (the 'missing money problem') and the introduction of capacity remuneration mechanisms to counter the issue. The authors propose a unilateral climate and supply market model to address these aforementioned concerns. The model's core component is an introduction of a CO_2 levy on final electricity consumption. Exemptions

[23] In 2017, EnBW and Dong Energy were awarded the right to build wind farms in the North Sea with submitted bids of 0 EUR/MWh. It should be noted that they will still receive some form of subsidy as they gained the right to operate those parks for 25 years and network charges for electricity consumers finance the costly grid connection(s) (Bundesnetzagentur 2017, p. 2).

from the levy are offered for electricity produced from CO_2-neutral sources. Suppliers have to provide guarantees of origin from CO_2-neutral production for every MWh delivered or pay the CO_2 levy. The government sets the amount of the levy, and the time period guarantees of origin remain valid for. This way the climate and supply market model incentivises the power production from CO_2-neutral sources during times when it does not cover demand or the target share of production.

The model can be used to implement a cross-sector carbon price as the analysis of the example of Switzerland illustrates. The example also shows that the model can be realised at an acceptable level of costs for consumers even though the exact costs depend on a number of the factors discussed, first and foremost the set CO_2 levy.

CO_2 pricing schemes targeting the power sector have to be carefully designed in order to comply with international and EU law. Some models, such as the British model, therefore exclude electricity imports from their CO_2 price floor. In other countries with greater transfer capacity to grids abroad, excluding imports would disproportionally disadvantage domestic power producers, which is why the British model only has limited adaption potential abroad. A legal assessment of the proposed climate and supply market model came to the conclusion that the suggested design complies with relevant legal frameworks.

The model offers various steering instruments for policymakers, and it can be altered to suit a given country. Aside from setting the level of the CO_2 levy, the government can specify the time period for which guarantees of origin remain valid and impose the number of guarantees of origin required for suppliers per MWh delivered. Together, these mechanisms enable the government to work towards a targeted share of CO_2-neutral production during any time period and incentivise the deployment of additional CO_2-neutral generation capacity.

Finally, even though the model has been developed in a European context, the general principle could also be used elsewhere. Within Europe, there are several countries for which the proposed climate and supply market model might represent a potential policy option.

References

Abbott, M. (2001). Is the security of electricity supply a public good? *The Electricity Journal, 14* (7), 31–33.

Abrell, J., Rausch, S., & Yonezawa, H. (2016). *Higher price, lower costs? Minimum prices in the EU emissions trading scheme* (SSRN Scholarly Paper No. ID 2764155). Rochester, NY: Social Science Research Network. Retrieved from https://papers.ssrn.com/abstract=2764155

Amt für Statistik. (2016). *Energiestatistik 2015*. Vaduz: Fürstentum Liechtenstein. Retrieved from http://www.llv.li/files/as/energiestatistik-2015.pdf

Ares, E., & Delebarre, J. (2016). *The carbon price floor*. London: House of Commons Library.

Aurora Energy Research. (2016). *Dash for interconnection: The impact of interconnectors on the GB market*. Oxford: Aurora Energy Research. Retrieved from https://www.auroraer.com/wp-content/uploads/2016/10/Dash-for-Interconnectors-Aurora-Energy-Research-February-2016.pdf

Boffey, D. (2017). Reform of EU carbon trading scheme agreed. *The Guardian*. Retrieved from https://www.theguardian.com/environment/2017/feb/28/reform-of-eu-carbon-trading-scheme-agreed

Bublitz, A., Keles, D., & Fichtner, W. (2017). An analysis of the decline of electricity spot prices in Europe: Who is to blame? *Energy Policy, 107*, 323–336.

Buchmüller, C. (2013). *Strom aus erneuerbaren Energien im WTO-Recht*. Baden-Baden: Nomos Verlagsgesellschaft mbH & Co. KG.

Bundesamt für Energie (BFE). (2017). *Schweizerische Elektrizitätsstatistik 2016*. Bern: Schweizerische Eidgenossenschaft. Retrieved from http://www.bfe.admin.ch/php/modules/publikationen/stream.php?extlang=de&name=de_306571764.pdf&endung=Schweizerische%20Elektrizit%E4tsstatistik%202016

Bundesnetzagentur. (2017). *Press release: Bundesnetzagentur announces successful bids in first auction for offshore wind farms*. Bundesnetzagentur. Retrieved from https://www.bundesnetzagentur.de/SharedDocs/Downloads/EN/BNetzA/PressSection/PressReleases/2017/13042017_windsee.pdf?__blob=publicationFile&v=2

Bundesrat. (2011). *Bundesgesetz über die Reduktion der CO2-Emissionen (CO2-Gesetz)*. Retrieved from https://www.admin.ch/opc/de/official-compilation/2012/6989.pdf

Carbon Market Watch. (2015). *Four magic potions to turn the EU ETS into an effective climate mitigation tool* (Policy Brief) (pp. 1–8). Brussels. Retrieved from http://carbonmarketwatch.org/four-magic-potions-to-turn-the-eu-ets-into-an-effective-climate-mitigation-tool/

Clark, P. (2017). *UK carbon emissions fall to late-19th century levels*. Retrieved from https://www.ft.com/content/2bc62cb8-004f-11e7-8d8e-a5e3738f9ae4

Committee on Assessing Approaches to Updating the Social Cost of Carbon, Board on Environmental Change and Society, Division of Behavioral and Social Sciences and Education, & National Academies of Sciences, Engineering, and Medicine. (2017). *Valuing climate changes: Updating estimation of the social cost of carbon dioxide*. Washington, D.C.: National Academies Press. https://doi.org/10.17226/24651.

Cottier, T., Delimatsis, P., Malumfashi, G., Matteotti-Berkutova, S., Nartova, O., De Sépibus, J., & Bigdeli, S. Z. (2011). Energy in WTO law and policy. In *The prospects of international trade regulation: From fragmentation to coherence* (pp. 211–244). Cambridge: Cambridge University Press.

Cottier, T., Espa, I., Hirsbrunner, S., Holzer, K., & Payosova, T. (2014a). *Differentiating energy tax on electricity: An assessment of the compatibility with WTO law and EU law* (Legal Opinion submitted to the Swiss Federal Financial Administration (FFA), the Swiss State Secretariat for Economic Affairs (SECO) and the Swiss Federal Office of Energy (SFOE)). Bern: University of Bern.

Cottier, T., Espa, I., Hirsbrunner, S., Holzer, K., & Payosova, T. (2014b). *CO_2 levies and tariffs on imported electricity*. Bundesamt für Energie BFE.

Cottier, T., Espa, I., Hirsbrunner, S., Holzer, K., & Payosova, T. (2014c). *Differential taxation of electricity: Assessing the compatibility with WTO law, EU law and the Swiss-EEC free trade agreement* (Legal Opinion submitted to the Swiss Federal Financial Administration (FFA), the Swiss State Secretariat for Economic Affairs (SECO) and the Swiss Federal Office of Energy (SFOE)). Bern: Swiss Federal Office of Energy (SFOE). Retrieved from https://www.seco.admin.ch/dam/seco/de/dokumente/Publikationen_Dienstleistungen/Publikationen_Formulare/Umwelt_Energie/Vereinbarte%20differenzierte%20Stromsteuer.pdf.download.pdf/ES2050%20zweite%20Etappe.%20differenzierte%20Stromsteuern.pdf

Crampton, P., & Stoft, S. (2006). *The convergence of market designs for adequate generating capacity*. White paper for the California Electricity Oversight Board. Retrieved from http://www.cramton.umd.edu/papers2005-2009/cramton-stoft-market-design-for-resource-adequacy.pdf

De Vries, L. J., & Hakvoort, R. A. (2003). *The question of generation adequacy in liberalized electricity markets*. Presented at the 26th Annual Conference International Association Energy Economics, Prague: IAEE. Retrieved from https://www.ecn.nl/fileadmin/ecn/units/bs/INDES/indes-ldv_paper.pdf

Delimatsis, P., & Mavromati, D. (2009). GATS, financial services and trade in renewable energy certificates (RECs) – Just another market based solution to cope with the tragedy of the commons? In *International Trade Regulation and the Mitigation of Climate Change*. Cambridge: Cambridge University Press.

Doukas, H., Patlitzianas, K. D., Kagiannas, A. G., & Psarras, J. (2008). Energy policy making: An old concept or a modern challenge? *Energy Sources, Part B: Economics, Planning, and Policy, 3*, 362–371.

European Commission. (2014). *Proposal for a decision of the European Parliament and of the Council concerning the establishment and operation of a market stability reserve for the Union greenhouse gas emission trading scheme and amending Directive 2003/87/EC /* COM/2014/020final - 2014/0011 (COD) */.* Brussels: European Commission. Retrieved from http://eur-lex.europa.eu/legal-content/EN/TXT/HTML/?uri=CELEX:52014PC0020&from=en

European Commission. (2016). *Final report of the sector inquiry on capacity mechanisms*. Brussels: European Commission. Retrieved from http://ec.europa.eu/energy/sites/ener/files/documents/com2016752.en_.pdf

European Commission. (2017). *Energy statistics*. Brussels: European Commission. Retrieved from https://ec.europa.eu/energy/sites/ener/files/documents/countrydatasheets_march2017.xlsx

European Council. (2017). *Revision of the emissions trading system: Council agrees its position – Consilium*. Retrieved from http://www.consilium.europa.eu/en/press/press-releases/2017/02/28-ets-revision/

European Court of Justice. Case C-204/12 Essent Belgium. (2014a). Retrieved from http://curia.europa.eu/juris/document/document.jsf;
jsessionid=9ea7d2dc30d642600f97ae054509900d9ede49291c2f.
e34KaxiLc3qMb40Rch0SaxyLbN50?text=&docid=157525&pageIndex=0&doclang=EN&mode=lst&dir=&occ=first&part=1&cid=242997

European Court of Justice. Case C-573/12 Åland Vindkraft, C-573/12 §. (2014b). Retrieved from http://curia.europa.eu/juris/document/document.jsf?text=&docid=154403&pageIndex=0&doclang=EN&mode=lst&dir=&occ=first&part=1&cid=125045

European Environment Agency. (2016). *Overview of electricity production and use in Europe*. Retrieved from https://www.eea.europa.eu/data-and-maps/daviz/co2-emission-intensity-3

Everts, M., Huber, C., & Blume-Werry, E. (2016). Politics vs markets: How German power prices hit the floor. *The Journal of World Energy Law & Business, 9*(2), 116–123.

Federal Office for the Environment. (2016). *Imposition of the CO_2 levy on thermal fuels*. Retrieved from https://www.bafu.admin.ch/bafu/en/home/topics/climate/info-specialists/climate-policy/co2-levy/imposition-of-the-co2-levy-on-thermal-fuels.html

Finon, D., & Pignon, V. (2008). Electricity and long-term capacity adequacy: The quest for regulatory mechanism compatible with electricity market. *Utilities Policy, 16*(3), 143–158.

Hancher, L., de Hauteclocque, A., & Sadowska, M. (2015). *Capacity mechanisms in the EU energy market: Law, policy, and economics*. Oxford: OUP.

Hepburn, C., Neuhoff, K., Acworth, W., Burtraw, D., & Jotzo, F. (2016). The economics of the EU ETS market stability reserve. *Journal of Environmental Economics and Management, 80*, 1–5.

Hirth, L. (2016). *What caused the drop in European electricity prices?* (SSRN Scholarly Paper No. ID 2874841). Rochester, NY: Social Science Research Network. Retrieved from https://papers.ssrn.com/abstract=2874841

HM Revenues & Customs. (2014). *Carbon price floor: Reform and other technical amendments*. Retrieved from https://www.gov.uk/government/uploads/system/uploads/attachment_data/file/293849/TIIN_6002_7047_carbon_price_floor_and_other_technical_amendments.pdf

HM Treasury. (2016). *Autumn statement 2016*. Retrieved from https://www.gov.uk/government/publications/autumn-statement-2016-documents/autumn-statement-2016

Holzer, K. (2014). *Carbon-related border adjustment and WTO Law*. Cheltenham: Edward Elgar.

Holzer, K., Espa, I., & Payosova, T. (2017). Promoting Green electricity through differentiated electricity tax schemes. In Cottier & Espa (Eds.), *International trade in sustainable electricity*. Cambridge: Cambridge University Press.

Horn, H., & Mavroidis, P. C. (2011). To B(TA) or Not to B(TA)? On the legality and desirability of border tax adjustments from a trade perspective. *The World Economy, 34*(11), 1911–1937.

Howard, R. (2016). *Next steps for the carbon price floor*. London: Policy Exchange. Retrieved from https://policyexchange.org.uk/wp-content/uploads/2016/11/Carbon-Price-Floor-Research-Note-FINAL3.pdf

Howse, R. (2009). *World Trade Law and renewable energy: The case of non-tariff measures*. New York and Geneva: United Nations Conference on Trade and Development. Retrieved from http://unctad.org/en/Docs/ditcted20085_en.pdf

Intercontinental Exchange (ICE). (2017). *EUA Futures*. Atlanta. Retrieved from https://www.theice.com/products/197/EUA-Futures/data

Ismer, R., & Neuhoff, K. (2007). Border tax adjustment: A feasible way to support stringent emission trading. *European Journal of Law and Economics, 24*(2), 137–164. https://doi.org/10.1007/s10657-007-9032-8.

Joskow, P. L. (2008). Capacity payments in imperfect electricity markets: Need and design. *Utilities Policy, 16*(3), 159–170.

Koch, N., Fuss, S., Grosjean, G., & Edenhofer, O. (2014). Causes of the EU ETS price drop: Recession, CDM, renewable policies or a bit of everything?—New evidence. *Energy Policy, 73*, 676–685.

Kreiser, L., Andersen, M. S., Olsen, B. E., Speck, S., Milne, J. E., & Ashiabor, H. (2015). *Environmental pricing: Studies in policy choices and interactions*. Cheltenham: Edward Elgar.

Kuik, O., & Hofkes, M. (2010). Border adjustment for European emissions trading: Competitiveness and carbon leakage. *Energy Security – Concepts and Indicators with Regular Papers, 38* (4), 1741–1748. https://doi.org/10.1016/j.enpol.2009.11.048.

Lu, B., Ma, X., & Feng, C. (2015). Theory and model of energy investment and financing. In J. Yan (Ed.), *Handbook of clean energy systems, 6 Volume Set*. New York: Wiley.

Marcu, A., Elkerbout, M., & Wijnand, S. (2016). *2016 State of EU ETS Report*. Brussels: Centre for European Policy Studies (CEPS). Retrieved from http://www.ceps-ech.eu/sites/default/files/State%20of%20EU%20ETS%20v16_0.pdf

Newbery, D. (2016). Missing money and missing markets: Reliability, capacity auctions and interconnectors. *Energy Policy, 94*, 401–410.

Nikas, A., Doukas, H., & López, L. M. (2018). A group decision making tool for assessing climate policy risks against multiple criteria. *Heliyon, 4*.

Norwegian Water Resources and Energy Directorate (NVE). (2017). *Norway and the European power market*. Retrieved from https://www.nve.no/energy-market-and-regulation/wholesale-market/norway-and-the-european-power-market/

Ofgem. (2017). *Electricity interconnectors*. Retrieved from https://www.ofgem.gov.uk/electricity/transmission-networks/electricity-interconnectors

Orkustofnun. (2017). *Development of electricity production in Iceland 2015* (Orkustofnun Data Repository). Reykjavik: Orkustofnun. Retrieved from http://os.is/gogn/Talnaefni/OS-2017-T005-01.pdf

Panezi, M. (2015). When CO_2 goes to Geneva - Taxing carbon across borders – Without violating WTO obligations. *CIGI Papers, 83*. Retrieved from https://www.cigionline.org/sites/default/files/cigi_paper_no.83.pdf

Petsonk, A. (1999). The Kyoto Protocol and the WTO: Integrating greenhouse gas emissions allowance trading into the global marketplace. *Duke Environmental Law & Policy Forum, 10*, 185.

Rattay, W. (2017). Emissionshandel: EU-Parlament stimmt für strengere Regeln. *Die Zeit*. Retrieved from http://www.zeit.de/wirtschaft/2017-02/emissionshandel-eu-parlament-zertifikate-umweltschutz

Rogge, K. S., Schneider, M., & Hoffmann, V. H. (2011). The innovation impact of the EU Emission Trading System — Findings of company case studies in the German power sector. *Ecological Economics, 70*(3), 513–523.

Swissgrid. (2017). *Energieübersicht Schweiz 2016*. Swissgrid. Retrieved from https://www.swissgrid.ch/dam/dataimport/energy_data/de/EnergieUebersichtCH_2016.xls

Sync Ni. (2013). *Foster welcomes Chancellor's decision to exempt Northern Ireland generators from Carbon Tax*. Retrieved from https://syncni.com/news/2/1459/foster-welcomes-chancellor-s-decision-to-exempt-northern-ireland-generators-from-carbon-tax/tab/1356

World Bank, Ecofys, & Vivid Economics. (2016). *State and trends of carbon pricing 2016*. Washington, DC: World Bank. Retrieved from https://openknowledge.worldbank.org/handle/10986/25160

The manufacturer's authorised representative in the EU is Springer Nature Customer Service Centre GmbH, Europaplatz 3, 69115 Heidelberg, Germany. If you have any concerns regarding our products, please contact ProductSafety@springernature.com

Printed and bound by CPI Group (UK) Ltd, Croydon, CR0 4YY
30/09/2026
02225651-0008